Frontiers in Particle Science and Technology 2016

Mitigation & Application of Particle Attrition

Houston, Texas, USA
11-13 April 2016

ISBN: 978-1-5108-2259-7

Frontiers in Particle Science and Technology 2016

Mitigation & Application of Particle Attrition

Houston, Texas, USA
11-13 April 2016

AIChE
The Global Home of Chemical Engineers

Printed from e-media with permission by:

Curran Associates, Inc.
57 Morehouse Lane
Red Hook, NY 12571

Some format issues inherent in the e-media version may also appear in this print version.

Printed by Curran Associates, Inc. (2016)

For permission requests, please contact AIChE
at the address below.

AIChE
120 Wall Street, FL 23
New York, NY 10005-4020

Phone: (800) 242-4363
Fax: (203) 775-5177

www.aiche.org

Additional copies of this publication are available from:

Curran Associates, Inc.
57 Morehouse Lane
Red Hook, NY 12571 USA
Phone: 845-758-0400
Fax: 845-758-2633
Email: curran@proceedings.com
Web: www.proceedings.com

TABLE OF CONTENTS

ACADEMIC KEYNOTE

ATTRITION & BREAKAGE FUNDAMENTALS

APPLICATION OF THE FUNDAMENTALS

BREAKAGE MODELING I

BREAKAGE MODELING II

TESTING & MEASUREMENTS OF PARTICLE ATTRITION I

POSTER SESSION & RECEPTION

INDUSTRIAL KEYNOTE

TESTING & MEASUREMENTS OF PARTICLE ATTRITION II

BREAKUP OF AGGLOMERATES

CASE STUDIES

Multiscale Strategy to Describe Breakage and Attrition Behavior of Agglomerates

Heinrich, S., University of Magdeburg

Originally presented on: 4/12/2016 8:15:00 - 9:00:00

During transportation, processing, storage and handling particles are mechanically stressed due to particle-particle and particle-wall contacts, which can lead to unwanted breakage or attrition phenomena and formation of dust that may reduce the product quality. In this contribution a coupling between different time and length scales within a multiscale strategy for the prediction of agglomerates deformation, breakage and attrition behaviour is presented [1]. This approach is based on modern experimental and numerical techniques by taking into account various mechanical loading conditions, such as compression, tension, shearing, bending and impact. By using different experimental setups, e.g. single particle compression tester, pneumatic impact gun, free fall apparatus, shear and attrition tester, the basic mechanical properties of single particles and particle collectives under static and dynamic as well as for dry and wet conditions are obtained. These are namely the modulus of elasticity (Young's modulus), stiffness, yield point, hardness, strength, cohesion, friction, restitution coefficient [2], breakage energy, breakage function and flowability. Based on the experimental data contact models to describe the mechanical interaction of solids particles sized from few micrometers to the millimeter range are developed. Moreover, the breakage criteria of the particles and agglomerates can be described depending on the microstructure (porosity, packing structure and adhesion mechanisms of primary particles) and material parameters (elastic, elastic-plastic or plastic behavior, viscoelasticity and glass transition temperature). The numerical calculations on the microscale, the lowest hierarchy scale, are performed with the in-house developed simulation framework MUSEN, which is based on the Discrete Element Method (DEM). The DEM has been extended with Bonded Particle Model (BPM) [3], where agglomerates have been represented as a set of primary particles connected with solid bonds (Fig. 1). After model validation and estimation of material parameters, the analysis of the agglomerates breakage behavior under different impact conditions is performed. On the mesoscopic scale the dynamics of particles in the apparatus is calculated with coupled DEM+CFD [4] approaches (Fig. 1). The coupling between meso and micro scales allows approximating parameters for the Population Balance Model (PBM) applied to simulate processes on the macro scale. Finally, we will show the applicability of our model strategy for fluidized bed systems with special focus on fluidized bed spray granulation and coating.

NETL

Driving Innovation ♦ Delivering Results

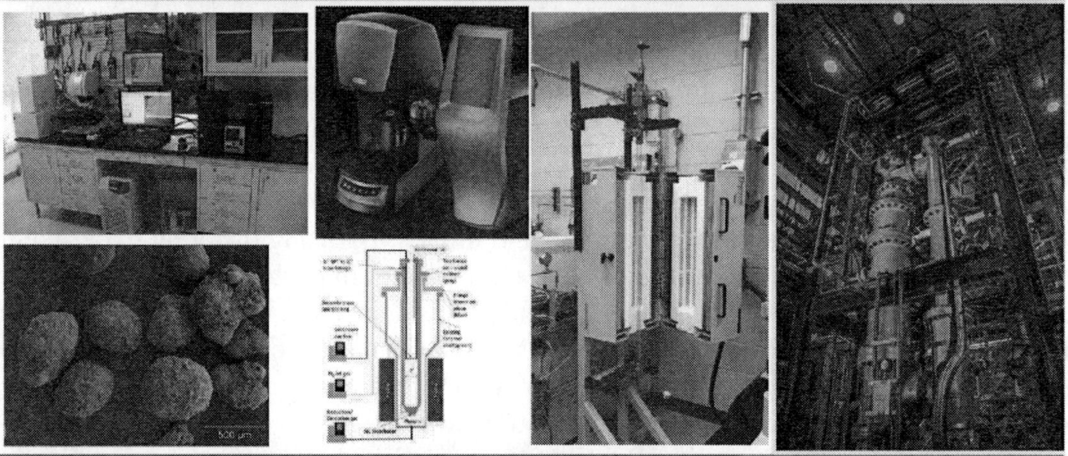

Development of First Principles Attrition Model to Determine Solid Makeup Cost of Novel CFB Technologies at NETL

Ronald W. Breault, Samuel Bayham and Esmail R. Monazam

April 12, 2016

U.S. DEPARTMENT OF ENERGY | National Energy Technology Laboratory

Motivation: Chemical Looping Combustion

NETL

- Chemical Looping Combustion is an advanced oxycombustion technology that utilizes a metal oxide to provide oxygen to the fuel.
- Metal oxide is reduced in fuel reactor and is oxidized in a separate air reactor. The CO_2 is kept separate from the air in the air reactor.
- Energy is released upon combustion of the reduced metal oxide
- **Favorable economics** of scaled-up process highly dependent on **cost** and **makeup rate** of metal oxide
- Makeup rate dependent on metal oxide **attrition** rate

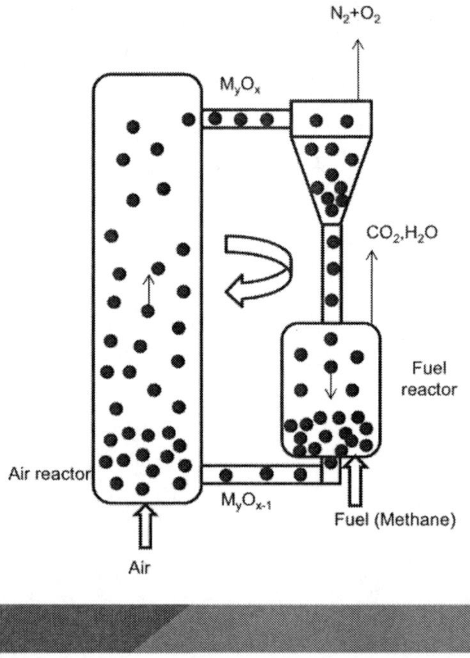

U.S. DEPARTMENT OF ENERGY | National Energy Technology Laboratory 2

Chemical Looping versus Fluidized Catalytic Cracking

Changes to FCC particle

No structural changes as material circulates through process, coking of surface

Changes to CLC particle

Significant structural changes occur as material circulates through the process, taking form of worst structure for each processing step

Goal and approach of attrition research at NETL

1. Development of **simple** (population-balance type) model to predict attrition losses based on fundamental erosion and abrasion mechanisms
   - Combine with technoeconomic modeling
2. Determine functionality of material parameters relevant to attrition as the carrier progresses through various oxidation and reduction states while thermally cycling.
3. Construct a high temperature attrition unit to validate spreadsheet model
   - Solids sampling system with bank of filters that can be switched in and out of service to capture fines as function of time
4. Utilize attrition data from NETL's 50 kW_{th} CLR unit as applicable

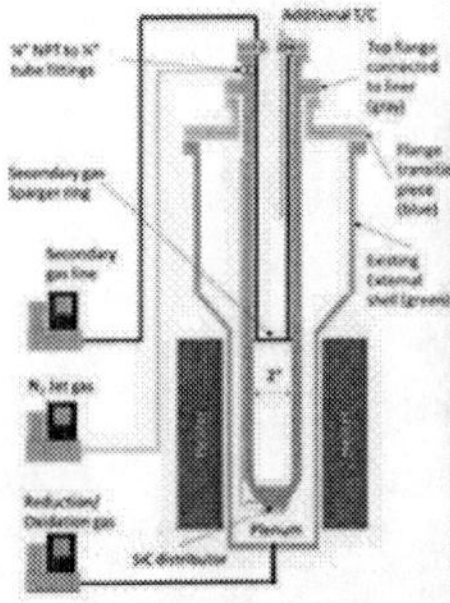

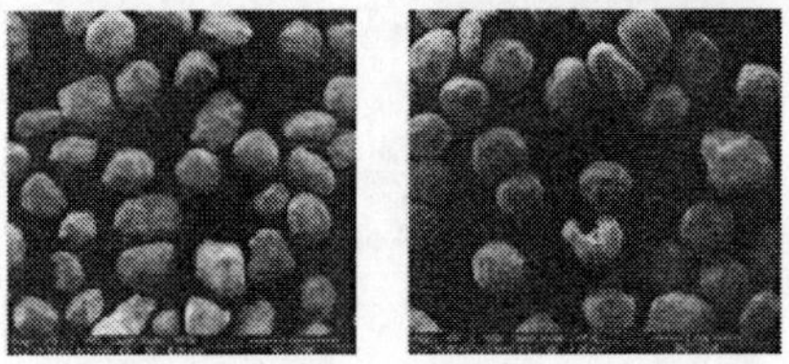

T.A. Brown et al. / Chemical Engineering Science 71 (2012) 449–467

Attrition of particulate solids

- **Attrition is the unintentional breakdown of solids due to stresses applied**
  - As opposed to comminution, where breakdown is intentional (e.g., coal pulverizer)
- **Three types of stress on the solid:**
  - Thermal:
    - Uneven expansion or contraction of the particle as it is heated or cooled, causing decrepitation
  - Chemical
    - internal stresses that change the lattice structure, weakening particles and produces surface features that can be easily abraded upon application of small external stresses
  - Mechanical
    - Abrasion (low velocity rubbing)
    - Fragmentation/chipping (high velocity impact)

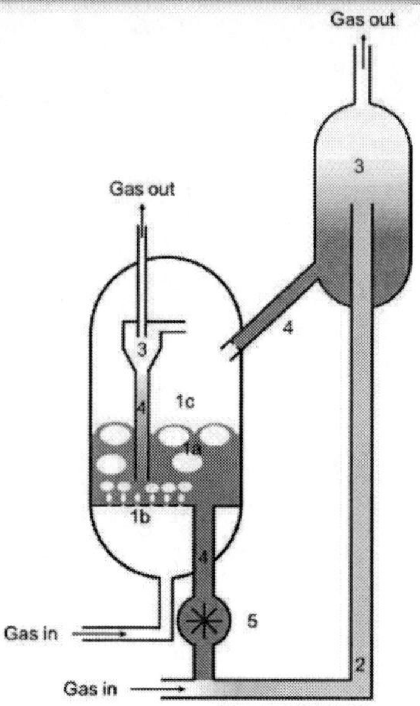

Mechanical attrition in CFB systems

- The rate of mechanical attrition is based on the velocity range and mode of contact.
- Abrasion is **low velocity** rubbing of particles against surfaces, producing very small fines
- Erosion consist of fragmentation and chipping as a result of **high velocity** impact
  - Does not occur below a certain threshold velocity

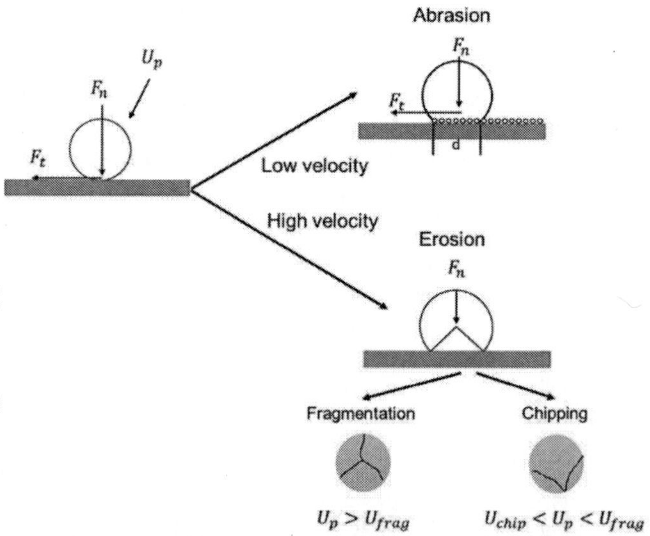

Traditional mechanical attrition models

Steady State Models

Jet attrition (Werther & Xi, 1993): $\dot{m}_{attr,jet} = C_{jet} n_{or} d_{or}^2 U_{or}^3$

Bed/bubble attrition (Ray et al., 1987): $\dot{m}_{attr,bed} = C_{bed} m_{bed} (U_g - U_{mf})$

Cyclone attrition (Reppenhagen & Werther, 2000): $\dot{m}_{attr,cyc} = \frac{C_{cyc} \dot{m}_{solids} U_{in}^2}{\sqrt{\mu_c}}$

Equations based on energy transfer method

These models capture the functionality of system parameters on attrition, but the attrition constant is difficult to know without running experiments at the required scale.

$$C_{\text{jet}} = \frac{\pi}{8} \frac{\eta_{j,abr}}{\sigma_p S_{mi}}$$

σ_p = Surface free energy (**measurable**)
S_{mi} = Specific surface energy of fines (**measurable**)
$\eta_{j,abr}$ = Energy transfer efficiency from jet to particle surface creation (**difficult to predict!**)

U.S. DEPARTMENT OF ENERGY | National Energy Technology Laboratory 7

Alternative approach to attrition modeling

Develop attrition models based on *fundamental material properties* as well as basic abrasion and erosion models.

Particulate wear: Archard's equation

$$\text{Volume worn} = \frac{kPL}{H}$$

Fragmentation: Ghadiri equation

$$\begin{pmatrix}\text{Volume fragmented} \\ \text{from mother particle}\end{pmatrix} = \alpha \frac{(\rho_P U_p^2 d_p H)}{K_c^2}$$

Given knowledge of the **forces** and **sliding distances** on the particles in the CFB system, the extent of attrition can be deduced if the following material properties are known.

Parameter	Defined as	How obtained
H	Material hardness	Material property test
K_c	Crack toughness	Material property test ($K_c = \phi\sigma\sqrt{a}$)
α and k	Fragmentation and wear constants	Attrition experiments with controlled conditions

NEEDS: Particle Properties - α, H, K_c, k

U.S. DEPARTMENT OF ENERGY | National Energy Technology Laboratory 8

Parameter determination: Hardness

- **Hardness**: Defined as the resistance to localized plastic deformation
- Determined by applying a known load to a shaped indenter and measuring the projected area A_{proj}
- Hardness has units of *pressure* (Pa)

$$H = \frac{P_{max}}{A_{proj}} [=] \frac{N}{m^2}$$

- Examples of hardness testing for non-particulate samples:
  - Brinell, Rockwell, Vickers, Knoop
- Hardness is usually greater than the material yield stress (e.g., $H \approx 3\sigma_y$) due to the surrounding material constraining motion of the indented material

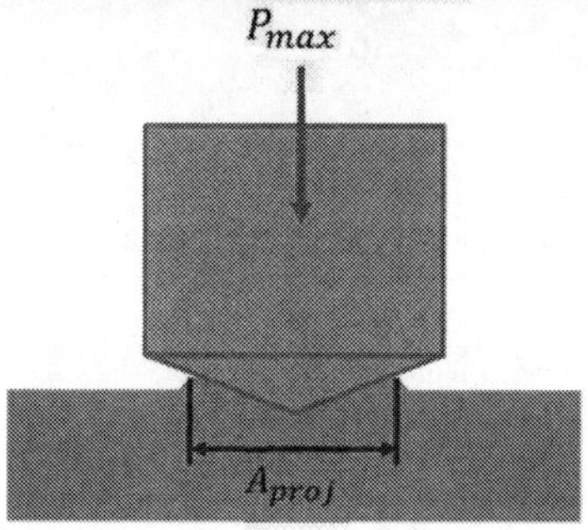

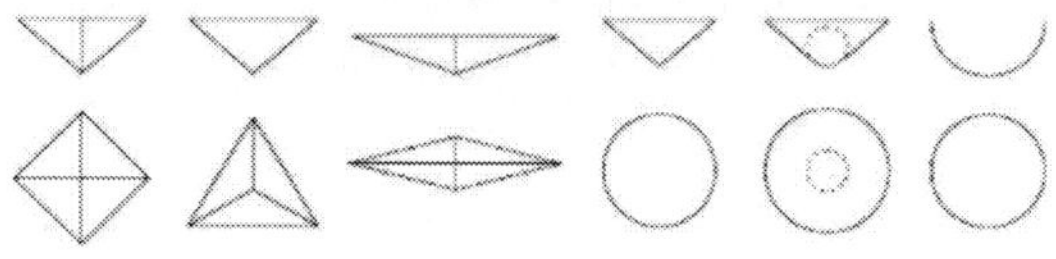

Ortiz, C. "Theoretical aspects of nanoindentation", Lecture notes, 2007. http://ocw.mit.edu/courses/materials-science-and-engineering/3-052-nanomechanics-of-materials-and-biomaterials-spring-2007/lecture-notes/lec22.pdf

Parameter determination: Fracture toughness

- **Fracture toughness**: Defined as the resistance of a material to brittle fracture
- Most materials consist of flaws such as voids, inclusions, and low angle grain boundaries. These act as "stress raisers", causing the local stress to increase beyond the yield stress.
- Studied by applying known stress to pre-cracked plate of known dimensions

$$K_{Ic} = \phi\sigma\sqrt{a}$$

- Where σ is applied stress, a is initial crack length, and ϕ is factor for certain geometries
- Materials that are more likely to undergo ductile fracture have greater fracture toughness

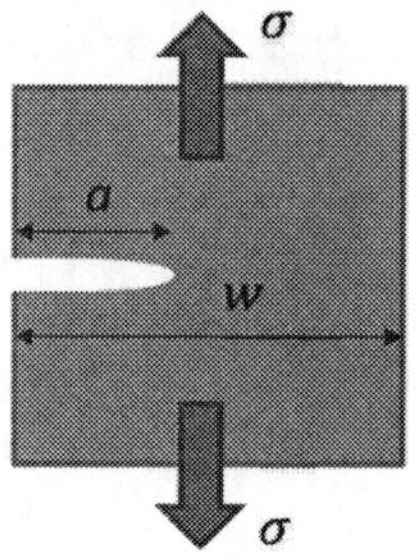

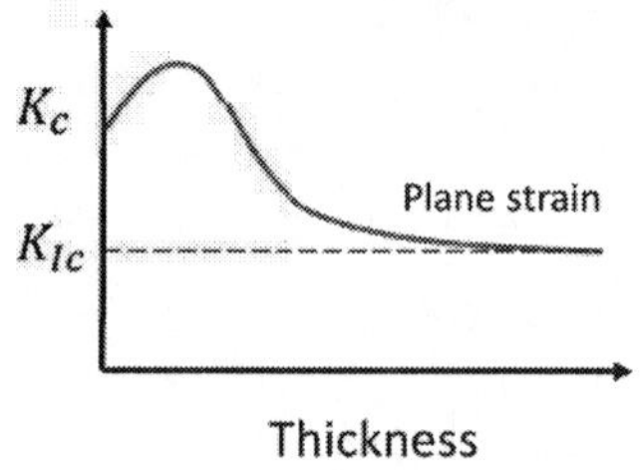

Parameter determination: Nanoindentation

- Hardness and fracture toughness are easy to obtain for non-particulate materials, ***but what about small particles?***
- Nanoindentation common technique to ascertain
  - Young's modulus (E)
  - hardness (H) and
  - fracture toughness (K_c)
- Ability to perform on small particles (100 microns) at high temperatures under oxidizing or reducing conditions

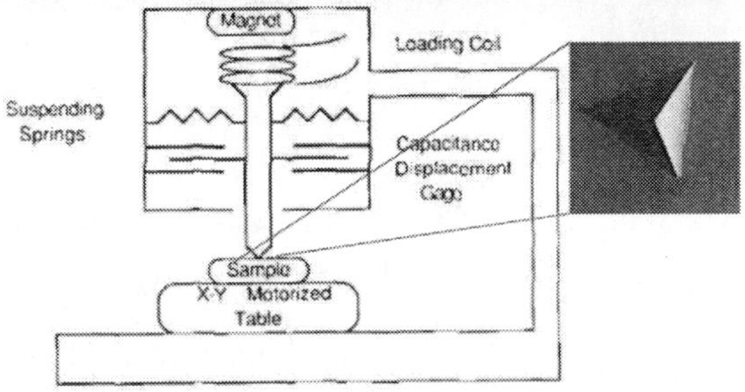

Doerner, M.F.; Nix, W. D. *J. Mater. Res.* 1 (1986), 601-609

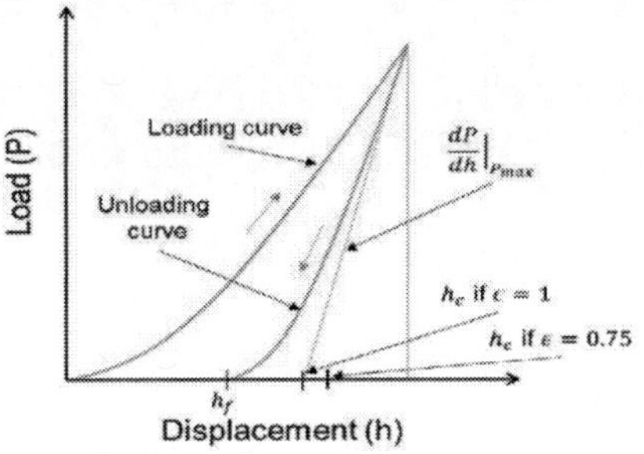

$$H = \frac{P_{max}}{A(h_c)}$$

P_{max} = Maximum load
h_c = actual depth of probe in material
$A(h_c)$ = calibrated function of depth

$$K_c = 0.016\sqrt{\frac{E}{H}}\frac{P_{max}}{c^{\frac{3}{2}}}$$

c = size of crack
E = Young's modulus
H = Hardness

Nanoindentation of hematite oxygen carrier

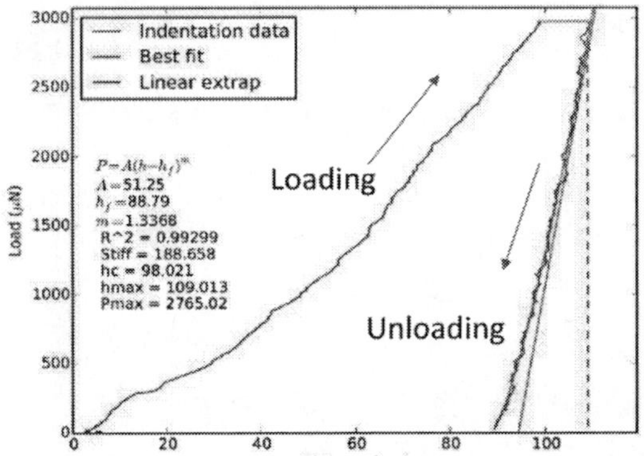

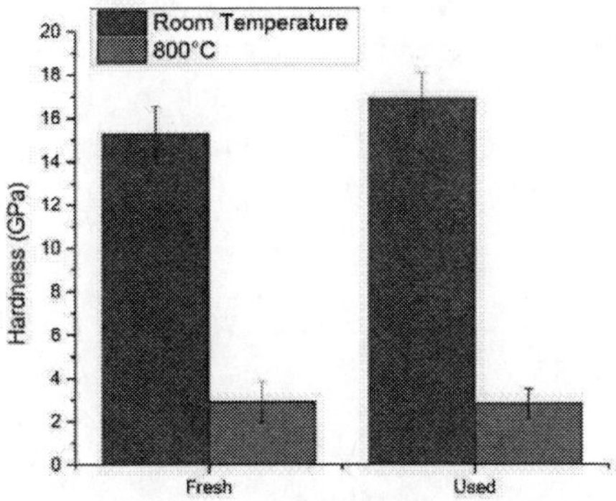

- Hematite tested at room temperature and at 800°C
- Hardness significantly greater at room temperature than at 800°C
- Data courtesy of Hysitron Inc., Minneapolis, MN.

Modeling approach at NETL

For a specific unit operation, mechanical attrition model can be developed using this procedure:

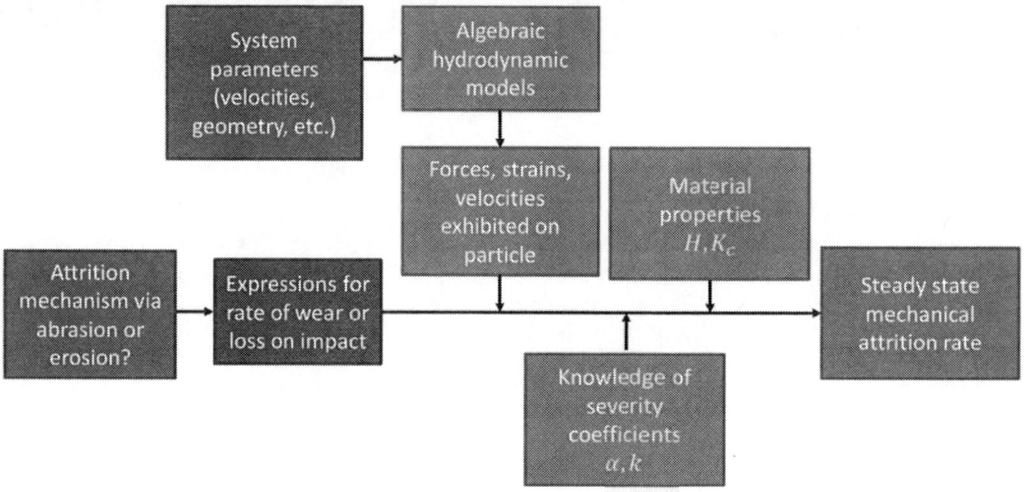

Example 1: Simple standpipe model

- Starting with Archard's wear equation, and applying it to Janssen's standpipe model, the following equation can be derived for the wear rate

$$\dot{m}_{attr} = \frac{k}{H}\left(\frac{\rho_s^2(1-\epsilon_g)gD_s}{4f_w}(\pi D_s L)\right)U_s$$

$D_s =$ Standpipe diameter
$L =$ Standpipe length
$U_s =$ Solids velocity
$g =$ Acceleration of gravity

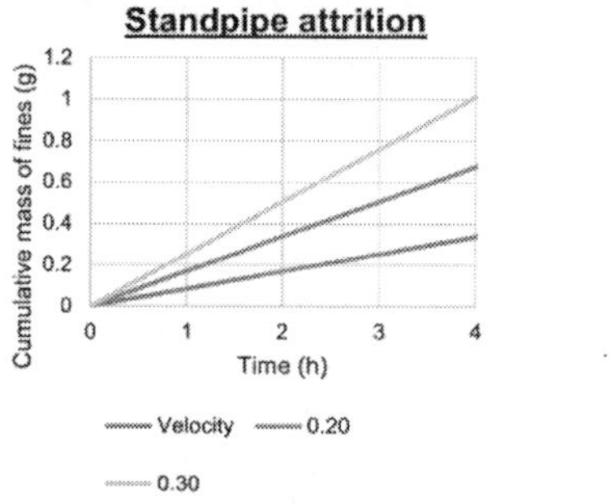

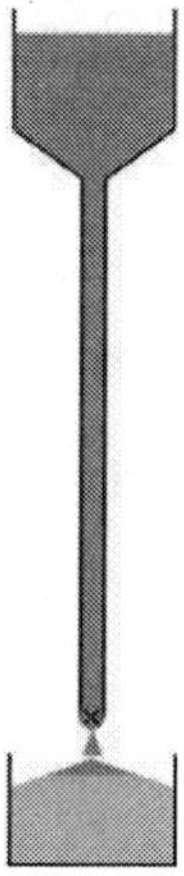

Parameter	Measurement Technique
Wear constant (k)	Standpipe experiment
Hardness (H)	Nanoindentation (Hysitron)
Particle-wall friction coefficient (f_w)	Annular shear cell, Jenike shear box, angled plate, or literature
Particle Density (ρ_s)	He Pycnometer
Gas holdup (ϵ_g)	Back calculate from bulk density of powder

Example 2: Cyclone

- Starting with Archard's wear equation, and applying it to the path that a particle might traverse in a cyclone, the wear equation can be derived as

$$\boxed{\dot{m}_{attr} = K_{cyc}\dot{m}_{c,in}u_{c,in}^2}$$

"Average" particle path

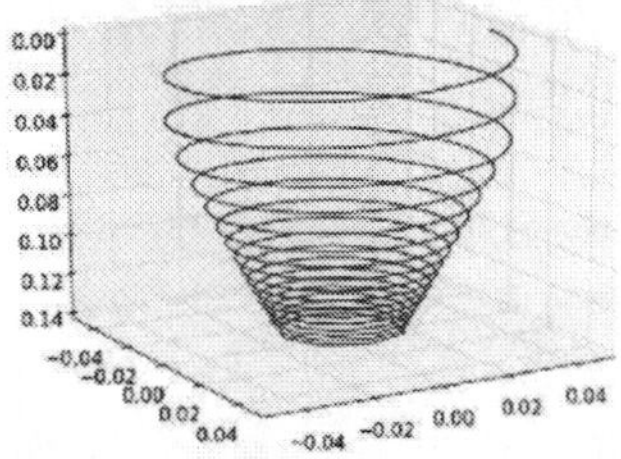

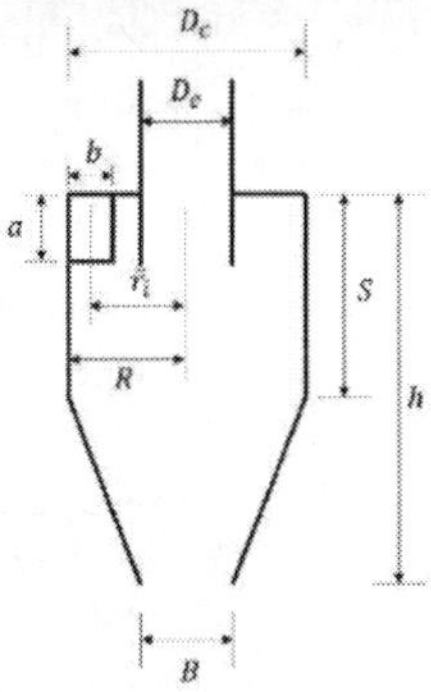

$$K_{cyc} = \begin{cases} 2\frac{k\rho_s}{H}\cdot\left(\frac{\bar{r_i}^2}{D_c^3}\frac{\pi(D_c^2 - D_e^2)S}{A}\sqrt{4\left(\frac{\bar{r_i}}{D_c}\right)^2 + \left(\frac{4A}{\pi(D_c^2 - D_e^2)}\right)^2}\right) & \text{Cylindrical cyclone only} \\ 2\frac{k\rho_s}{H}\cdot f(\bar{r_i}, D_c, D_e, B, h, A, S) & \text{Cyclone with conical barrel} \end{cases}$$

Compare to equation derived by Reppenhagen and Werther [1]

$$\dot{m}_{attr} = C\dot{m}_{c,in}\mu^n u_{c,in}^2 \qquad \mu = \frac{\dot{m}_{c,in}}{\rho_f u_{c,in} A} \qquad C = \text{material and efficiency constant}$$

U.S. DEPARTMENT OF ENERGY | National Energy Technology Laboratory

1. Reppenhagen, J.; Werther, J. *Powder Technol.* **113** (2000) 55-69.

15

Design of high temperature jet cup

- Construct a high temperature attrition unit.
- Plenum gas to reduce or oxidize sample material (no jet gas)
- Jet gas (N_2) to attrit particle in "off cycle"
- Secondary gas (N_2) to ensure consistent fines entrainment due to different experimental conditions and temperature decrease in top of liner
  - Designed to entrain particle of size <40 um

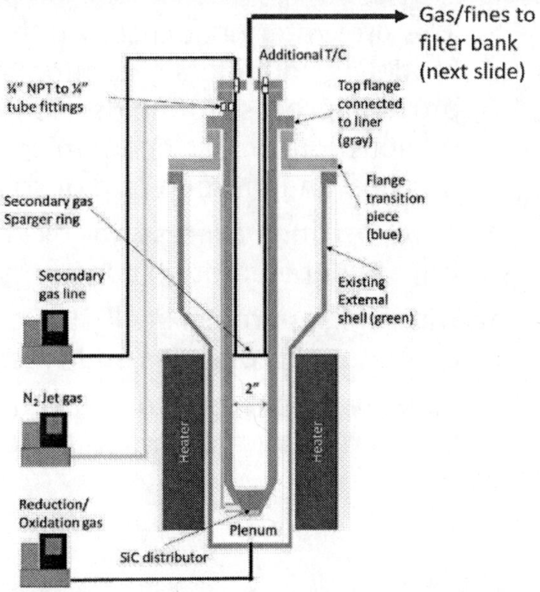

U.S. DEPARTMENT OF ENERGY | National Energy Technology Laboratory

16

Design of high temperature jet cup

- Build gas outlet to accommodate two sets of five filters separated by four-way valves
  - Reactor flow diverted between one bank or the other
  - Pre-purging of air from filter needed before introduction of reducing gas!
- Solids sampling system with bank of filters that can be switched in and out of service to capture fines as function of time
- Operator changes hand valves for filters that do not have reactor gas outlet flow

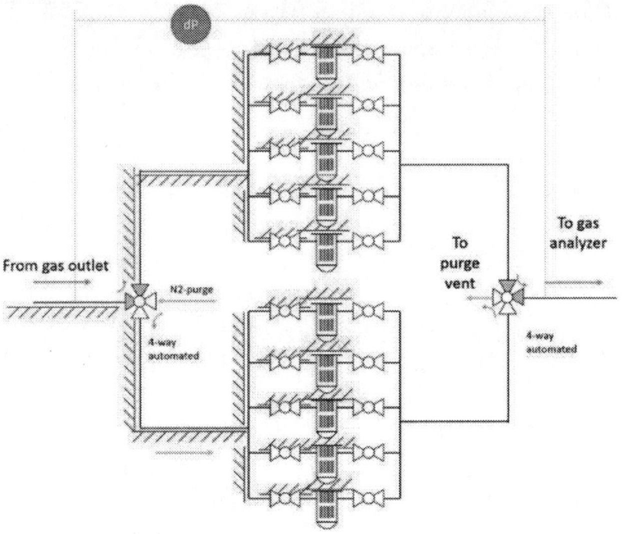

U.S. DEPARTMENT OF ENERGY | National Energy Technology Laboratory

Concluding remarks

- This presentation focuses on the need for mathematical modeling and offers recommendations on how to model the attrition process for fields such as chemical looping combustion, whose economics are highly dependent on attrition and makeup rate of solid oxygen carrier.
- Many attrition expressions exist for different areas of the CFB but are dependent on constants that are unknown.
- Modeling approach at NETL: Develop modeling scheme based on material properties and wear/fragmentation rate laws
- Validate model using data from high-temperature attrition unit and 50 kW_{th} Chemical Looping Combustion unit at NETL

U.S. DEPARTMENT OF ENERGY | National Energy Technology Laboratory

Exploring the Interplay Between Attrition and Separation

Ben J. Freireich

The Dow Chemical Company

Core R&D – Solids Processing

Outline

- Problem Definition
- Model Description
- Model Variations
- Model Results
- Conclusions

2

Problem Definition

- Separation
- Attrition
  - Fragmentation
  - Abrasion
- Inlet velocity…
  - Better separation
  - More attrition

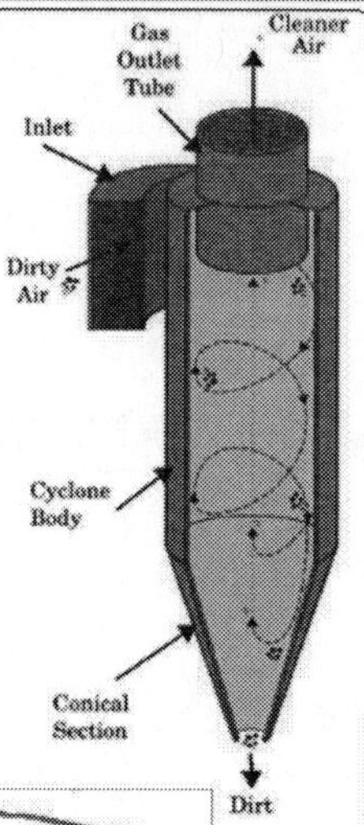

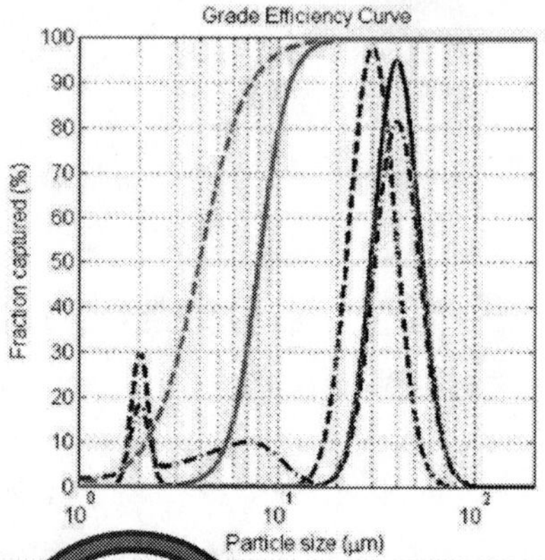

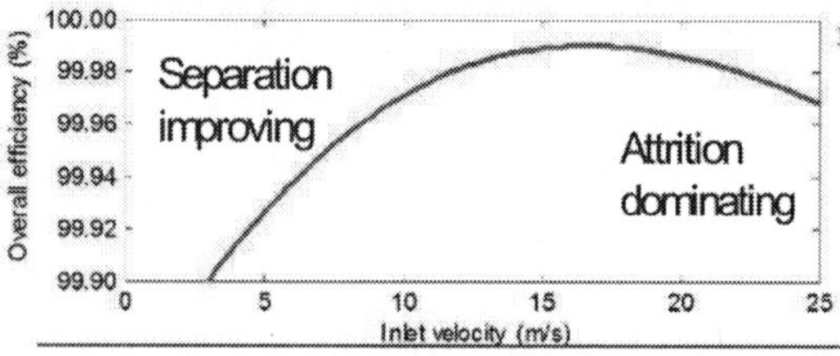

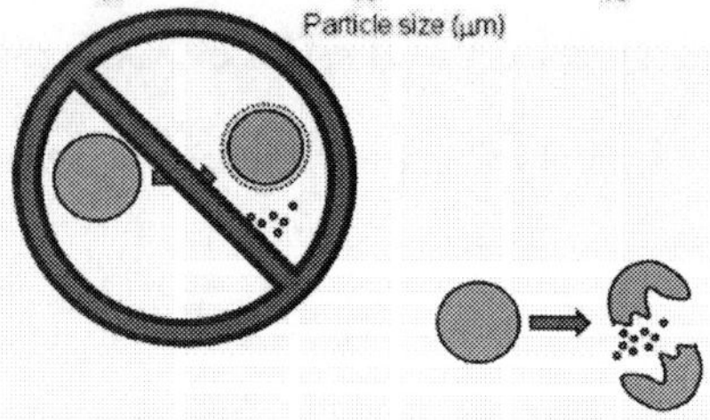

http://en.wikipedia.org/wiki/File:Cyclone_separator.svg

Model Description

- Single close coupled cyclone
  - Avoids classification due to entrainment or subsequent cyclones
  - Similar to single "effective" GEC
- Overall efficiency
  - $\eta = \int G y dx = \dfrac{\dot{M} - L}{\dot{M}}$
  - $y dx$ = mass fraction with size x to $x+dx$
  - $G(x)$ = probability of retaining size x on single pass
- Attrition (only in cyclone)
  - $s(x)$ = fraction of particles fragmented
  - $b(x|x')$ = PSD of daughters from parent x'

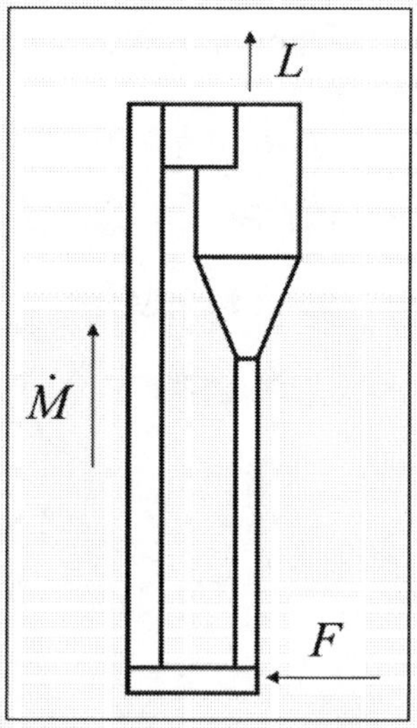

Model Description cont'd…

- Mass balance per size

Accumulation → $\frac{\partial m}{\partial t} = Fy_F - \dot{M}(1-G)y - s\dot{M}y + \dot{M}\int_{x'=x}^{x'\to\infty} b(x|x')s(x')y(x')dx$

(Annotations: Feed → Fy_F; Cyclone losses → $\dot{M}(1-G)y$; Mother fractures → $s\dot{M}y$; Daughters appear → $\dot{M}\int_{x'=x}^{x'\to\infty} b(x|x')s(x')y(x')dx$)

- Normalize steady state model
  - $(1-G+s)y = (1-\eta)y_F + \int_{x'=x}^{x'\to\infty} b(x|x')s(x')y(x')dx'$
- Overall balance
  - $\eta = \int_{x=0}^{x\to\infty} G(x)y(x)dx$

(Annotations: Overall efficiency → η; Grade efficiency → $G(x)$; Steady state PSD (solve model) → $y(x)$)

Sub-models

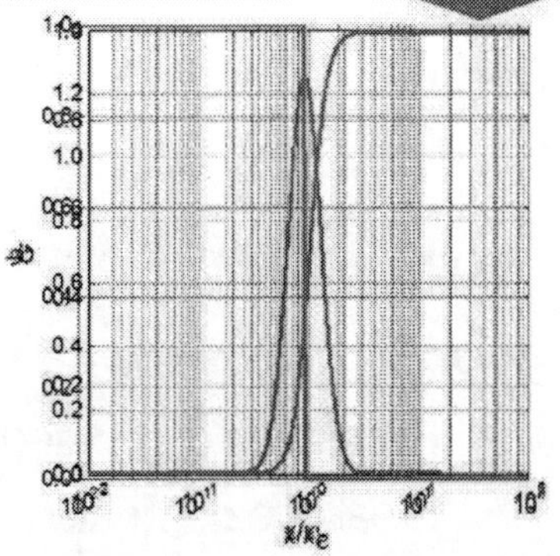

- Feed
  - Lognormal PSD, $d_{50} = x_F$, $\sigma_g = 1.4$
- Cyclone
  - Dirgo and Leicht, $G = \left[1 + \left(\frac{x_C}{x}\right)^{\beta}\right]^{-1}$
- Attrition
  - Fracture probability, $s = s_0\left(\frac{x}{x_F}\right)^{n_s}$
    - » Start with $n_s = 0$
  - Daughter PSD: Uniform $b(x|x') = n_b\left(\frac{x}{x'}\right)^{n_b-1}$
    - » Start with $n_b = 1$

Model Variations

$$(1-G+s)y=(1-\eta)y_F+\int_{x'=x}^{x'\to\infty}b(x|x')s(x')y(x')dx'$$

- What we want
  - Steady state PSD
  - Overall efficiency

$$\eta=\int_{x=0}^{x\to\infty}G(x)y(x)dx$$

- How we get it
  - Model 0: Just use the feed PSD ← $\eta_0=\int_{x=0}^{x\to\infty}Gy_F dx$
  - Model 1: Assume daughters vanish ← $(1-\eta_1)^{-1}=\int_{x=0}^{x\to\infty}(1-G+s)^{-1}y_F dx$
  - Model 2: Solve for steady state PSD
  - Model 3: Size dependent fragmentation ← $s=s_0\left(\frac{x}{x_F}\right)^{n_s}$
  - Model 4: Size biased fragments ← $b(x|x')=n_b\left(\frac{x}{x'}\right)^{n_b-1}$
- So what?

Model 0 - 2 results

- Model 0: Naïve
- Model 1: Vanishing daughters
  - Separation dominated
  - Breakage dominated
- Model 2: Finite daughters
  - Separation dominated
  - ~Breakage dominated
  - Generally lower losses than Model 0
    daughters can be captured

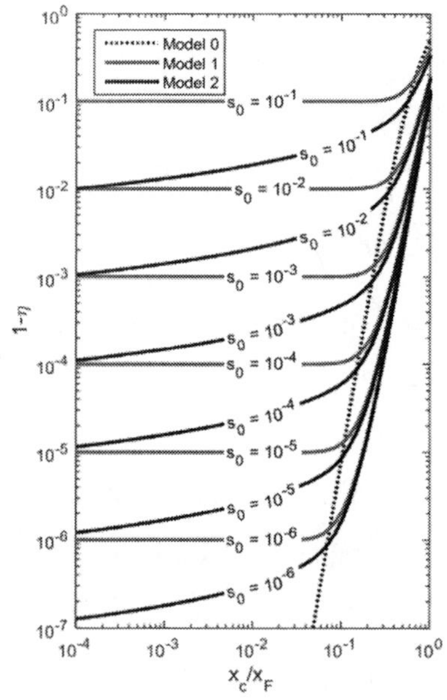

(Model 1 and 2 are surprisingly similar for $x_C > 0.1\ x_F$)

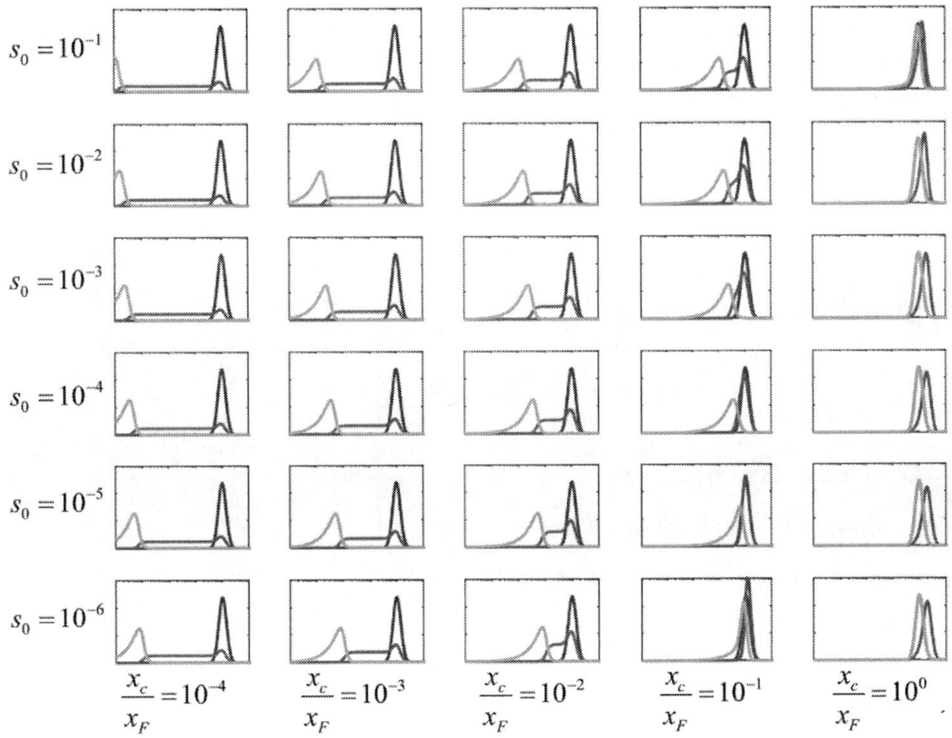

$s_0 = 10^{-1}$
$s_0 = 10^{-2}$
$s_0 = 10^{-3}$
$s_0 = 10^{-4}$
$s_0 = 10^{-5}$
$s_0 = 10^{-6}$
$\frac{x_c}{x_F} = 10^{-4}$
$\frac{x_c}{x_F} = 10^{-3}$
$\frac{x_c}{x_F} = 10^{-2}$
$\frac{x_c}{x_F} = 10^{-1}$
$\frac{x_c}{x_F} = 10^{0}$

9

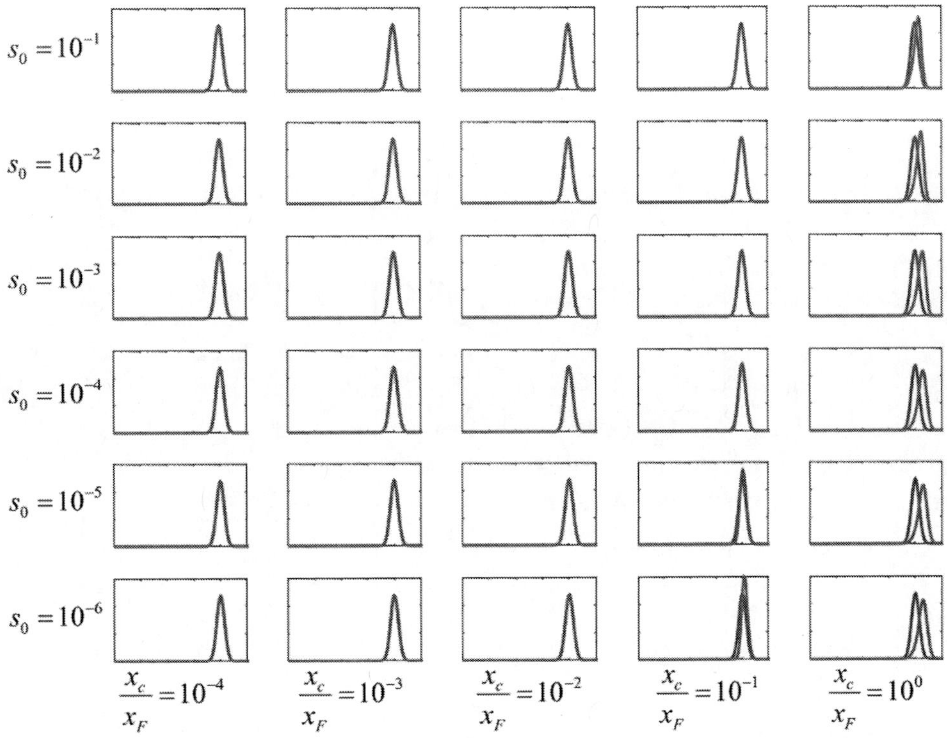

$s_0 = 10^{-1}$
$s_0 = 10^{-2}$
$s_0 = 10^{-3}$
$s_0 = 10^{-4}$
$s_0 = 10^{-5}$
$s_0 = 10^{-6}$
$\frac{x_c}{x_F} = 10^{-4}$
$\frac{x_c}{x_F} = 10^{-3}$
$\frac{x_c}{x_F} = 10^{-2}$
$\frac{x_c}{x_F} = 10^{-1}$
$\frac{x_c}{x_F} = 10^{0}$

10

Model 2 and 3 results

- Model 2: Size independent
  - $n_s = 0$
- Model 3: Size dependent
  - $n_s = 1$
  - $n_s = 2$

$$s(x) = s_0 \left(\frac{x}{x_F} \right)^{n_s}$$

- Higher size dependence
  - slower fine particle attrition
  - greater classification dependence
  - loss rate $\propto x_c^{n_s}$ for small cut sizes

(All still surprisingly similar for $x_C > 0.1\ x_F$)

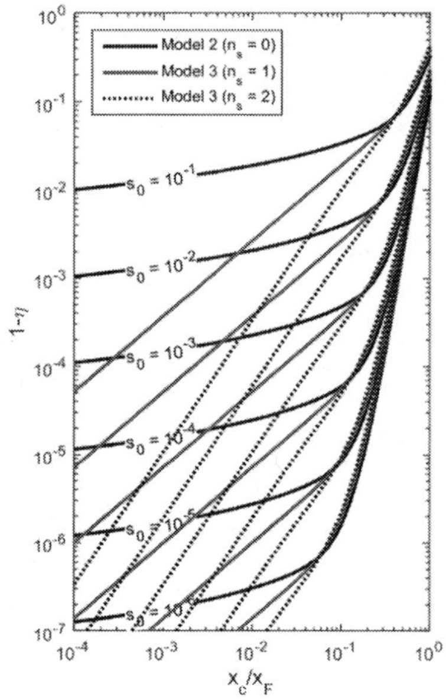

(Opposite trends for higher cut sizes)

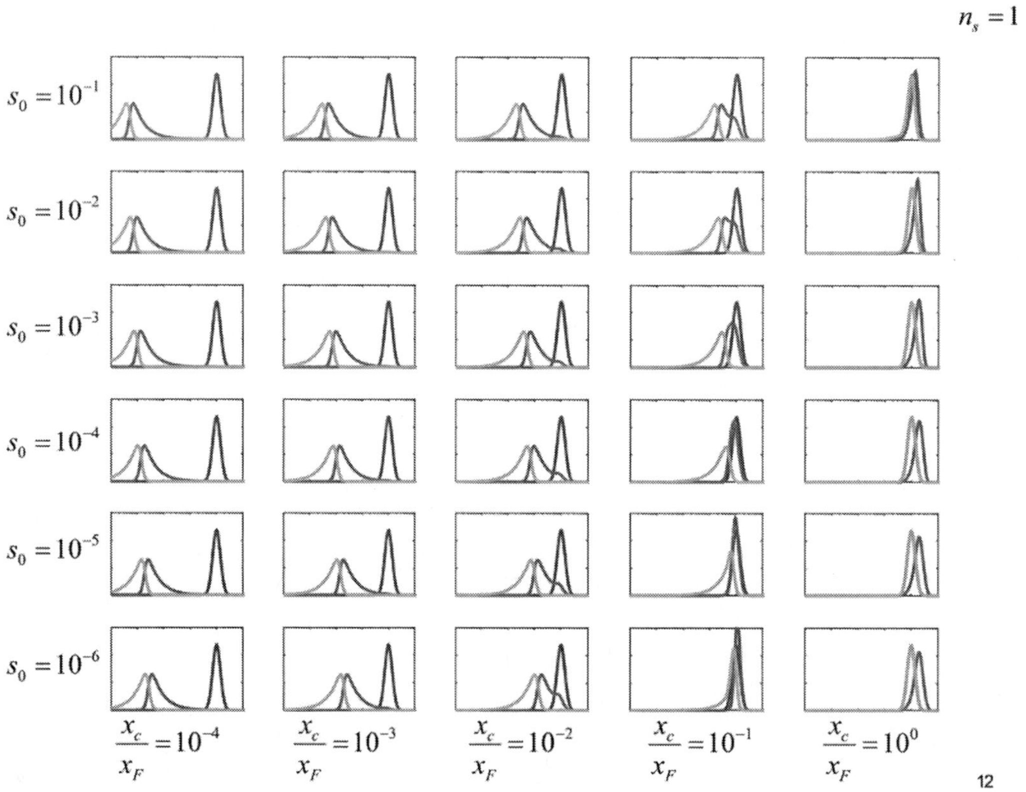

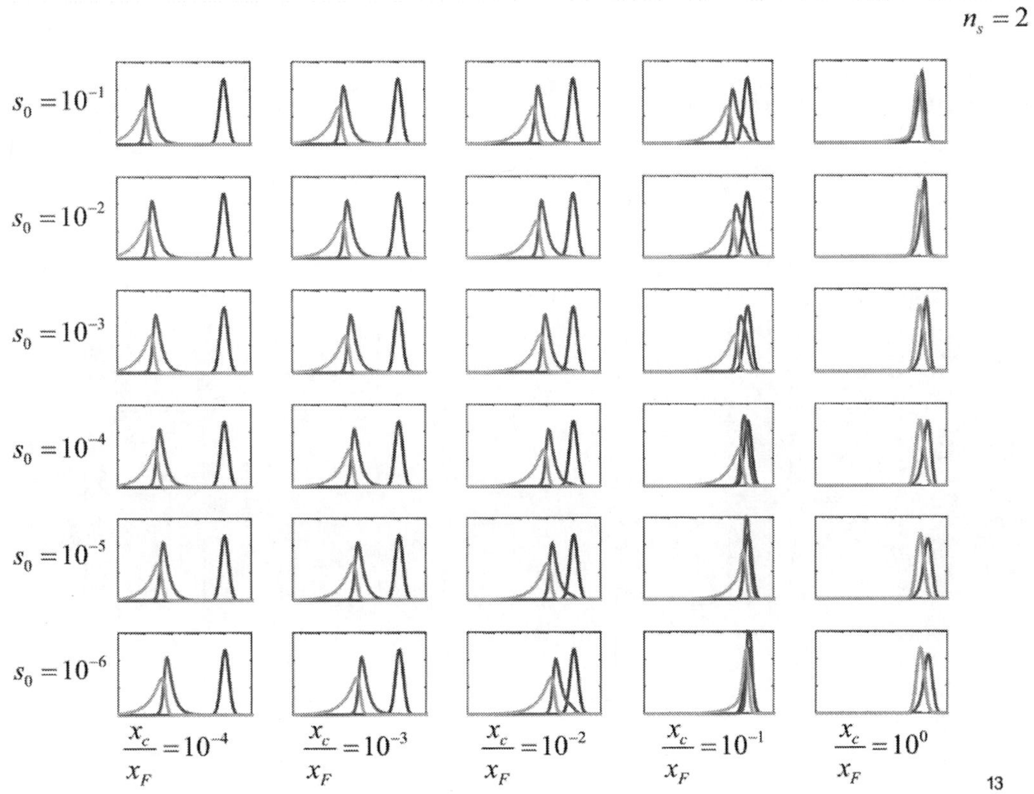

13

Model 2 and 4 results

$$b(x \mid x') = n_b \left(\frac{x}{x'} \right)^{n_b - 1}$$

- Model 2: Uniform
  - $n_b = 0$
- Model 4: Skewed
  - $n_b = 1/2$
  - $n_b = 2$
- High n_b, more fines, more loss
- Low n_b, less fines, less loss
- Loss rate ∝ n_b

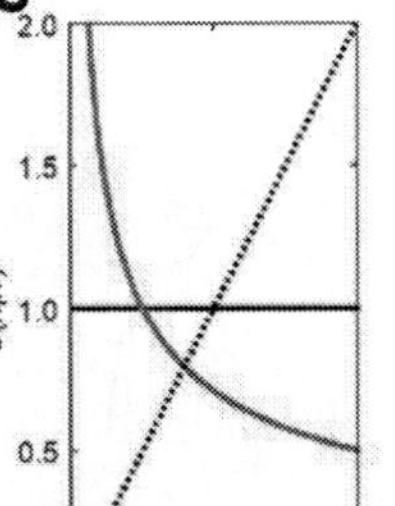

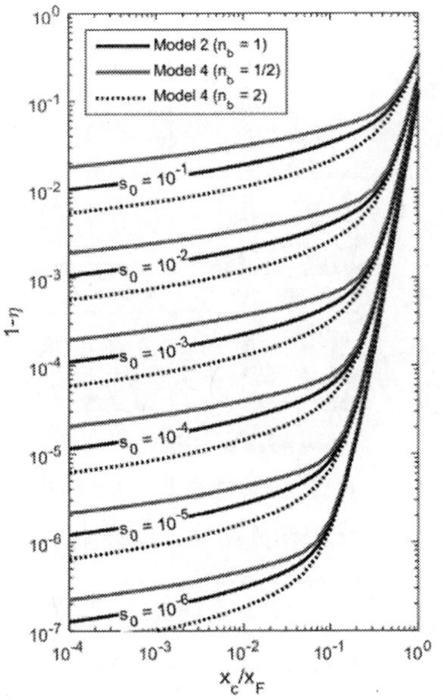

(All still surprisingly similar for $x_C > 0.1\ x_F$)

14

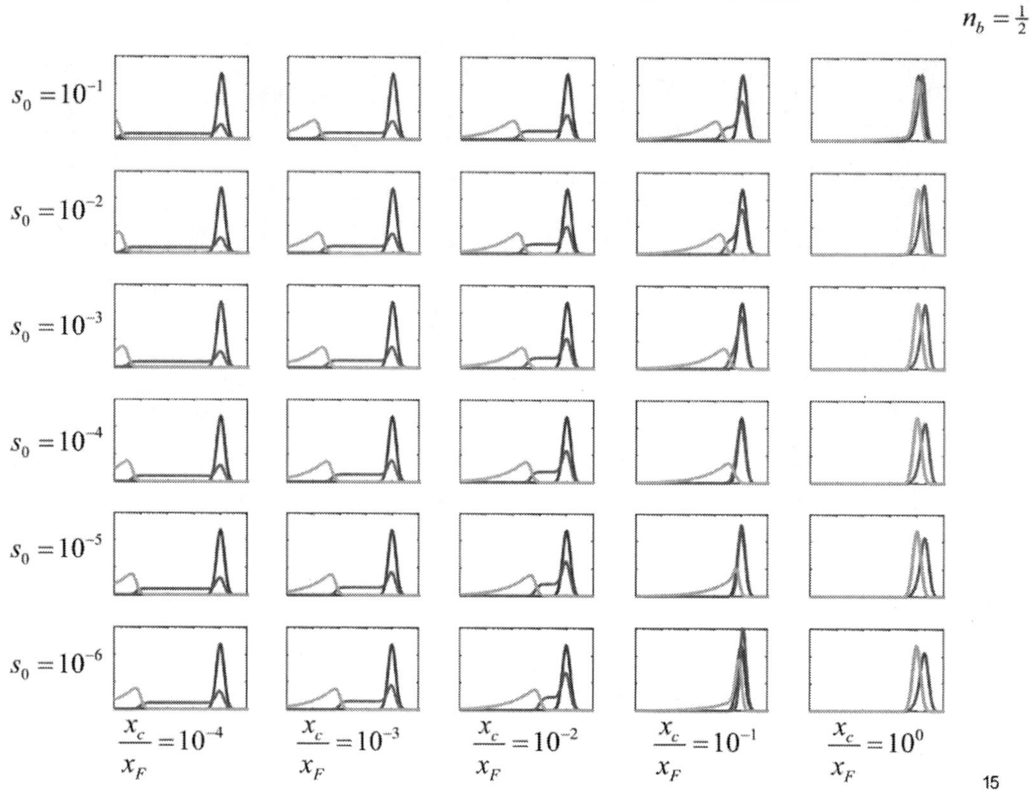
$n_b = \frac{1}{2}$
$s_0 = 10^{-1}$
$s_0 = 10^{-2}$
$s_0 = 10^{-3}$
$s_0 = 10^{-4}$
$s_0 = 10^{-5}$
$s_0 = 10^{-6}$
$\frac{x_c}{x_F} = 10^{-4}$
$\frac{x_c}{x_F} = 10^{-3}$
$\frac{x_c}{x_F} = 10^{-2}$
$\frac{x_c}{x_F} = 10^{-1}$
$\frac{x_c}{x_F} = 10^{0}$
15

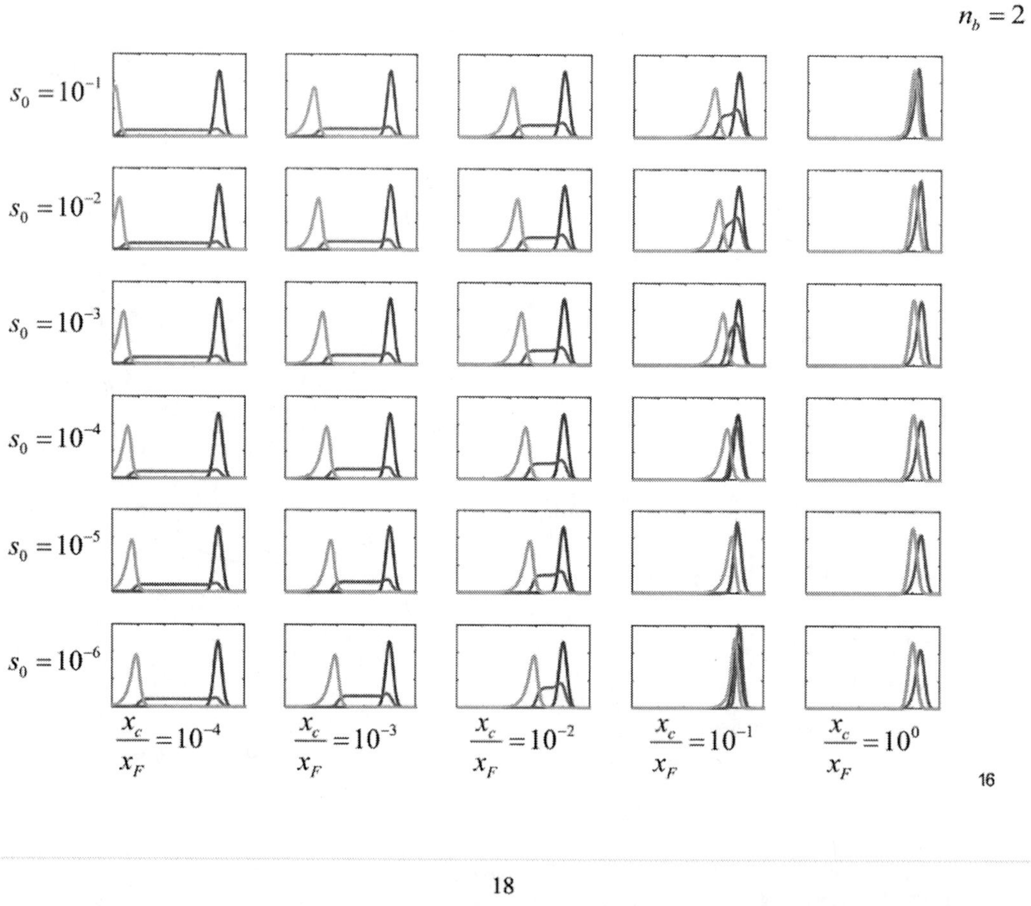
$n_b = 2$
$s_0 = 10^{-1}$
$s_0 = 10^{-2}$
$s_0 = 10^{-3}$
$s_0 = 10^{-4}$
$s_0 = 10^{-5}$
$s_0 = 10^{-6}$
$\frac{x_c}{x_F} = 10^{-4}$
$\frac{x_c}{x_F} = 10^{-3}$
$\frac{x_c}{x_F} = 10^{-2}$
$\frac{x_c}{x_F} = 10^{-1}$
$\frac{x_c}{x_F} = 10^{0}$
16

Conclusions

- Maps of fragmentation dominated attrition were generated as a function of attrition rate and cyclone cut size
- Influence of various model assumptions were investigated
  - Mode 0 (just separation, no attrition) much too optimisitic
  - Model 1 (vanishing fines)
    - » works reasonably well for high cut sizes
    - » over predicts losses (10x) for low cut sizes, PSD too coarse
  - Model 3 (size dependent fragmentation)
    - » loss rate $\propto x_c^{n_s}$ for small cut sizes
    - » PSD always finer than Model 2
  - Model 4 (size biased fragments)
    - » Loss rate $\propto n_b$
    - » PSD similar to Model 2, but loss PSD has shorter/longer tail

17

Back up slides

18

Model 0 - 2 results

- Model 0: Naïve
- Model 1: Vanishing daughters
  - Separation dominated
  - Breakage dominated
- Model 2: Finite daughters
  - Separation dominated
  - ~Breakage dominated
  - Generally lower losses than Model 0
    daughters can be captured

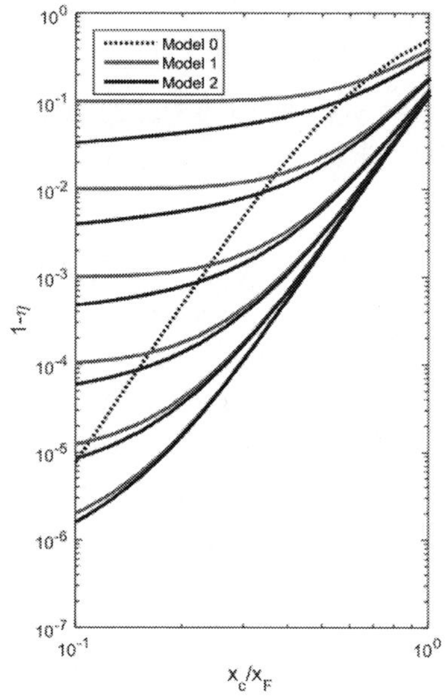

(Model 1 and 2 are surprisingly similar for $x_C > 0.1\ x_F$)

19

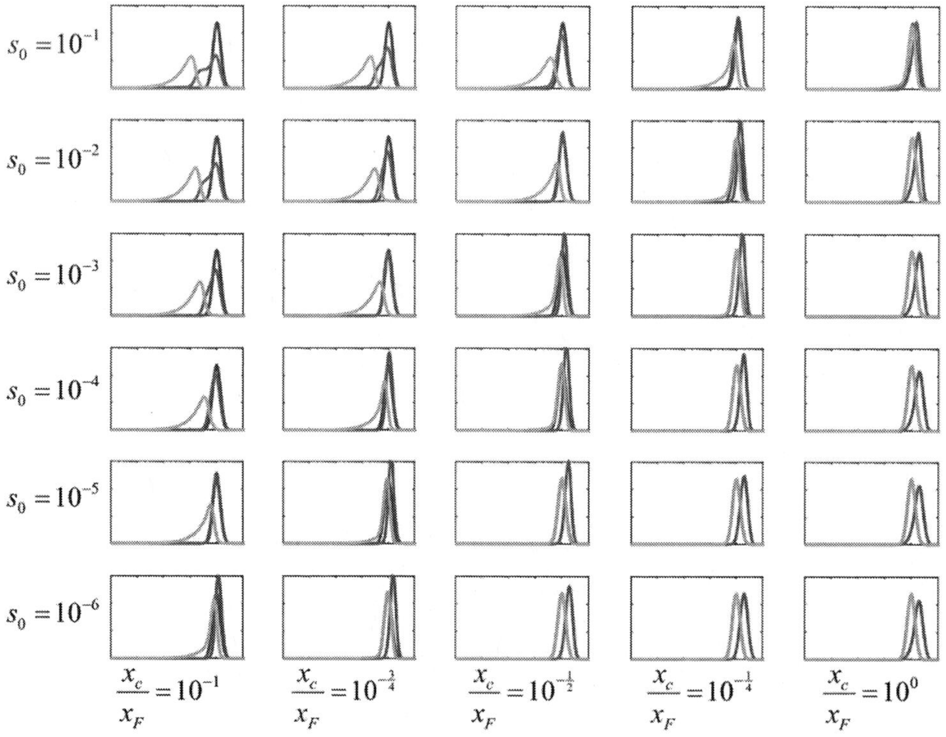

20

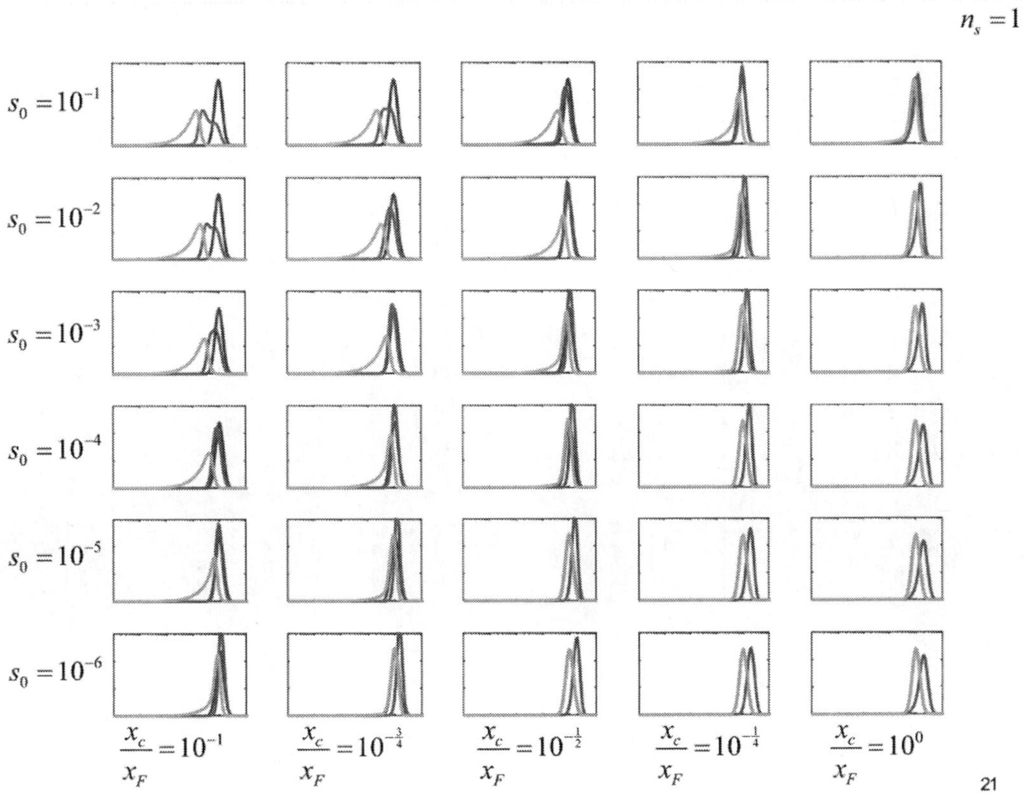

21

Model 2 and 3 results

- Model 2: Size independent
  - $n_s = 0$
- Model 3: Size dependent
  - $n_s = 1$
  - $n_s = 2$

$$s(x) = s_0 \left(\frac{x}{x_F} \right)^{n_s}$$

- Higher size dependence
  - slower fine particle attrition
  - greater classification dependence
  - loss rate $\propto x_c^{n_s}$ for small cut sizes

(All still surprisingly similar for $x_C > 0.1\ x_F$)

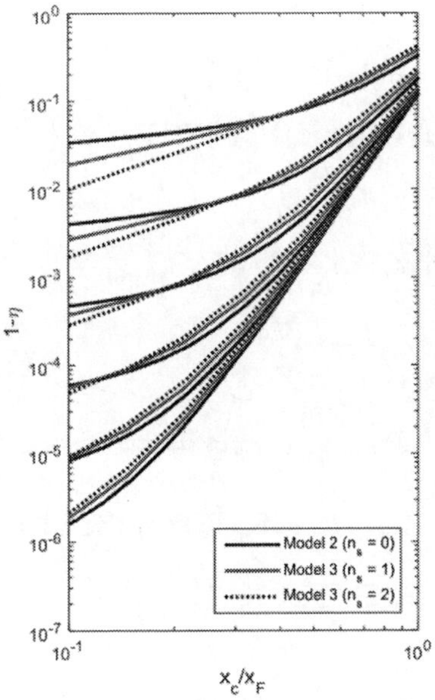

(Opposite trends for higher cut sizes)

22

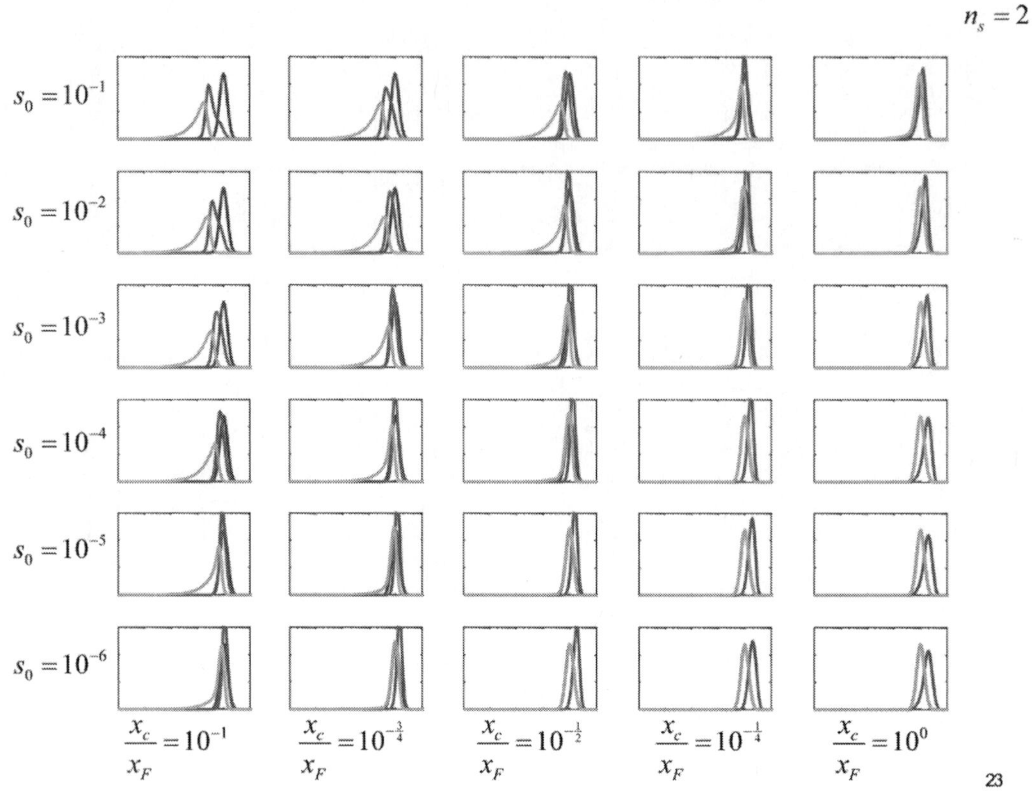

23

Model 2 and 4 results

$$b(x \mid x') = n_b \left(\frac{x}{x'} \right)^{n_b - 1}$$

- Model 2: Uniform
  - $n_b = 0$
- Model 4: Skewed
  - $n_b = 1/2$
  - $n_b = 2$
- High n_b, more fines, more loss
- Low n_b, less fines, less loss
- Loss rate $\propto n_b$

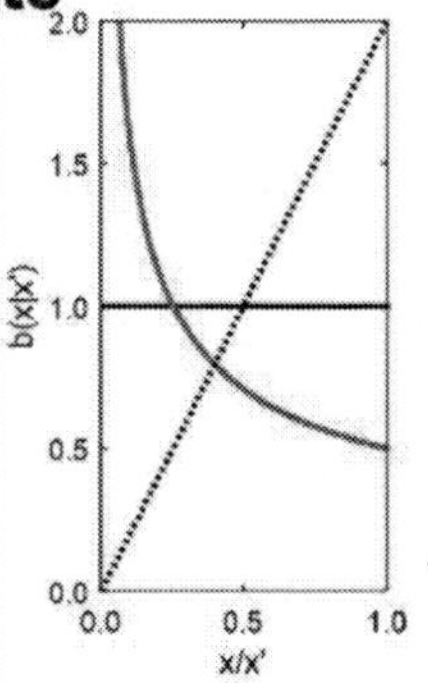

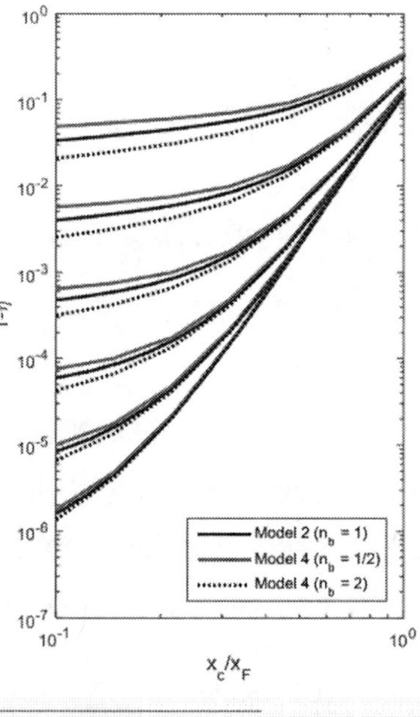

(All still surprisingly similar for $x_C > 0.1\ x_F$)

24

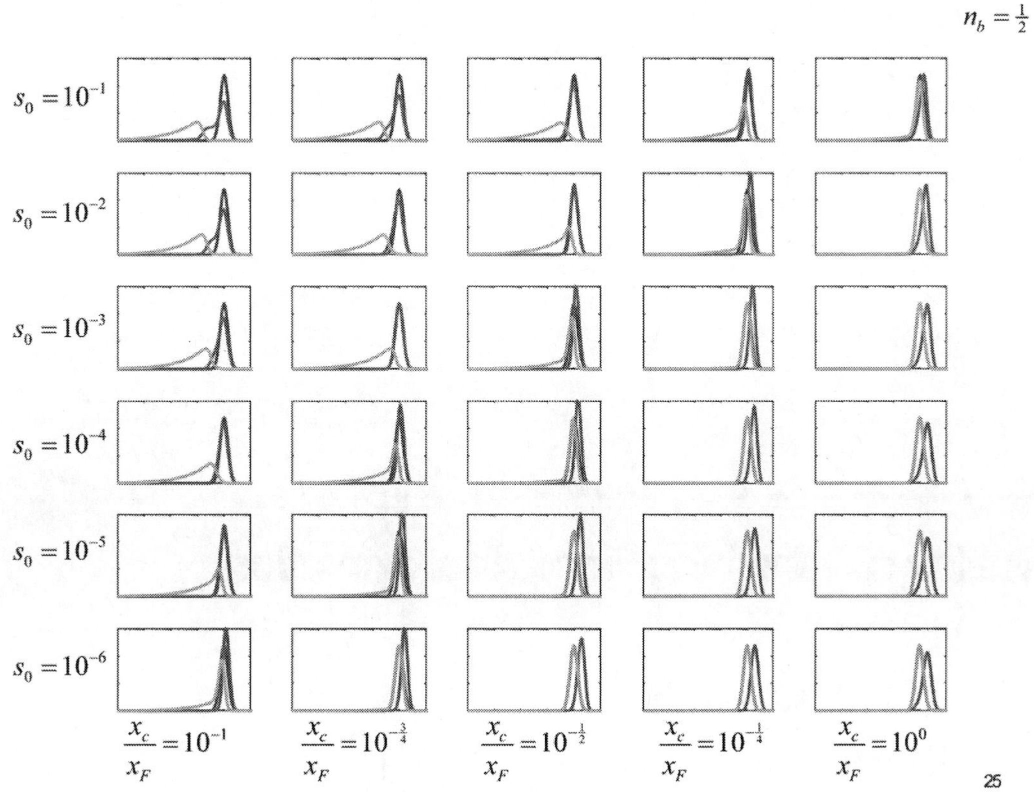
$n_b = \frac{1}{2}$
$s_0 = 10^{-1}$
$s_0 = 10^{-2}$
$s_0 = 10^{-3}$
$s_0 = 10^{-4}$
$s_0 = 10^{-5}$
$s_0 = 10^{-6}$
$\frac{x_c}{x_F} = 10^{-1}$
$\frac{x_c}{x_F} = 10^{-\frac{3}{4}}$
$\frac{x_c}{x_F} = 10^{-\frac{1}{2}}$
$\frac{x_c}{x_F} = 10^{-\frac{1}{4}}$
$\frac{x_c}{x_F} = 10^{0}$

25

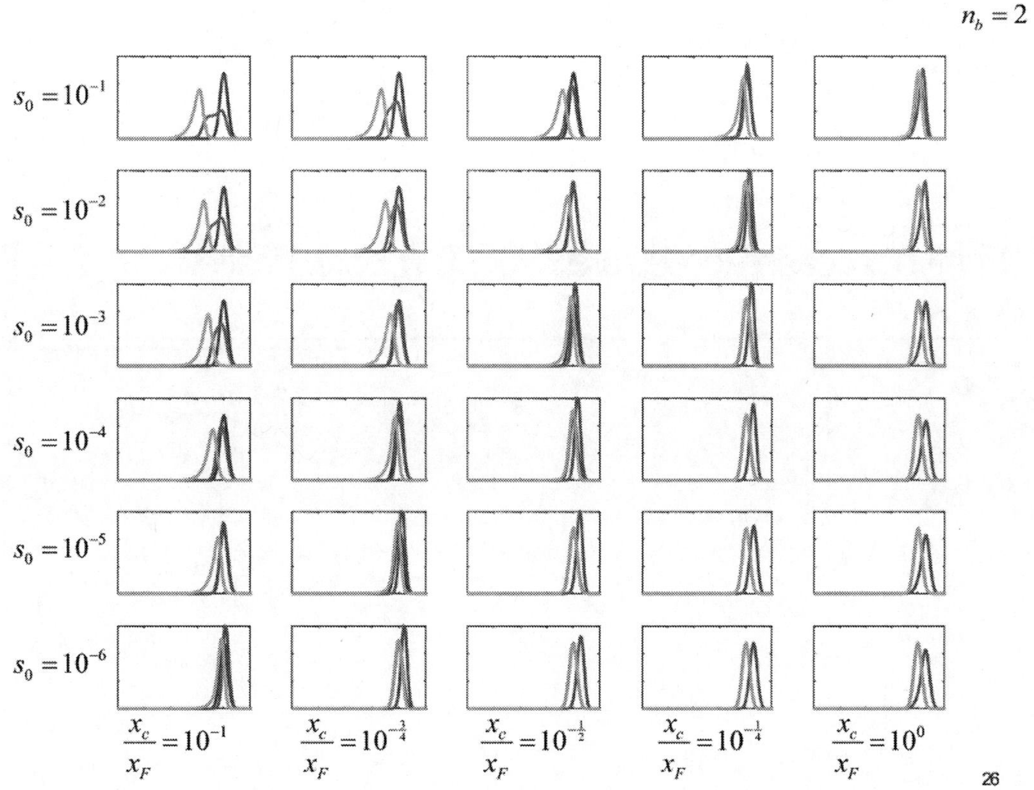
$n_b = 2$
$s_0 = 10^{-1}$
$s_0 = 10^{-2}$
$s_0 = 10^{-3}$
$s_0 = 10^{-4}$
$s_0 = 10^{-5}$
$s_0 = 10^{-6}$
$\frac{x_c}{x_F} = 10^{-1}$
$\frac{x_c}{x_F} = 10^{-\frac{3}{4}}$
$\frac{x_c}{x_F} = 10^{-\frac{1}{2}}$
$\frac{x_c}{x_F} = 10^{-\frac{1}{4}}$
$\frac{x_c}{x_F} = 10^{0}$

26

Quality of Vanishing Fines Approximation

- Model 1: Vanishing fines
  - Losses always higher
- Model 3: Uniform finite fines

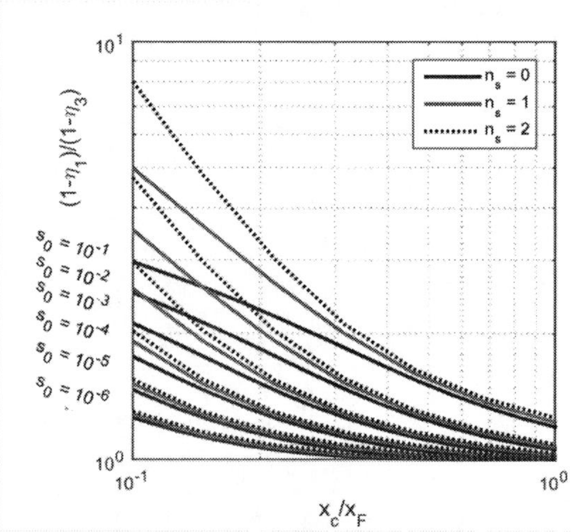

- Approximation is better at
  - higher cut sizes
  - lower size dependence

- For typical cases losses are < 4x with approximation

Towards Better Size Reduction Flow Sheets: How the Application of Some Simple, yet Powerful Philosophies and Principles Can Improve Both Process Operation and Product Quality

Karl Jacob
The Dow Chemical Company
Engineering & Process Sciences
Solids Processing Lab
Midland, Michigan

Dedication

This presentation is dedicated to the late R.R. "Dick" Klimpel

- As a grad student with L. Austin at PSU, Dick helped to define population balance methods for size reduction
- 30+ year career at Dow
- Wrote "Process Engineering of Size Reduction"
- Ardent and enthusiastic supporter of particle technology
- Gaudin Medal winner from the AIME

Size Reduction

- Applied to nearly every material in the world
- Size reduction is employed across a size range from 50 nanometers to a meter or more
- While it is desirable to "engineer" our particles, sometimes size reduction is unavoidable or absolutely necessary.

- YET…Problems abound in size reduction
  - Fines
  - Wear
  - Production rate
  - Wrong particle size
  - Heat
  - …

Often simple, yet powerful ideas are at the root of these issues

Inspirational Video

#1: Klimpel's Rule

- Quote from Dick, "Karl, never try to do more than a decade of size reduction in one step!"
- Simply put: $10 \times dp_{50,out} < dp_{50,in}$
- Why is this important??

–Fines!!

- The management of fines is absolutely critical to all manner of product quality issues across all kinds of materials: superabsorbent powders, pharmaceuticals, specialty chemicals, etc.
- Nearly always, we need to consider the fate of the fines – in many cases, they are simply discarded and represent lost yield.

#1: Klimpel's Rule

- Mechanistically, if we try for more than decade of size reduction, think about the particles here – they will exposed to more and more breakage events at the expense of increased fines production

- Does this always apply to my comminution process??
  - Not necessarily – example is mineral ores where we need to get to fine sizes to liberate the desired species
  - BUT, remember those increased fines will have other negative impacts like plant industrial hygiene and powder flow

#2: Ben's Rule

- You will make fines! Regardless of how much of an effort you make to prevent fines production, fines will still be created. A good example is tempered glass!

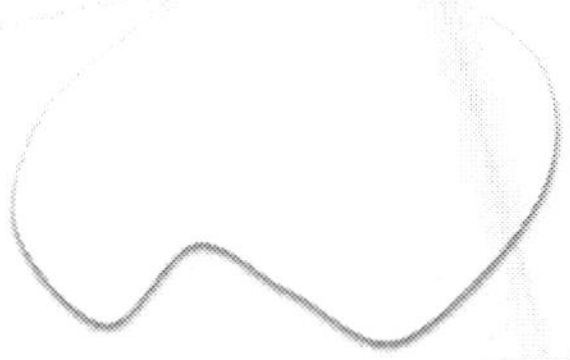

#2: Ben's Rule

- **Much like rule #1, we see that fines and fines management are crucial to successful operation of our size reduction (intended) and attrition (unintended) processes.**
- **This rule doesn't give you a directive as to how to not produce fines, but rather forces you to acknowledge that fines will be generated.**

#3: Karl's Rule

- Ignore everything you've been taught about energy!!
- You can use Kick's, Bond's or Rittinger's Laws to maybe predict how much energy might be used.
- BUT, I have yet to meet an engineer who says, "my mill is using too much energy!"
- Why? Most *CHEMICAL* products are high value added materials and energy isn't a significant consideration in the value equation. Tremendous premium on making the CORRECT particles!

#4: Importance of the Grinding System

- Remember that the mill is only one part of the overall system

Typical Grinding Circuit

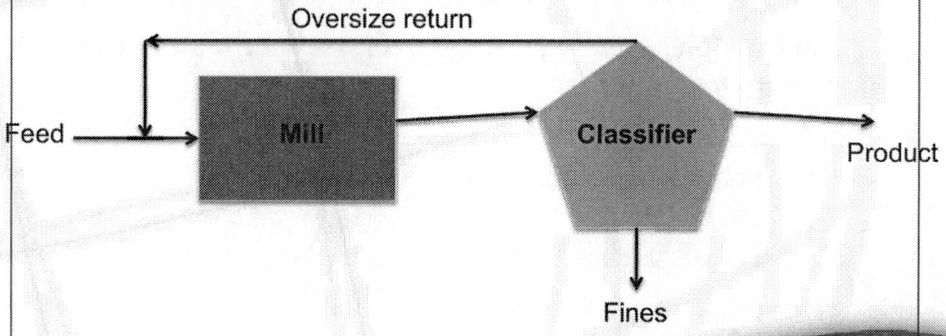

#4: Importance of the Grinding System

- Need to consider all the ancillaries as they are important factors in successful operation of the grinding circuit
  - Hoppers
  - Classifiers
  - Chutes
  - Conveyors
  - Dust collectors
  - Pumps
  - Feed tanks and agitators
  - Etc.

#4: Importance of the Grinding System

- **Absolutely crucial to the successful operation is the understanding of the interplay between ALL the variables in the system.**
- **Many people don't recognize the impact of feed rate on the probability that a particle will be selected for breakage and the resulting daughter particle size.**
- **Typical example: creation of high oversize return streams due to high feed rates**

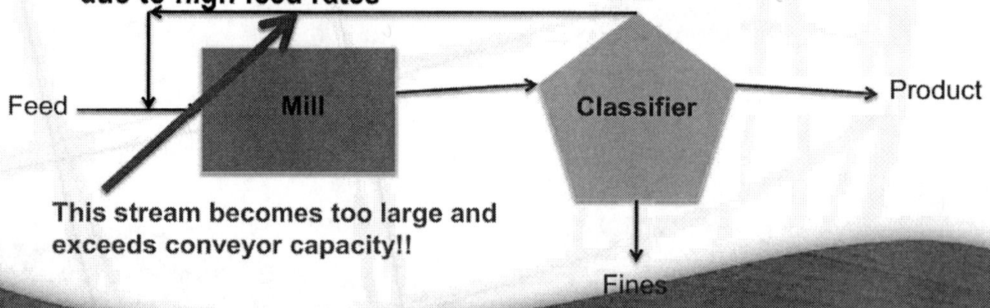

#5: Open Loop Grinding

- There is a strong tendency to do pilot testing in open loop for a wide variety of reasons
- I have seen many an engineer proudly exclaim that they've solved the plant's grinding problem by doing a vendor trial in open loop!
- DON'T DO THIS!
- Nearly all production plant situations are operated closed loop and the fact that the breakage selection and daughter fragment matrices (S_i and $B_{i,j}$) are a function of feed rate causes problems

#5: Open Loop Grinding

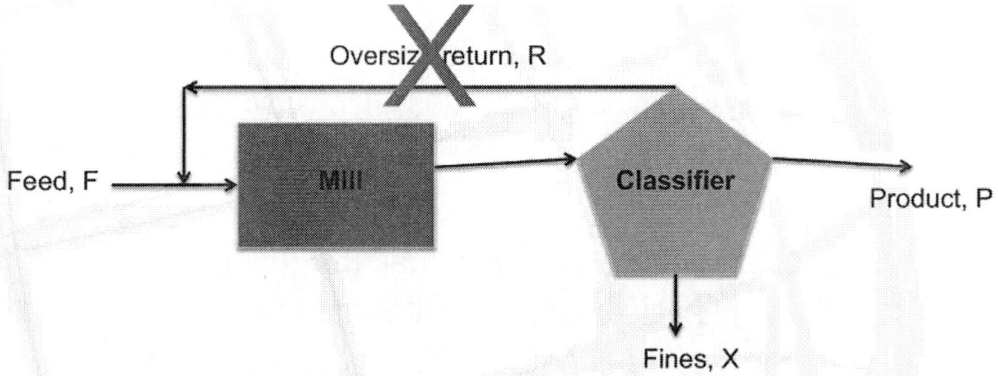

- **In many milling circuits, R can be roughly equal to F or even greater for that matter! With the dependence of mill performance on feed rate, then the output stream may not be as fines free as desired!!**

#5: Open Loop Grinding

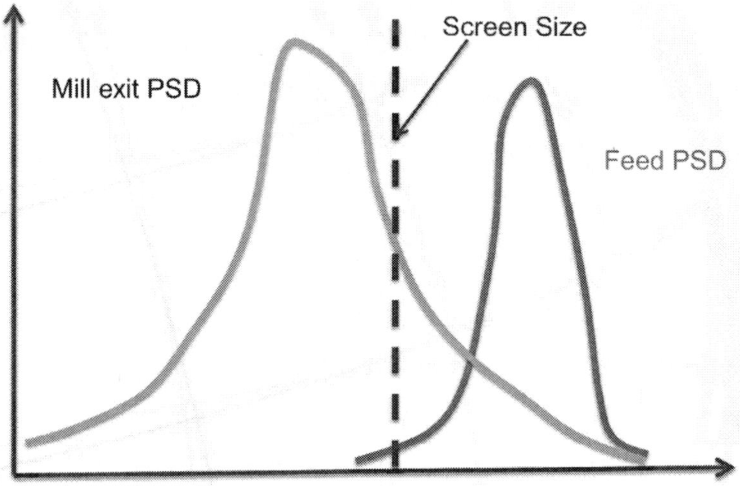

#6: Consider screen size in internally classified mills

- This is a subset of operating open loop and applies to mills that are usually internally classified (have some sort of device to limit particles larger than size x from leaving the mill)
- Many people will put a screen in their mill (hammer mill, Fitz mill, etc.) that is commensurate with the specification size (I need no more x % greater than y microns is a typical specification).
- This invariably results in ***<u>significant</u>*** overgrinding of the material. Plus the heat generated can result in product degradation.

#6: Screen Size in Internally Classified Mill

Screen prevents oversize from exiting the mill. Material in mill must endure repeated breakage events in order to exit.

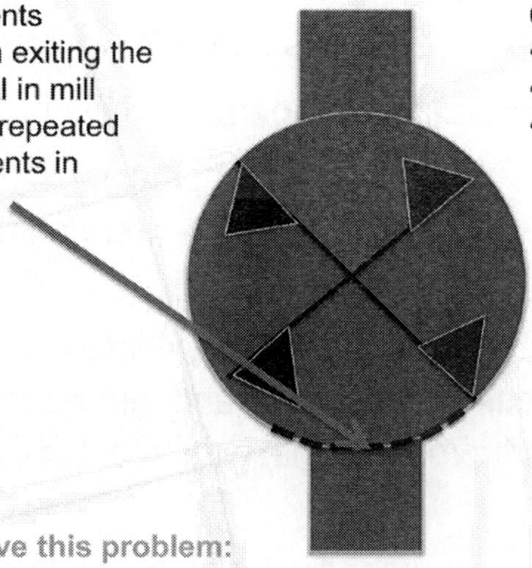

Consequences are usually dire:

- Degradation due to heat
- Excessive fines production
- Mill wear and screen breaks

How to solve this problem:

- Relax the specification (often not possible)
- Relax the screen size, use external screener to remove the oversize and add the oversize back to the feed

#7: Wear

- Every size reduction operation is subject to wear! Part of your mill or media will end up in the product. Even very hard media like yttria-stabilized zirconia beads wear after enough use! Example: a 5 micron loss on an 800 micron grinding bead means that this 2% loss in media is in your product
- Dealing with wear
  - Expect to replace grinding media
  - Recognize that mill wear will affect grinding behavior
  - Expect to replace wear parts frequently
  - Specialty materials (hard facing, polyurethanes, etc.) can improve part life

#8: Wet milling

- Many of the other seven previous concepts still apply
- Basic chemical engineering is of paramount importance
  - Need to strongly consider heat management – think about having a 3 or 5 HP motor on a 1 L mill – how do I remove that heat?
  - Fluid flow – when this system begins to thicken as a result of size reduction, can I both pump it and remove the heat – remember you are probably changing from turbulent to laminar flow
- Aggregation control is essential to successful wet milling

#8: Wet Milling

This is wet milling gone bad! In less than 3 minutes this system went from a thin to shaving cream.

#9: Use math to your advantage

- **When used in concert with the previous eight concepts, modeling your system mathematically can be powerful tool to help improve both product quality and process operation.**
- **Questions that you could answer might be…**
  - **What if I change my feed size?**
  - **What are the respective sizes of my streams?**
  - **What if I change my screen size?**
  - **How much turn up or turn down do does my process have?**
- **As Bert and Ben will tell us, we don't wade into these waters lightly. Many size reduction/attrition problems don't need the 16 lb sledge that population balance modeling provides, but there will be cases where it is very advantageous.**

#9: Use math to your advantage

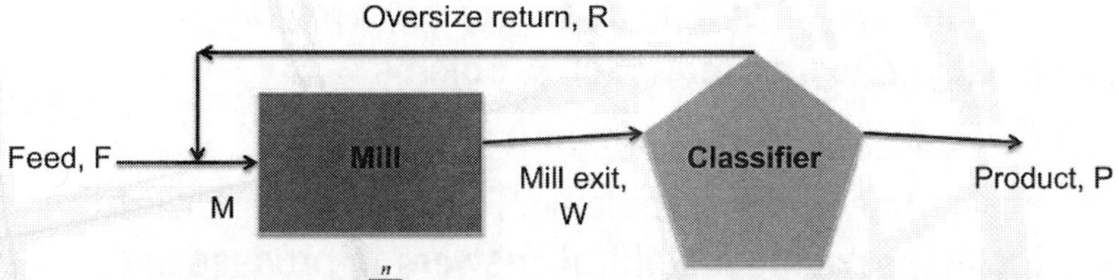

$$W_i = M_i(1-S_i) + \sum_{j=i}^{n} B_{i,j} S_j M_j \qquad R_i = G(x_i)\, W_i$$

- There are a number of mass balance equations which can be written in addition to the above which permit a closed form solution to this problem.
- In an ideal world, one can feed the mill tightly sized fraction and determine directly the Si and Bi,j but this is not routinely done.
- However, even a simple example can shed light on the grinding problem.
- Many of the grinding "rules" can find their basis in this kind of analysis.

#9: Simple example

$$F = \begin{matrix} 65 \\ 30 \\ 4 \\ 1 \end{matrix} \quad S = \begin{matrix} 0.95 \\ 0.70 \\ 0.40 \\ 0 \end{matrix} \quad G(x) = \begin{matrix} 0.97 \\ 0.75 \\ 0.50 \\ 0.01 \end{matrix} \quad B_{i,j} = \begin{matrix} 0 & 0 & 0 & 0 \\ 0.25 & 0 & 0 & 0 \\ 0.50 & 0.50 & 0 & 0 \\ 0.25 & 0.50 & 1 & 0 \end{matrix}$$

Results:

$$R = \begin{matrix} 3.3 \\ 22.4 \\ 38.0 \\ 0.5 \end{matrix} \quad P = \begin{matrix} 0.1 \\ 9.6 \\ 38.0 \\ 52.3 \end{matrix}$$

$$\sum R_i = 64.2$$

Takeaways from this analysis:

- **Note size of recycle stream**
- **Creation of significant size 4 product**
- **Once framework is created, this allows us to investigate easily process changes**
- **We still need A LOT more work to understand effect of mill parameters on both selection and breakage.**

Final Thoughts…

- In many cases, simple concepts can provide solutions to size reduction problems
- Fines control is of paramount importance
- For more complex problems, PBM can be a powerful tool in providing insightful answers to process and product problems

- QUESTIONS??

Industrial Challenges in Particulate Flow and Attrition

Paul Mort, Research Fellow,
Procter & Gamble Company
Member of IFPRI

P&G

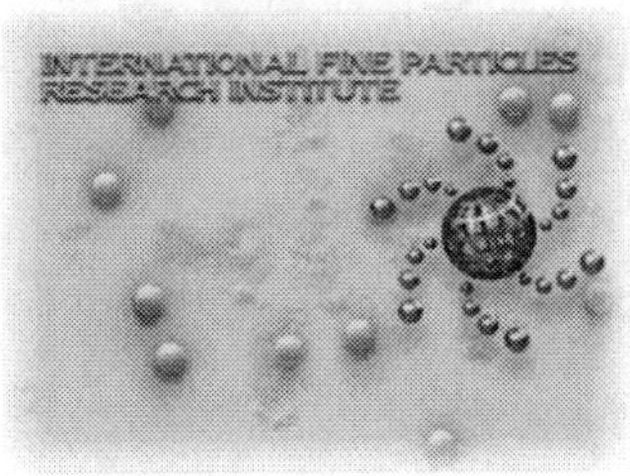

AIChE Frontiers of Particle Science & Technology, April 2016

IFPRI work in Size Reduction, including Attrition

- Size Reduction of Particles; van Brakel (1981)
- Attrition Studies; Beddow (1983)
- Studies on Abrasive and Erosive Wear of Materials by Particles; Hashimoto (1983)
- Mechanical Properties of Powder Bed Processes and Application to Particulate Processes; Makino (1984)
- Particle Attrition; Vervoon (1986)
- Attrition of Bulk Particulate solids, Bridgewater (1989)
- Comminution by Impact; Leschonski (1989)
- Fracture Mechanics; Isherwood (1990)
- The Rate and the Limit Particle Size of Ultrafine Grinding of Hard Materials in Liquid; Jimbo (1993)
- Experimental Simulation of Processes in Ball Mills by Fragmentation of Particle Assemblies.; Weichert (1994)
- Test Devices for Comminution; Schonert (1995)
- Increasing Steepness of the Particle Size Distribution in Grinding of Chemicals &Pharmaceuticals; Klimpel (1996)
- Computer Simulation of Particle Breakage; Campbell (1998)
- Impact Attrition of Particulate Solids; Ghadiri (1998)
- Experimental Study of Fragmentation by High Velocity Impacts and by Air Jet Milling; Dodds, Guigon (1998)
- Achievements in Size Reduction; de Jong (1998)
- Parameters of Controlled Grinding; Leschonski (2000)
- Contact/Impact Charging of a Single Particle; Yamamoto (2002)
- Mechanochemistry of Materials; Saito (2003)
- Sub Micron Particle Production in Stirred Media Mills; Schwedes, Kwade (2006)
- Nanomilling in Organic Solvents in Stirred Media Mills and Microstructural Analysis; Peukert (2006)
- Milling of Organic Solids; Ding (2009)
- Characterisation of Particle Grindability by Indentation and Impact; Hare (2014)
- Grindability Test Modelling, Measurement and Mill Fingerprinting; Ooi (2015)
- Review on Non-Mechanical Approaches to Milling; Tavares (2016)

Test Devices for Comminution; Schonert (1995)

In a mill the particles are stressed and broken only within a small part of the entire mill volume, e.g.:

- within the contact volumes of the balls in grinding media mills,
- contact volumes between rollers and the table in vertical spindle mills, or
- at the beater surfaces in an impact mill.

The particles have to flow into these active spots, to be stressed and to leave afterwards. A comminution process, therefore, is controlled by

1. particle motion inside the mill chamber,
2. stress intensity and the particle arrangement in the active spots, and
3. breakage behavior of the material.

↓

Focus of this presentation:

Compressive-shear mode of attrition,
Flow and stress fields in dense bulk flows

Motivation – optimizing processes that handle particulates either as intermediates or end products

- We desire efficient processes:
  - Maximize production throughput ← Dense flows are of interest
  - Minimize the spent energy. ← Optimize the use of energy to achieve desired product transformation
- Two aspects in this optimization:
  1. Process equipment and its integration within a system; and
  2. Properties of the particulate material that are relevant to the processing conditions, specifically the material's response to flow and stress fields that are imposed by the process equipment.
- Industrially-relevant materials have a complex coupling among:
  - imposed flow field and boundary conditions,
  - constitutive properties of the material
    function of shear rate, packing density...
  - resultant stress & energy consumption.

$$\int_{time} \frac{\text{stress} \cdot \text{flow}}{\text{mass}} = \text{specific energy}$$

- Attrition ≡ unwanted particle size reduction ← ~ specific energy or power input
- Process optimization requires understanding of flow and stress fields.

Multi-scale Approach to Particulate Flow

Is there an optimal particle flow operating regime?

- **Fluid-like;**
- **Dense flow;**
- **Efficient particle interactions:**
  - **Boundaries,**
  - **Particles,**
  - **added Liquid,**
  - **Gas.**
- **Low E/M.**

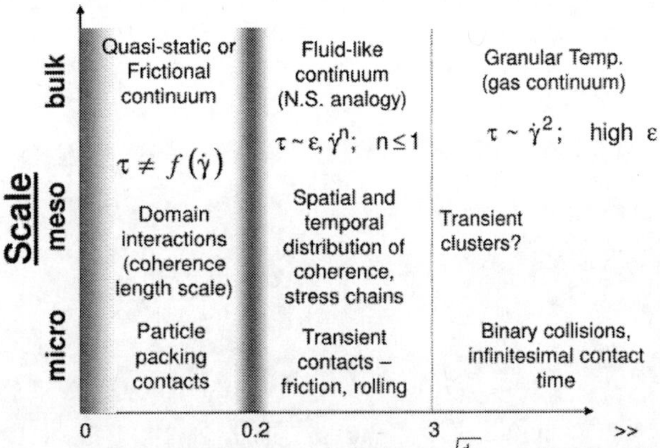

Dimensionless shear rate, $\dot{\gamma}^* = \dot{\gamma}\sqrt{\frac{d_p}{g}}$ (Tardos, Savage)

Inertia / confinement,	$I = \frac{\dot{\gamma} d_p}{\sqrt{P \cdot \rho_0}}$				(GDR Midi)
Contact time	t_c / t_{bc}	high	~2	1	(Campbell)
Stiffness/shear	$\frac{k}{\dot{\gamma} \rho \cdot d^3}$	high		low	
Stokes #	$\frac{m\dot{\gamma}}{6\pi\eta d}$	0	~10	high	(Brady)

Coupling of particulate flow and stress fields – a multi-scale challenge

micro…

- Fundamental understanding of particulate flow & stress fields and their material coupling.
- Particle-particle / particle-boundary interactions:
  - particle characteristics & properties: size, size distribution, shape, surface characteristics, complex modulus.
- Flow and stress in critical sections of the process, e.g., gap between an impeller and a process boundary.
- Unit operation: flow fields, residence time distributions and scale-up.
- System scale of ancillary process interactions:
  - feeders, intermediate storage, product and intermediate handling, recycle integration, etc.

…macro…

1) What can we infer about stress based on power and/or torque measures?
   - Granular rheology

3) Effect of size and shape distributions on jamming onset

2) Implication of transient jamming on stress and stress fluctuations

Opportunities for optimization and control:

- Systems approach,
- Modeling,
- Open questions…

1) What can we infer about stress fields in a well-characterized dense flow?

Granular rheology

- Instrumented axial Couette
  - Constitutive characterization of dense flows in the intermediate regime (IFPRI, Tardos & Kheiripour-Langroudi, 2011)
- Collaboratory: Model the 3D Couette flow
  - Wassgren, Campbell (IFPRI / NSF, 2013)

Granular Rheology – stress analysis in Couette flow

Tardos and Kheiripour-Langroudi, CCNY, IFPRI

Interrogate stress field in a drained vertical Couette flow

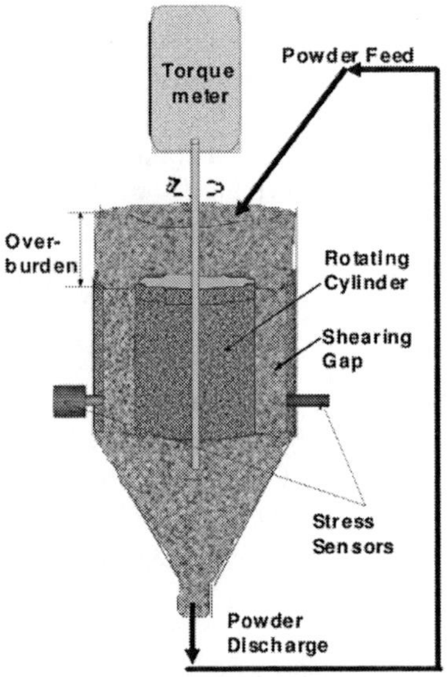

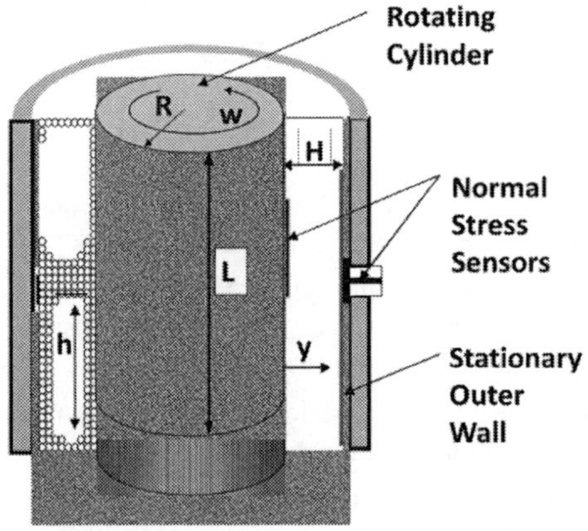

Experimental options:

➢closed-bed (Batch, → close packed);

➢open-bed (Continuous, → allow dilation)

Shear / normal stress ratio as a function of boundary conditions

Tardos and Kheiripour-Langroudi, CCNY, IFPRI

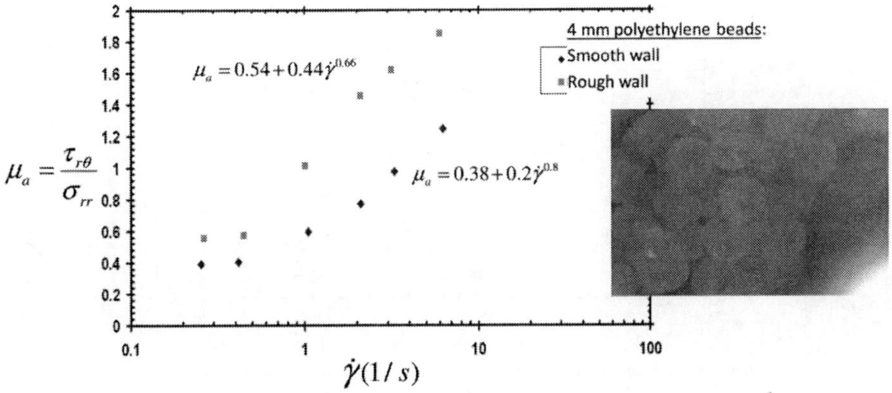

Determine Constants in Constitutive Equation from Rheometer Experiments in Couette Device

Tardos and Kheiripour-Langroudi, CCNY, IFPRI

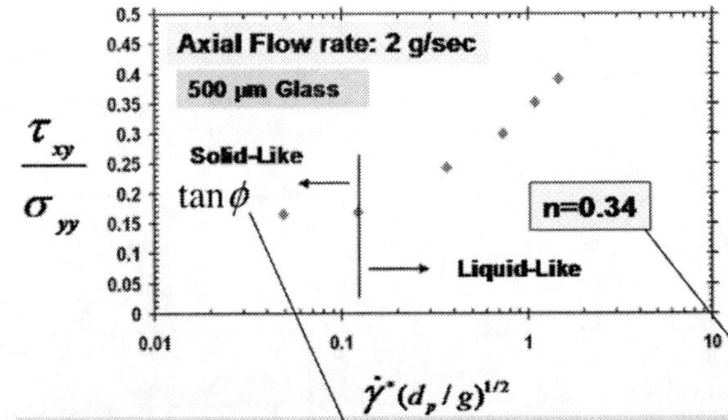

$$T_{ij} = p\,\delta_{ij} + \sqrt{2}\,p\,\sin(\phi)\frac{\dot{\gamma}_{ij}}{|\dot{\gamma}_{ij}|} + \sqrt{2}\,ap\,\cos(\phi)\frac{\dot{\gamma}_{ij}}{|\dot{\gamma}_{ij}|}\dot{\gamma}_{ij}^{n}$$

Schaeffer Equation (solid- like behavior) — Liquid-like behavior

Growing interest in applying intermediate regime rheology to <u>continuum</u> flow solvers

Effect of particle shape

Tardos and Kheiripour-Langroudi, CCNY, IFPRI

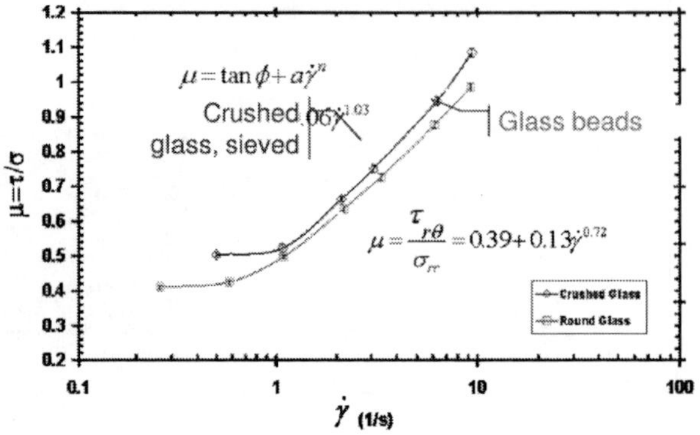

Figure 3: Measured ratio of shear to normal stresses (apparent friction coefficient) as a function of shearing rate; 1 mm in diameter, round and crushed (odd-shaped) glass. Fitted curve also shown in the figure.

Coupling of stress and dilation

Tardos and Kheiripour-Langroudi, CCNY, IFPRI

Capacitance sensor measures solids fraction in Couette gap.

- Various elastomeric materials (A,B,C) show markedly different stress and packing profiles
- Power law rheology is stronger for materials that retain packing state with increasing shear.
  - Analogy w/ incompressible fluids?
- Materials that dilate with large fluctuations in packing can maintain relatively stable stress ratios at higher strain rates.
  - Average can be steady, local highly fluctuating.

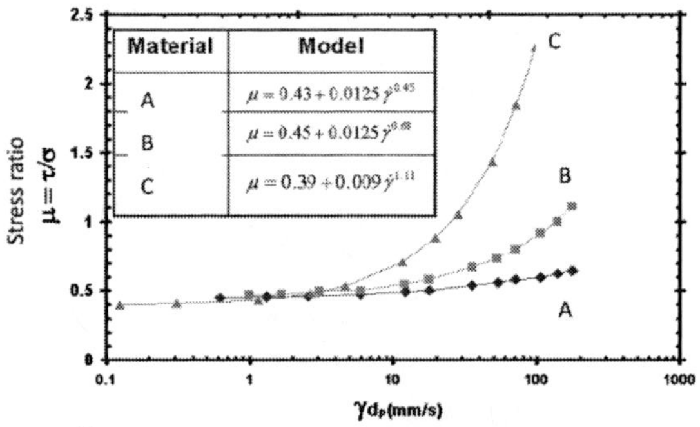

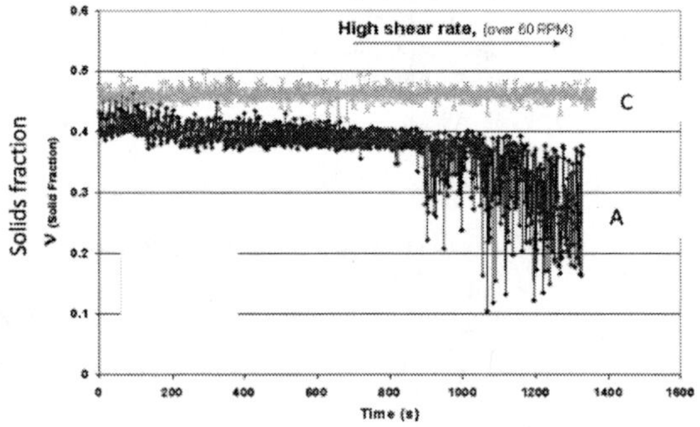

3D Couette, Experimental (Tardos, Kheiripour-Langroudi)

1 mm dia glass beads

Normal stress & Torque

Data measured over a series of rotational speeds in the Axial Couette cell operating in continuous flow mode. The torque measurement is averaged over the full volume of material in the shear gap. The normal stress sensor is localized, and its acquisition speed is much faster than the torque sensor.

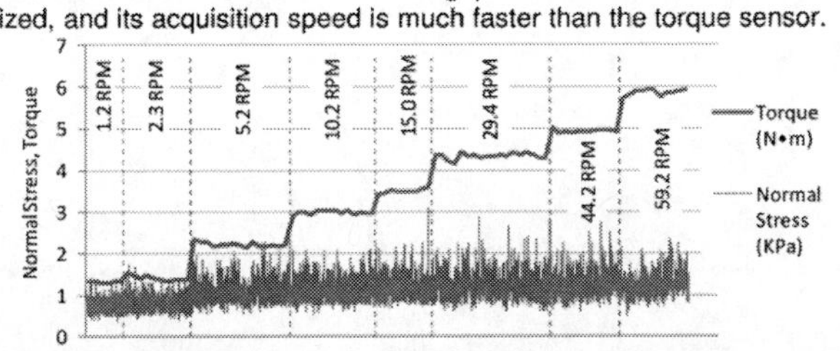

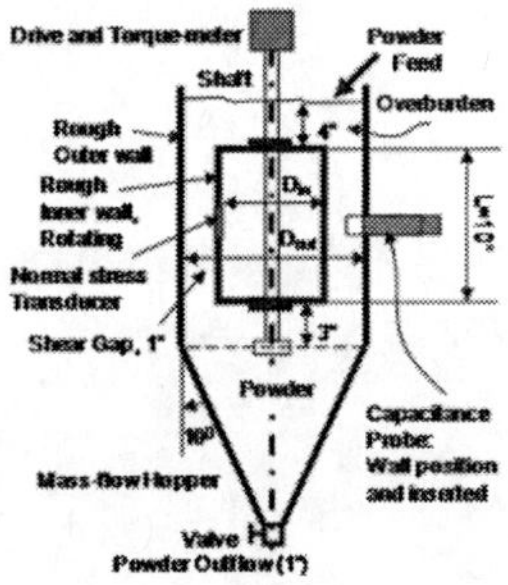

Solids packing fraction

Measured as a function of rotational shear, continuous flow in Axial Couette (25.4 mm gap). The graph is an overlay of separate measurements collected over sequential time series at rotational speeds noted.

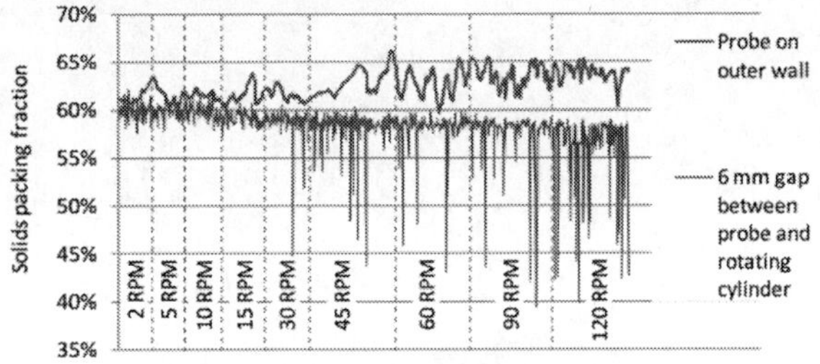

Dense Granular Flow – A Collaborative Study, Powder Technol, 2015.

3D Couette, Granular Rheology

1 mm dia glass beads

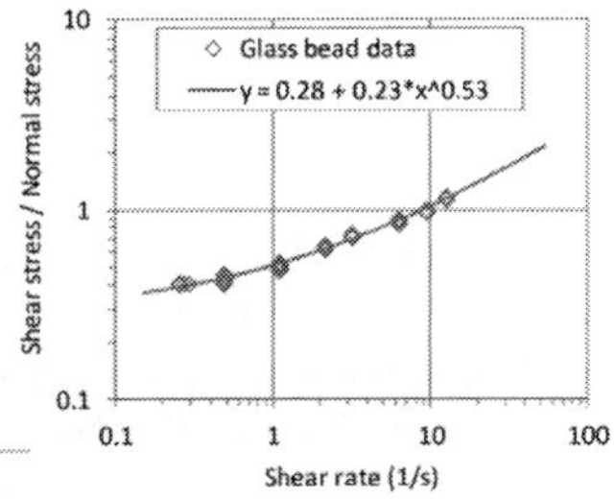

Experimental (Tardos, Kheiripour-Langroudi)

Rheological power-law fit derived from average-value torque and normal-stress data, 1 mm glass beads.

Simulation (Campbell, Wassgren)

Parameter study using constant volume boundary conditions (Wassgren)

a) Calibrating simulations to experimental data at 15 RPM (shear rate ~3.14 /s), best convergence is obtained when particles do not roll against each other ($\mu_r/d_p \rightarrow \infty$);

b) Predicted shear/normal stress ratio plotted as a function of the shear rate for a range of solid fractions without rolling interactions, i.e., $\mu_r \rightarrow \infty$. The dashed line is the experimental fit, from above.

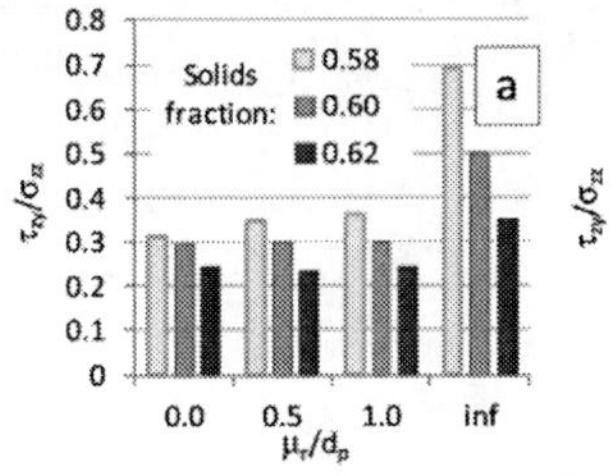

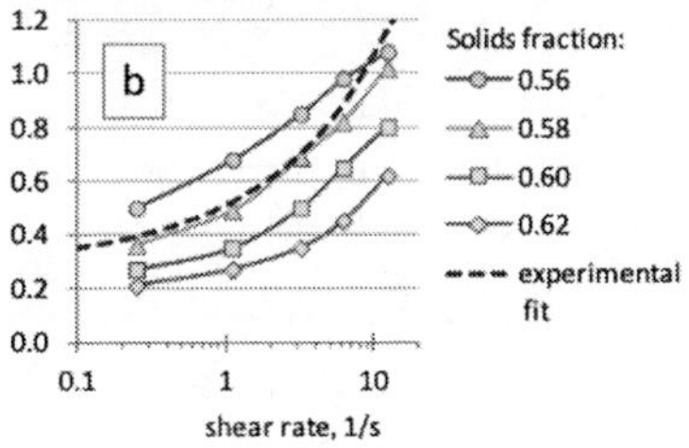

Dense Granular Flow – A Collaborative Study, Powder Technol, 2015.

2) Local stress chains and stress fluctuations

- Stress chain visualization and measurement
  - Behringer group, Duke University
    - Annular shear cell, 1996
    - 2D Couette, 2010
    - Dynamics and Rheology of 2D Hopper Flows (IFPRI, 2013)
- Collaboratory: Model the 2D Hopper Flow (IFPRI / NSF, 2013)
  - Kondic, Shattuck
- Attrition testing
  - Bridgwater, Cambridge, 1989
    - Annular shear cell
    - Stirred hopper attrition test (Cone Cell)

Stress Fluctuations for Continuously Sheared Granular Materials

Miller, O'Hern, Behringer (Phys Rev Lett, 1996)

In a confined flow, local stress fluctuations may be much higher than the average stress.

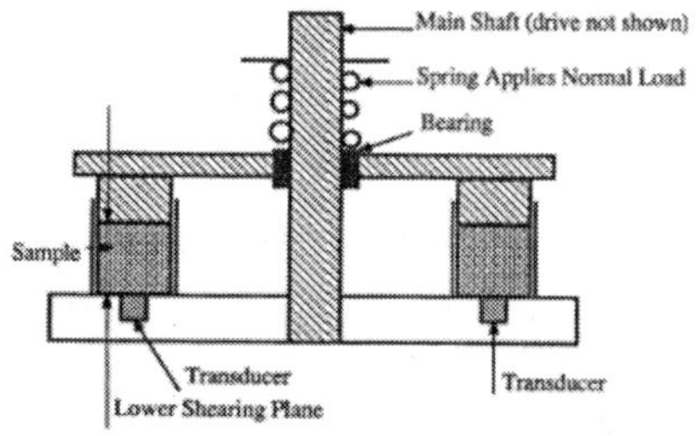

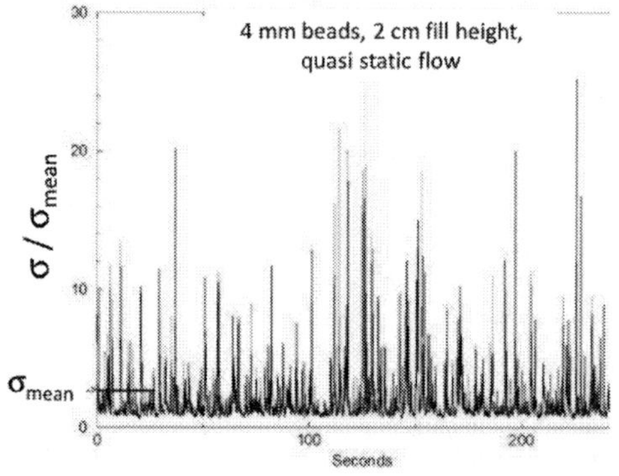

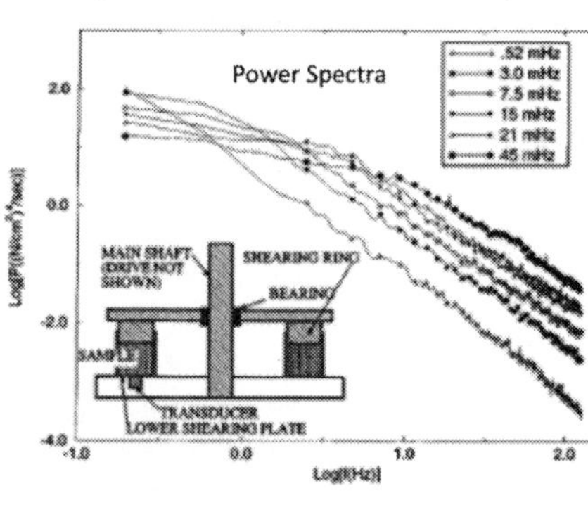

Spatial distribution of stress chains

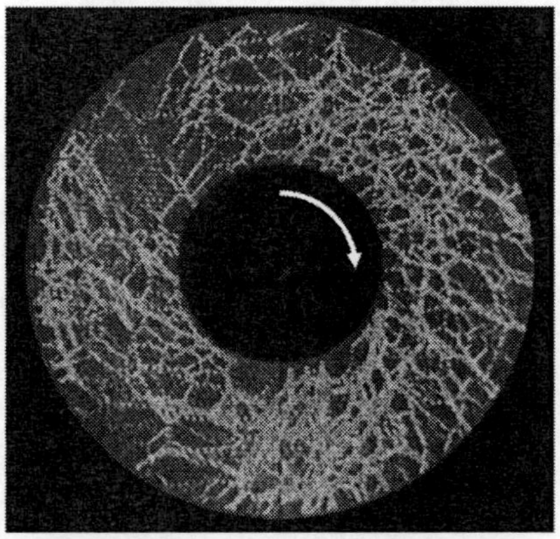

Iconic image of stress chains illuminated by photo-elastic particles in vertical axis 2-D Couette flow

Behringer group, Duke University

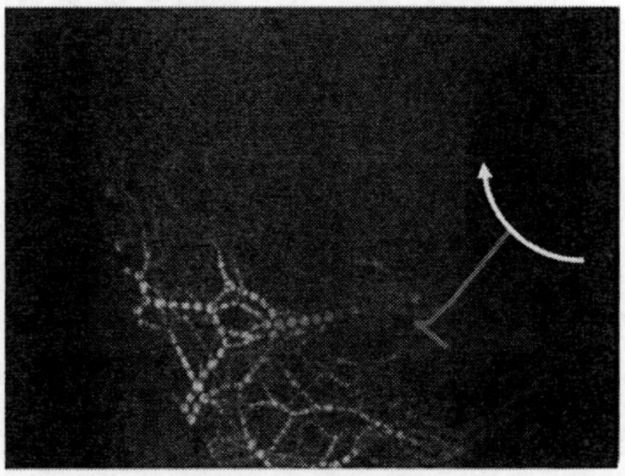

Horizontal axis, modified impeller "paddle" edge highlighted in red.

Stress chains extend from paddle tip to wall of mixer – implies opportunity for jamming.

Behringer, Clark, 2010

2D Hopper Flow

Experimental (Behringer, Tang)

2D hopper flow of polymer disks:

- average disk diameter D_p,
- disk thickness h,
- hopper orifice D (D»h), and
- 2D close packing, ν=0.84:

a) Flow rate data for hopper discharge at different half-angles (from vertical) plotted according to 2D Beverloo equation...

$Q^{2/3}/d_p = [(4/\pi)\ C\ v\ \rho_s\ g^{1/2}\ h]^{2/3}\ (D/d_p - k)$

b) Discharge coefficient, $C = C_{90}+A\bullet\cot^B(\alpha)$, as a function of hopper half angle, with k = 1.48.

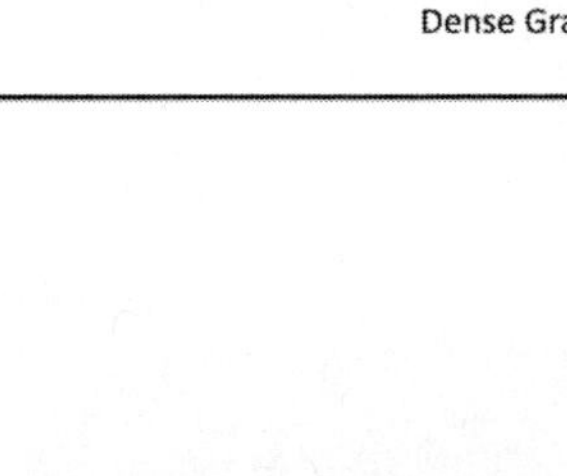

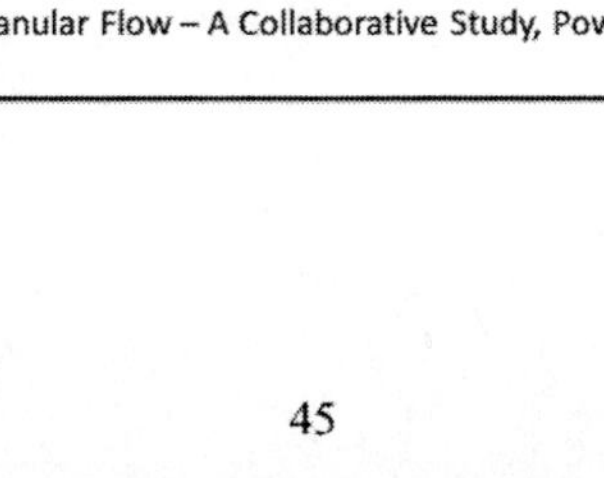

Simulation (Kondic, Shattuck)

Calibration of DEM models by adjusting particle friction and restitution parameters to match measured emptying time; used relatively wide orifice, D/d_p » 4.5, to avoid jamming.

Simulated times are marginally faster compared to experimental results: 15.87±0.05 s.

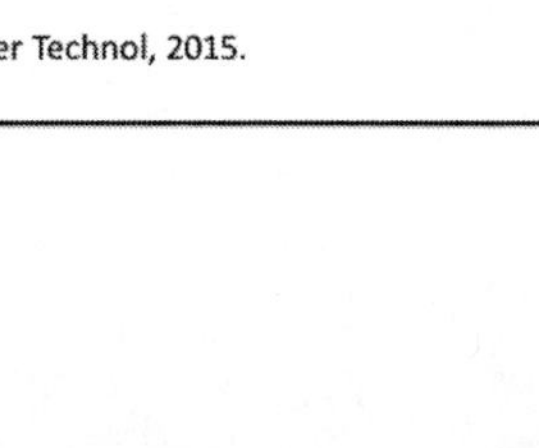

Dense Granular Flow – A Collaborative Study, Powder Technol, 2015.

2D Hopper: Jamming & Clogging

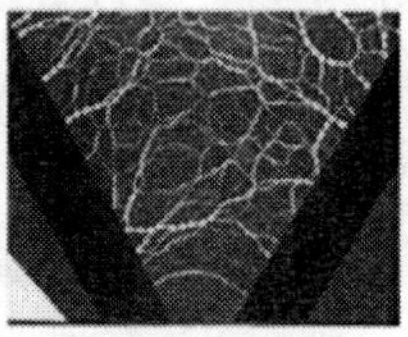

Experimental
(Behringer, Tang)

Jamming follows an exponential distribution with the mean time to clog (λ) being a strong function of dimensionless orifice size.

The best exponential fit was obtained relative to the square of the dimensionless orifice size, further adjusted by the boundary-layer dimension, k=1.48. Over the range tested, the hopper half-angle appears to have little effect on the clogging interval.

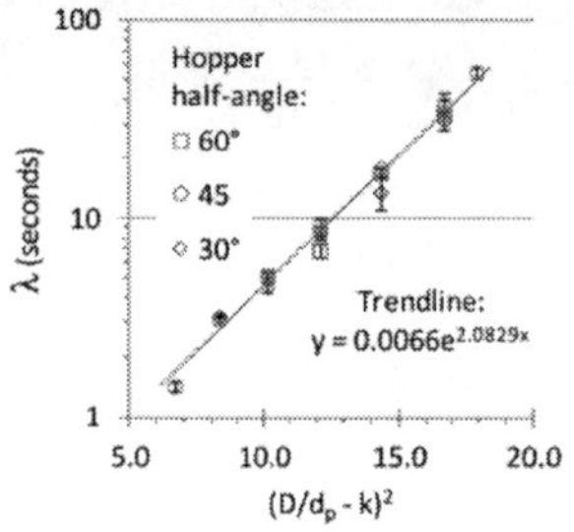

Simulation (Kondic, Shattuck)

Comparison of 2D hopper modeling results for the mean time interval (λ) between jams sufficient to clog the outlet as a function of dimensionless orifice size reduced by Beverloo "k" coefficient. For relatively small dimensionless orifices ($D/d_p < 4$) and large vertical half-angles ($a \geq 60°$), the survival time realizations are remarkably consistent with the exponential trend seen in experiments. On the other hand, simulations seem to under-predict survival times for steeper hoppers.

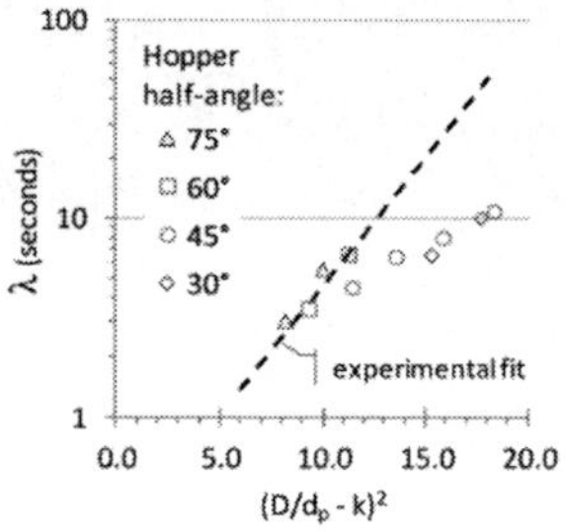

Dense Granular Flow – A Collaborative Study, Powder Technol, 2015.

2D Hopper, Flow and Stress Fields

Experimental (Behringer, Tang)

Coarse-grained velocity and stress fields in 2D hopper with half-angle 30° from vertical, i.e., a "mass flow" condition: a) average velocity; b) velocity fluctuation; c) average stress; d) stress fluctuation.

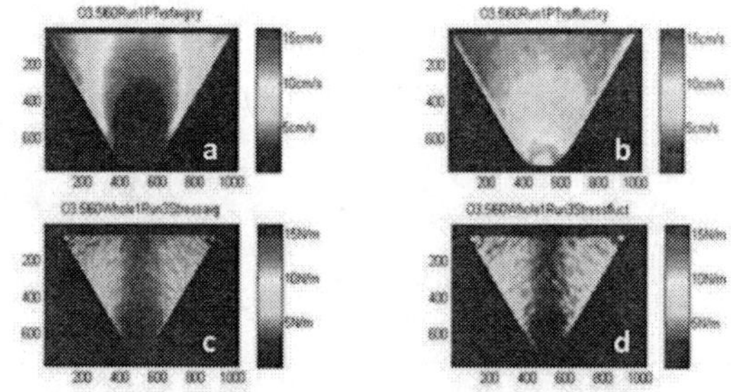

Simulation (Kondic, Shattuck)

Comparative simulations of velocity and force fields: a) average velocity; b) velocity fluctuation; c) average force on a particle; d) force fluctuation.

Velocity data have common units with experiments, cm/sec.

Force data have units of particle mass • gravity, where 1 unit = 1.46 mN.

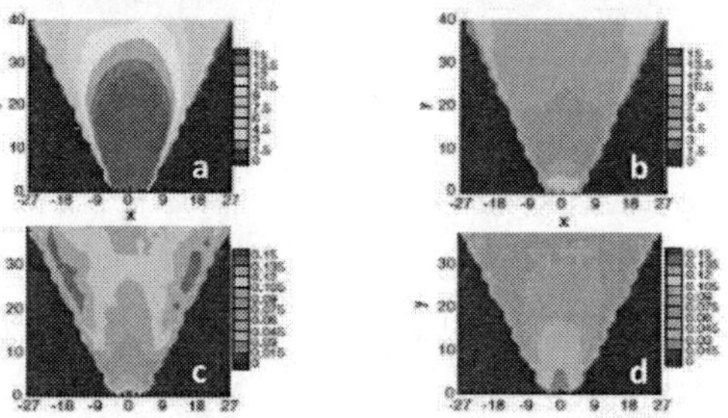

Dense Granular Flow – A Collaborative Study, Powder Technol, 2015.

Attrition in an annular shear cell, fitted with fins to simulate flow through gaps

Bridgwater, IFPRI 1989

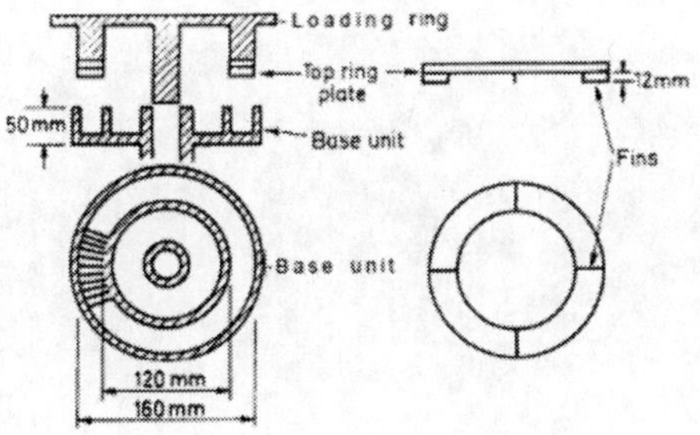

Modified annular shear cell design

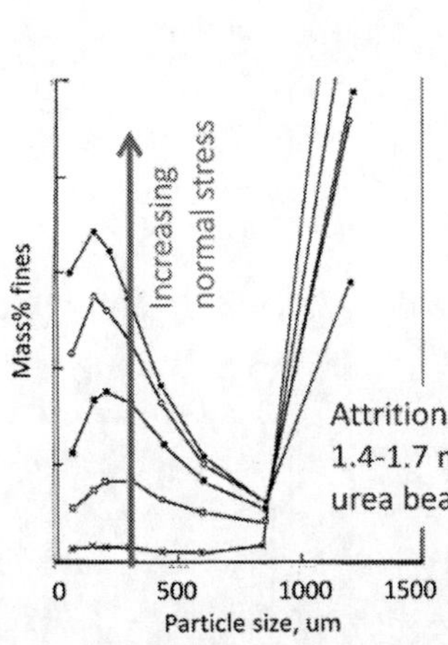

- Small gaps produce a higher degree of attrition than larger gaps.
- Higher stresses around the fins cause the production of more fines and less immediate undersize.

Attrition in a stirred hopper (Cone Cell)

Bridgwater, IFPRI 1989

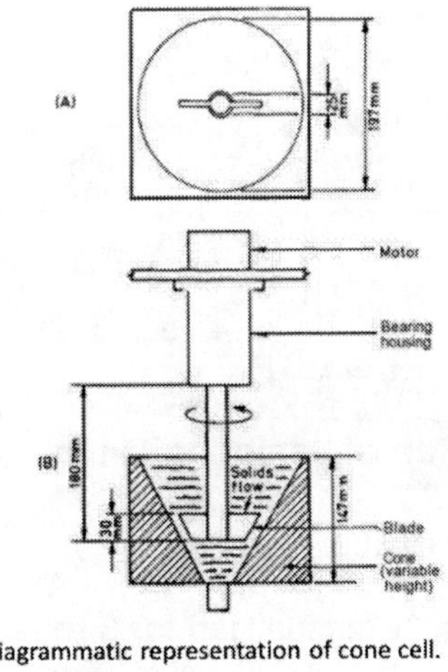

Diagrammatic representation of cone cell.
(A) Top view
(B) Front view, partially sectioned.

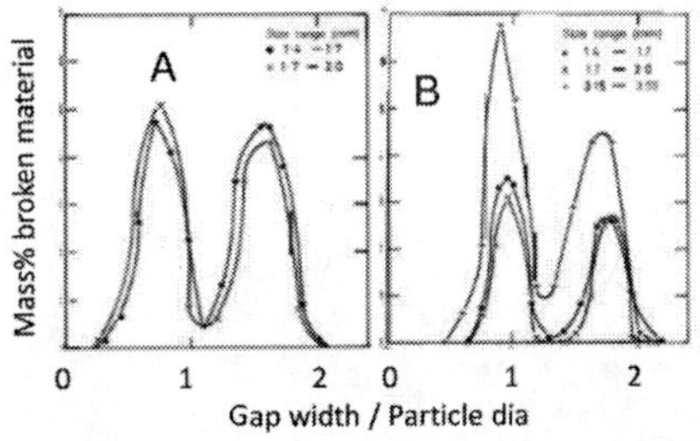

Percentage of broken materials (A, catalyst; B, urea) against dimensionless gap size.

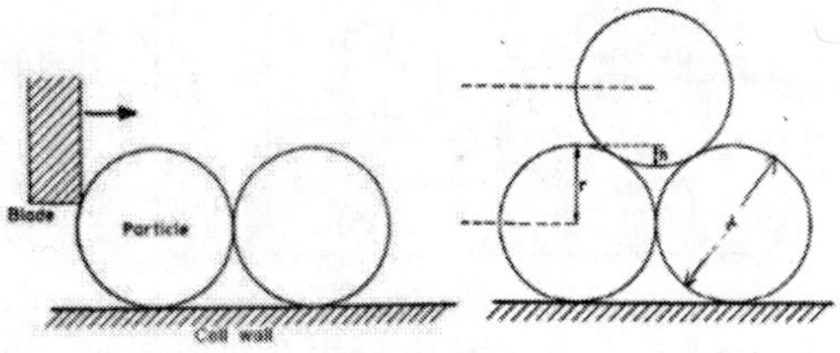

Attrition occurs in conditions of jamming, i.e., where impeller blade contacts particles in a way that creates a force chain against the wall, either through individual particles (LHS) or a stacked multilayer packing (RHS)

3) Effect of particle size and shape distributions on jamming

- Statistical approach: infer stress-chain dynamics from jamming statistics.
  - Mort, Geiger and Wandstrat (P&G), w/ Behringer as consultant*
- Vertical flow of granules through an orifice.
  - Jamming Onset measured as a function of dimensionless orifice size; jamming probability used to differentiate characteristics of otherwise free-flowing granules
  - Empirical correlation of flow behavior with granular characteristics including:
    - **Size and Shape distributions**
    - **Bulk Frictional properties**

*Granular flow through an orifice – effect of granule properties on the onset of jamming," Proceedings of the AIChE Annual Meeting, 2007.

"Characterizing flowability of granular materials by onset of jamming in orifice flows," Papers in Physics, V 7, 2015.

Cylindrical Hopper Plan - 1

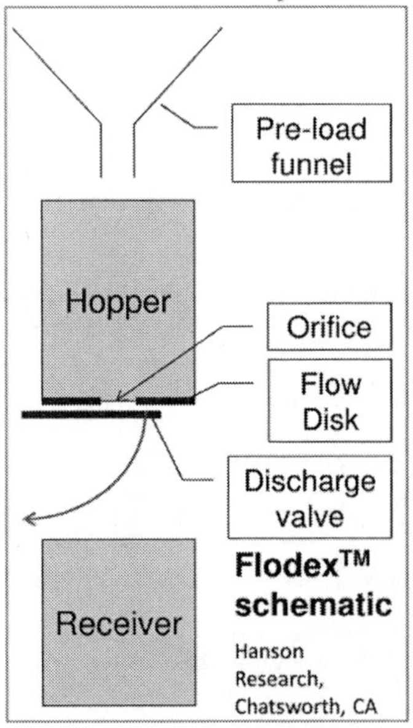

Flodex^{TM} schematic
Hanson Research, Chatsworth, CA

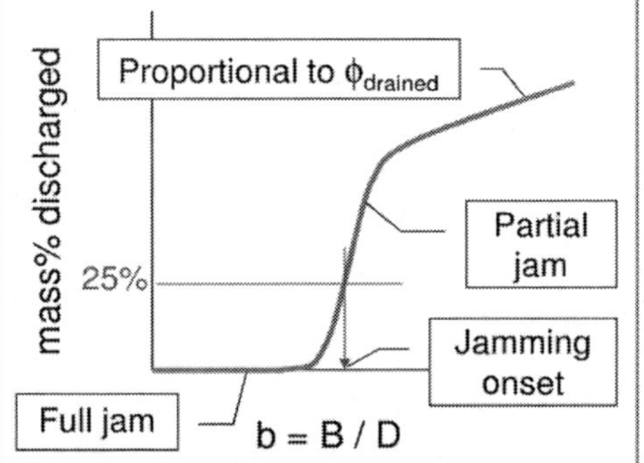

Interpolation of Jamming Onset

- Use interchangeable flow disks with pre-set orifice sizes (2.5, 3, 3.5, 4, 5,… mm).
- Measure mass% discharged for 3 reps per orifice size.
- Interpolate to find Jamming Onset (J).

Cylindrical Hopper Plan - 2

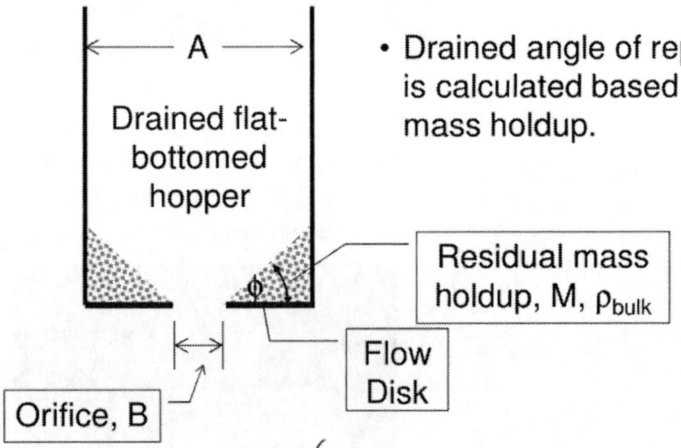

- Drained angle of repose, ϕ_{drain}, is calculated based on retained mass holdup.

$$\phi_{drain} = arctan\left(\frac{24 \cdot M_{retained}}{\pi \cdot \rho_{bulk} \cdot \left(2 \cdot A^3 - 3 \cdot B \cdot A^2 + A^3\right)}\right)$$

Cylindrical Hopper Results

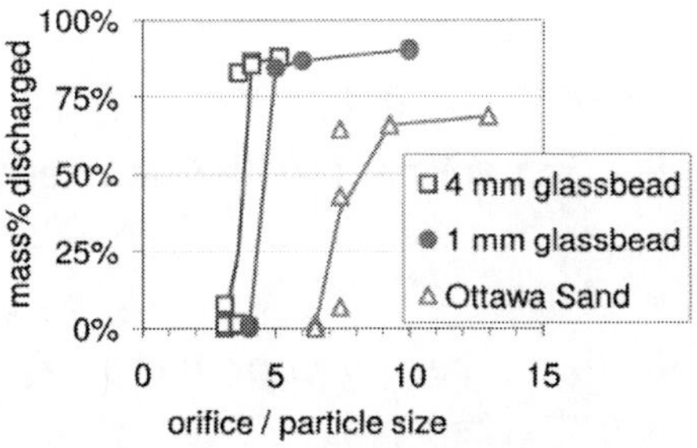

Sample	J	ϕ_{drain} (°)
4 mm glassbead	3.6	25.5
1 mm glassbead	4.3	19.0
Ottawa sand	7.1	38.6

- Cohesionless glassbeads follow theory (3-5 particle Jamming Onset).
- Ottawa Sand (washed and tightly classified) has a higher relative jamming onset and drained angle of repose.

Automated Image Analysis, Shape and Size Distribution

SolidSizer™, JM Canty Process Technologies, Buffalo, NY

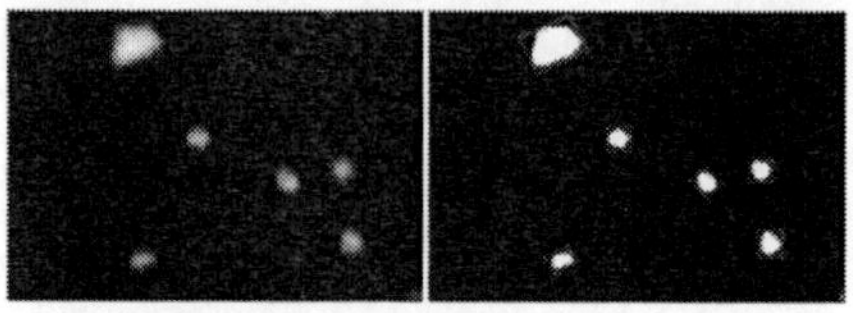

Particle Size Distribution:

- Particle size basis, $D_{particle} = \sqrt{(D1 \times D2)}$.
- Breadth of distribution quantified as geometric standard devation (σ_g).

Shape:

- Quantified by Aspect Ratio, AR = D1/D2.
- Median aspect ratio (AR_{50}) used as a characteristic.

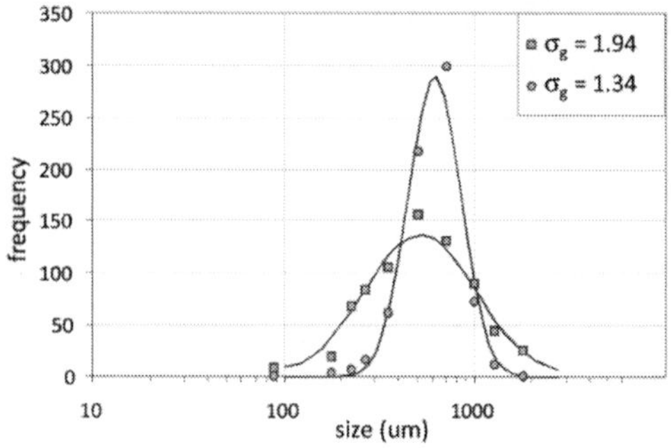

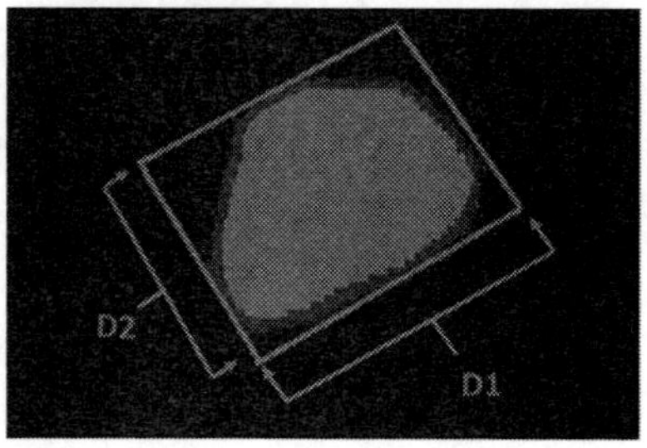

Friction

Frictional properties quantified by drained angle of repose (ϕ) measurement from orifice flow:

- Friction appears to be important when in excess of a certain value, ϕ^*, where $\phi^* \sim 45°$.
- Define Excess Friction factor, XF = MAX(1, tan(ϕ))

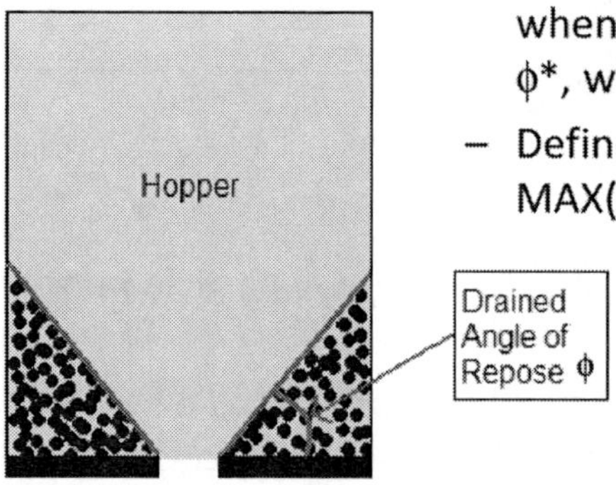

Jamming Onset: Multivariate regression result

Multivariate model of jamming onset (J) as a function of:

- particle aspect ratio (AR_{50}),
- size distribution (σ_g), and
- excess friction factor (XF).

An "Intrinsic Jamming Onset" is defined by the extrapolated intercept (k_j), i.e., k_j is the jamming onset for a "perfect" spherical, monodisperse, frictionless particle.

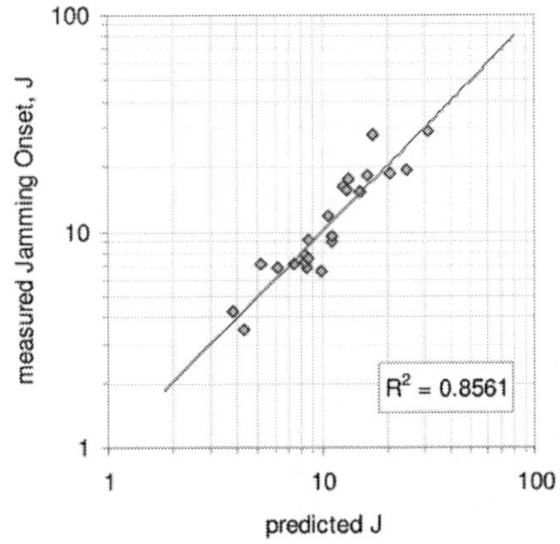

$$\ln(J) = \ln(k_j) + a\cdot\ln(AR_{50}) + b\cdot\ln(\sigma_g) + c\cdot\ln(XF)$$

$$k_j = 2.78,\ a = 2.7,\ b = 1.4,\ c = 1.2.$$

Summary: Open Questions on Attrition

- Optimization of unit operations based on granular rheology.
  - Stress field in dense flows, E/M, power spectra
  - Transient jamming
  - Boundary conditions
  - Systems integration
  - Implication of attrition on rheology
- IFPRI Legacy, van Brackel, 1981.

  "Investigations concerned with attrition (unwanted size reduction of particles) are scarce compared with literature on various aspects of comminution."
  - Catalyst attrition in fluid catalytic cracking systems
  - Dust formation in grinding
  - Miscellaneous fluid-particle systems
    - Packed and moving beds, spouted beds
    - Classification and heat exchange (grids)
    - Solids handling (bin filling, discharge, etc...)
- Way forward...
  - Industry / Academic collaboration
  - Adopt IFPRI model for "round table" workshop, develop strategy for size reduction program.

Fundamentals of Jet Milling and Hammermilling

Liu, G., DuPont

Originally presented on: 4/12/2016 11:30:00 - 12:00:00

Both jet milling and hammermilling are the most common size reduction operations in many industries. Fully understanding the fundamentals of these mills can assist the selection of the right type of milling equipment, as well as improving both their operability and energy efficiency. Here the milling mechanism, the importance of feeding, and the design of air systems for both mills are discussed. The selection of nozzle for jet mills, the classification principle, and the use of steam in jet milling are also deliberated.

1

EXAMPLES OF PARTICLE ATTRITION MODELING IN DEM SIMULATIONS

Carl Wassgren, Purdue University

Collaborators

2

- Yu Guo (U of FL)
- Xia Hua (Ford)
- Madhusudhan Kodam (Dow)
- Jennifer Curtis (UCD)
- Bruno Hancock (Pfizer)
- Bill Ketterhagen (Pfizer)

Three Examples

3

- Pharmaceutical tablet attrition
- Rigid fiber breakage
- Flexible fiber breakage

4

Tablet Attrition in a Rotating Drum

Tablet Attrition in a Rotating Drum

5

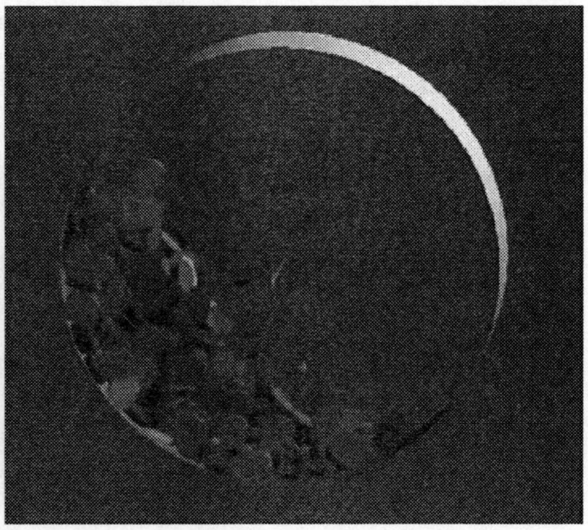

cylindrical MCC tablets

Kodam, M., 2010, *Attrition of Pharmaceutical Tablets*, Ph.D. Thesis, Purdue University.

Attrition Modeling Approach

6

- Use DEM to predict
  - impact frequencies and speed distributions
  - irreversible tangential work
- Implement sub-models to predict attrition
  - edge-chipping (Ghadiri and Zhang, 2002)
  - abrasive wear (Archard, 1953)

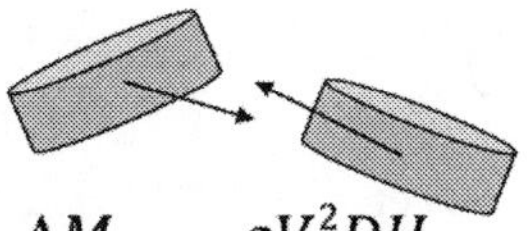

$$\frac{\Delta M}{M} = \alpha \frac{\rho V^2 D H}{K_{IC}^2}$$

edge-chipping

$$\Delta M = K_A \frac{\rho F_N \Delta s}{H}$$

abrasive wear

ρ	:	density
V	:	impact speed
D	:	tablet diameter
H	:	hardness
K_{IC}	:	critical stress intensity factor
F_N	:	normal force
Δs	:	tangential displacement
M	:	tablet mass
α, K_A	:	fitting constants

Attrition Modeling Approach...

7

- Tablet shape is assumed to remain unchanged
  - attrition calculations performed in DEM post-processing
- Parameters used in the models found from
  - literature (critical stress intensity factor)
  - measurement (size, density, hardness, edge-chipping and Archard wear constants)
  - found that mass loss due to Archard wear << edge-chipping

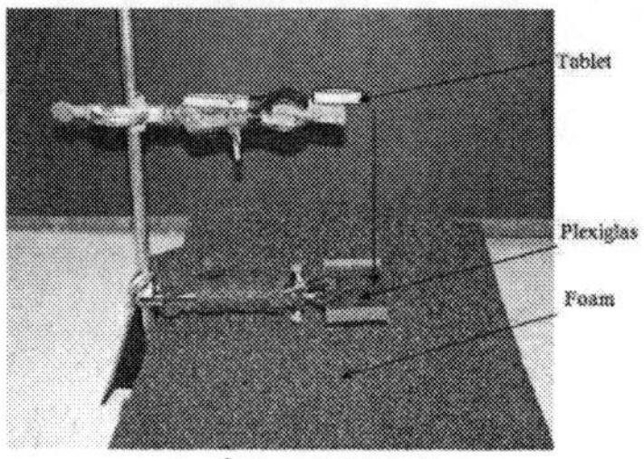

impact test

pin-on-disk test

Threshold Impact Speed, V^*

8

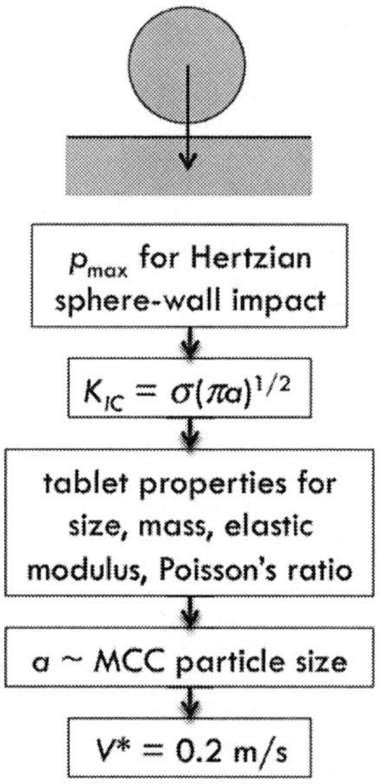

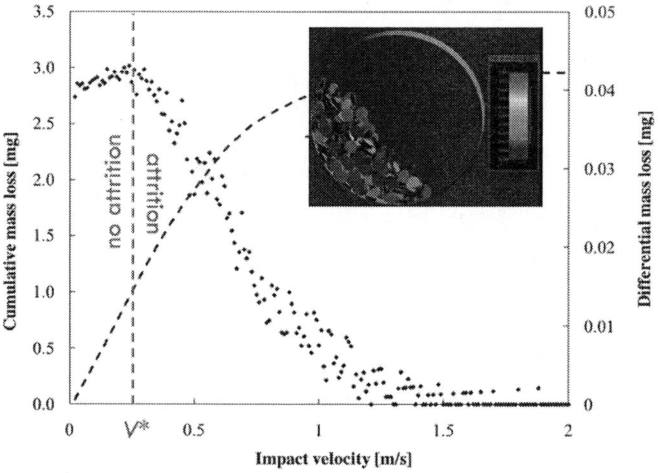

Results: Friability Tester

9

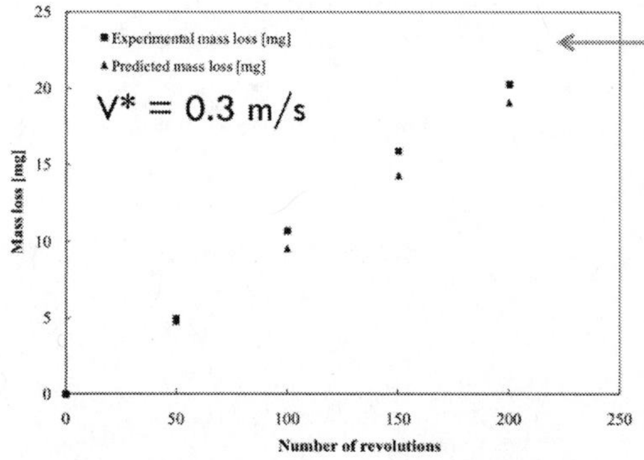

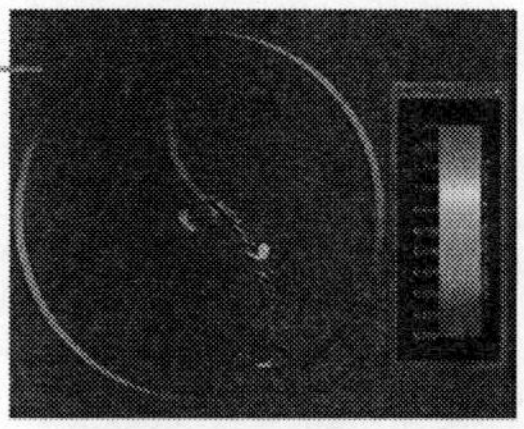

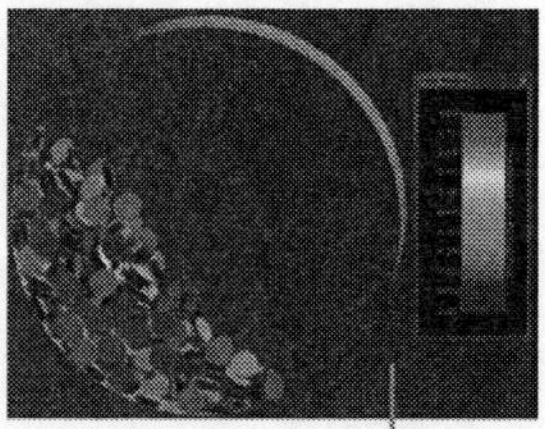

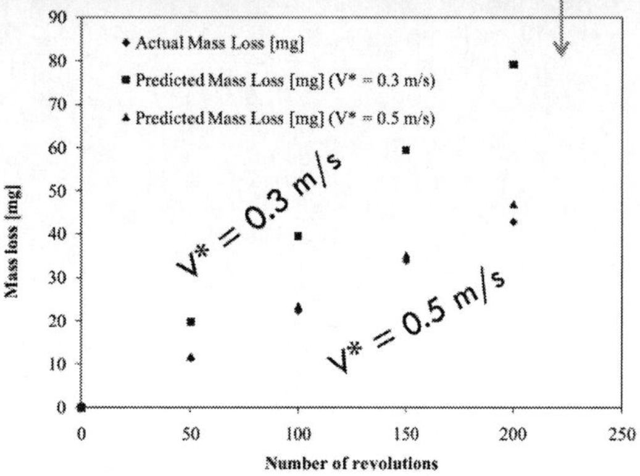

The fit value for V^* for the friability tester was a poor V^* for the rotating drum

Conclusions

10

- DEM + edge-chipping sub-model holds promise for predicting attrition in rotating drum systems
  - Should include a threshold impact speed
- Some validation and model issues to be resolved
  - Tablets were manufactured differently between friability tester and rotating drum
    - friability tester: wall lubrication => weaker tablet edges
    - rotating drum: no wall lubrication => stronger tablet edges
  - Tablet-tablet collision also involves KE transfer

11 Breakage of Rigid Fibers

System of Interest

12

- Model of an agitated filter dryer (AFD)
- Rod-shaped active pharmaceutical ingredient (API) particles
- Particles modeled as rigid, unbreakable sphero-cylinders

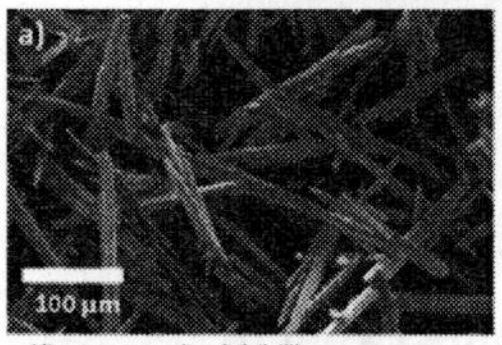

(Remy et al., 2015)

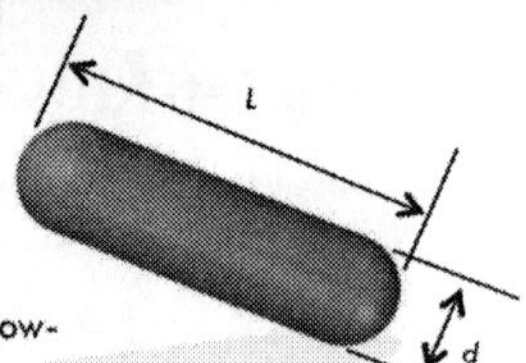

Hua et al., 2015, "The internal loads, moments, and stresses in rod-like particles in a low-speed, vertical axis mixer", *Chem Eng Sci*, Vol. 134, pp. 581 – 598.

Internal Particle Loads and Moments

13

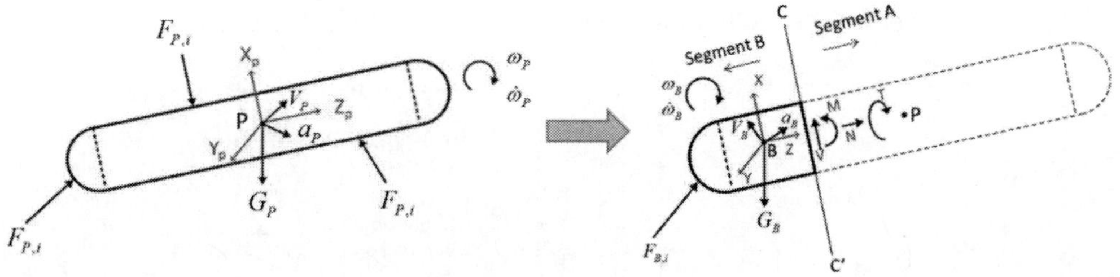

$$\mathbf{a}_B = \mathbf{a}_P + \boldsymbol{\omega}_P \times (\boldsymbol{\omega}_P \times \mathbf{r}_{B/P}) + \dot{\boldsymbol{\omega}}_P \times \mathbf{r}_{B/P}$$

$$\boldsymbol{\omega}_B = \boldsymbol{\omega}_P$$

$$\dot{\boldsymbol{\omega}}_B = \dot{\boldsymbol{\omega}}_P$$

translational acceleration; angular velocity and acceleration

$$\mathbf{N} + \mathbf{V} + \sum \mathbf{F}_{B,i} + \mathbf{G}_B = m_B \mathbf{a}_B$$

$$I_B \dot{\boldsymbol{\omega}}_B + \boldsymbol{\omega}_B \times (I_B \boldsymbol{\omega}_B) = \left(\sum \mathbf{r}_{i,B} \times \mathbf{F}_{B,i}\right) + \mathbf{M} + \mathbf{T} + \left(\mathbf{r}_{CC'/B} \times \mathbf{V}\right)$$

force and moment balance

$$\mathbf{N} = N\hat{\mathbf{e}}_z$$

$$\mathbf{V} = V_x \hat{\mathbf{e}}_x + V_y \hat{\mathbf{e}}_y$$

$$\mathbf{M} = M_x \hat{\mathbf{e}}_x + M_y \hat{\mathbf{e}}_y$$

$$\mathbf{T} = T\hat{\mathbf{e}}_z$$

forces and moments at cross-section

Internal Stresses

14

Beam model:

$$\sigma_{xx} = 0$$

$$\sigma_{yy} = 0$$

$$\sigma_{zz} = \frac{NR^2 + 4M_x y - 4M_y x}{\pi R^4}$$

$$\sigma_{xy} = \sigma_{yx} = 0$$

$$\sigma_{yz} = \sigma_{zy} = \frac{4V_y\left(R^2 - y^2\right) + 6Tx}{3\pi R^4}$$

$$\sigma_{xz} = \sigma_{zx} = \frac{4V_x\left(R^2 - x^2\right) - 6Ty}{3\pi R^4}$$

(stress predictions improve as AR ↑)

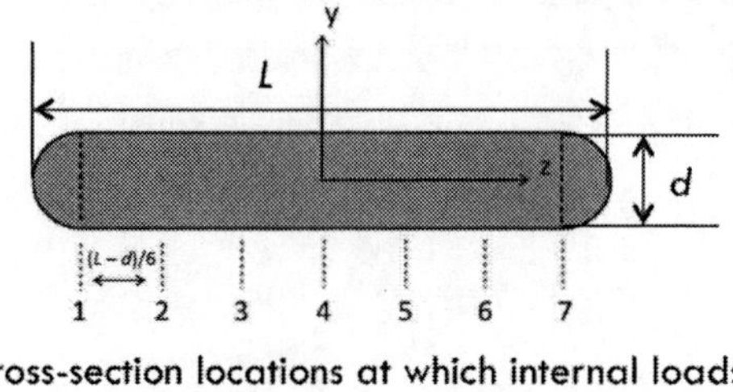

cross-section locations at which internal loads and moments are determined

Common failure models:

Brittle failure:

$$|\sigma_1| \text{ or } |\sigma_3| \geq S_u$$

Ductile failure:

$$\sigma_M = \sqrt{\tfrac{1}{2}\left[(\sigma_1 - \sigma_2)^2 + (\sigma_2 - \sigma_3)^2 + (\sigma_1 - \sigma_3)^2\right]}$$

$$\sigma_M \geq \sigma_y$$

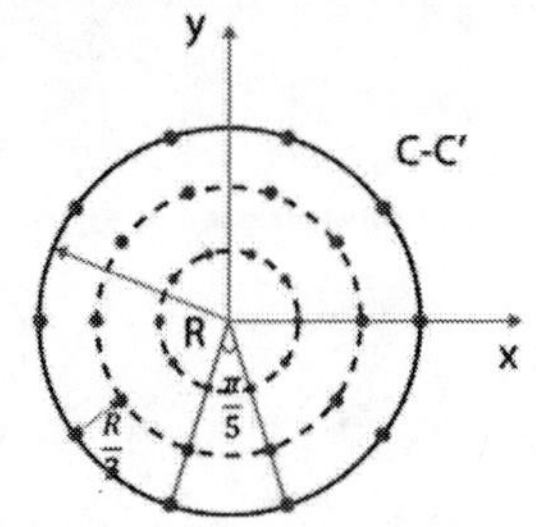

sample points (in red) at which stresses are calculated

Loads and Moments

15

- On average, max loads and moments at center-plane
- Loads and moments ↑ as AR ↑ (normal load insensitive to AR)
- Twisting moment << bending moment

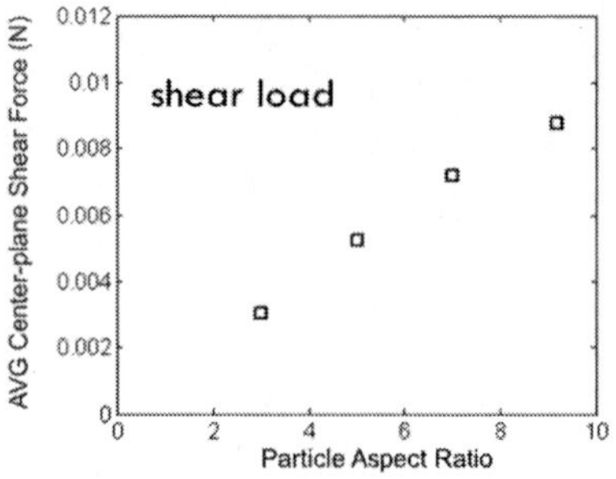

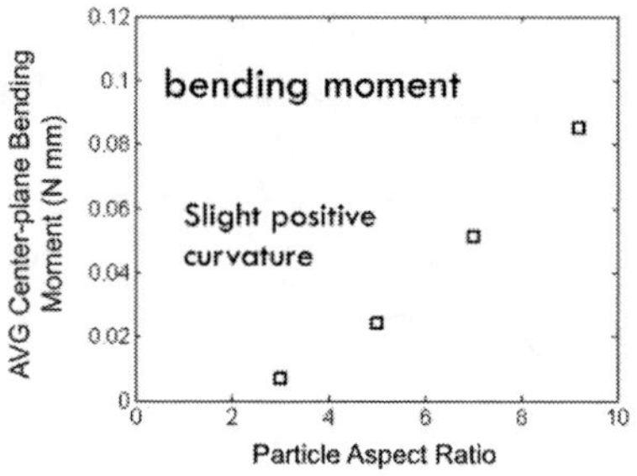

- Maximum principal stress and Mises stress have similar values
- Maximum principal stress located at circumference of center-plane

Maximum Principal Stress Distributions

16

- Larger max principal stresses follow a Weibull distribution (two fitting parameters)
- Complementary cumulative distribution can be used to estimate fraction of particles expected to break

$$f(x) = abx^{b-1}\exp(-ax^b)$$

$$1 - F(x) = \exp(-ax^b)$$

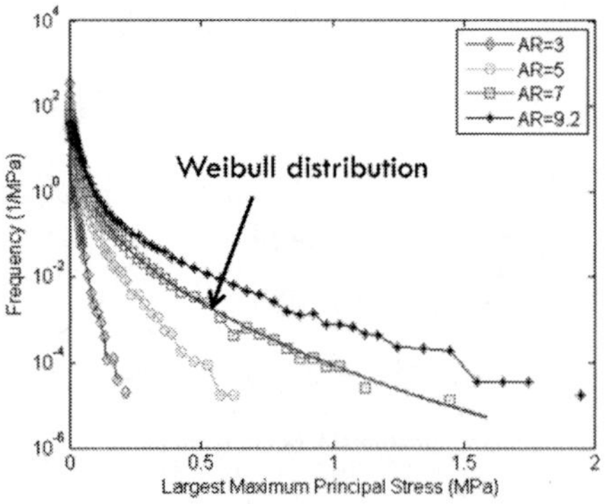

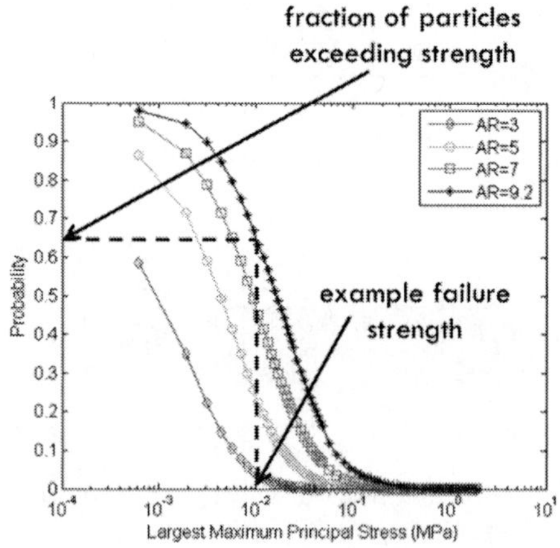

Conclusions

17

- Internal load and moment balance coupled with beam theory offers potential for predicting breakage of high aspect ratio, rigid, rod-like particles, e.g., API crystals
  - Potentially faster than bonded-element approach
  - Smoother particles than bonded-spheres
  - Fragments not necessarily discrete sizes, but the smallest size is limited by particle cross-sectional diameter (a sphere)
- In-simulation particle breakage can be included

18

Breakage of Flexible Fibers

System of Interest

19

- Attrition cell
- Rod-shaped API particles
- Particles modeled as flexible, breakable, bonded spheres

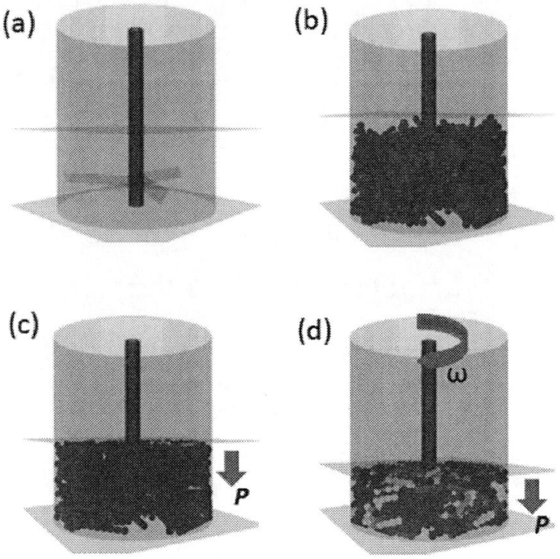

Guo et al., 2013, "Validation and time step determination of discrete element modeling of flexible fibers", *Powd. Tech.*, 249, 386 – 395.
Guo et al., 2015, "Computational study of granular shear flows of dry flexible fibres using the discrete element method", *JFM.*, 775, 24 – 52.

Fiber Model

20

- Bonded spheres (originally by Potyondy and Cundall, 2004)

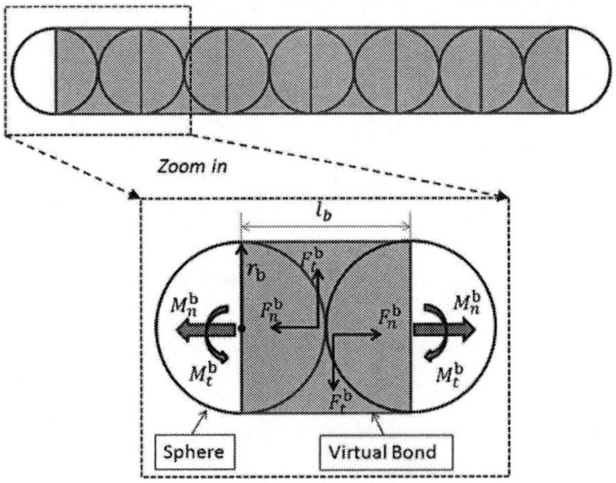

$$dF_n^b = K_n^b d\delta_n^b = \frac{E_b A}{l_b} d\delta_n^b = \frac{E_b A}{l_b} v_n^r dt$$

$$dF_t^b = K_t^b d\delta_t^b = \frac{G_b A}{l_b} d\delta_t^b = \frac{G_b A}{l_b} v_t^r dt$$

$$dM_n^b = K_{\text{tor}}^b d\theta_n^b = \frac{G_b I_p}{l_b} d\theta_n^b = \frac{G_b I_p}{l_b} \omega_n^r dt$$

$$dM_t^b = K_{\text{ben}}^b d\theta_t^b = \frac{E_b I}{l_b} d\theta_t^b = \frac{E_b I}{l_b} \omega_t^r dt$$

$$F_{dn}^b = \beta_b \sqrt{2 m_s K_n^b} v_n^r$$

$$F_{dt}^b = \beta_b \sqrt{2 m_s K_t^b} v_t^r$$

$$M_{dn}^b = \beta_b \sqrt{2 J_s K_{\text{tor}}^b} \omega_n^r$$

$$M_{dt}^b = \beta_b \sqrt{2 J_s K_{\text{ben}}^b} \omega_t^r$$

Fiber Breakage

21

- If (bond stress) > (bond strength) => breakage

$$\sigma^b_{\max} = \frac{F^b_n}{A} + \frac{|M^b_t| r_b}{I} > \sigma_{\text{ten}}$$

$$\tau^b_{\max} = \frac{|F^b_t|}{A} + \frac{|M^b_n| r_b}{I_p} > \tau_{\text{sh}}$$

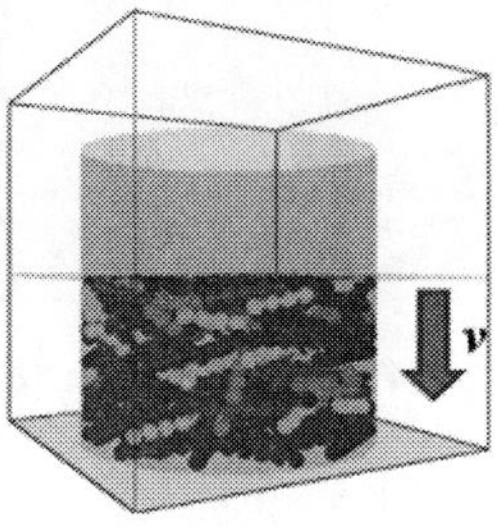

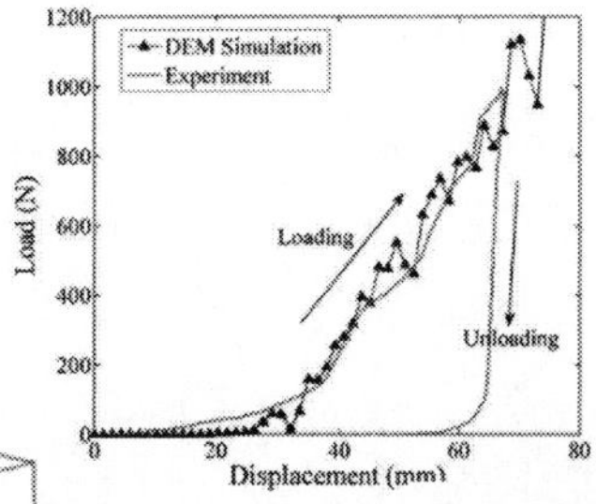

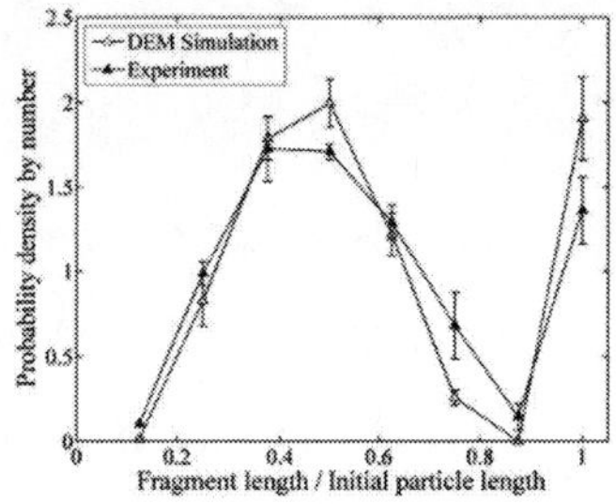

Particle Damage in Attrition Cell

22

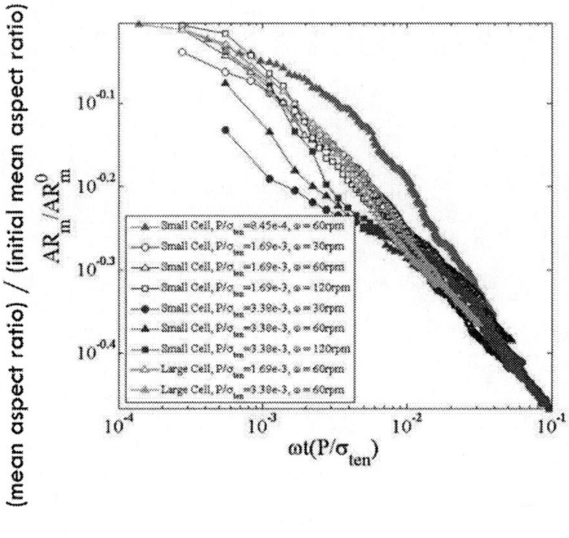

- (weight fraction of fragments) and (mean aspect ratio) are proportional to (total work done on particles)

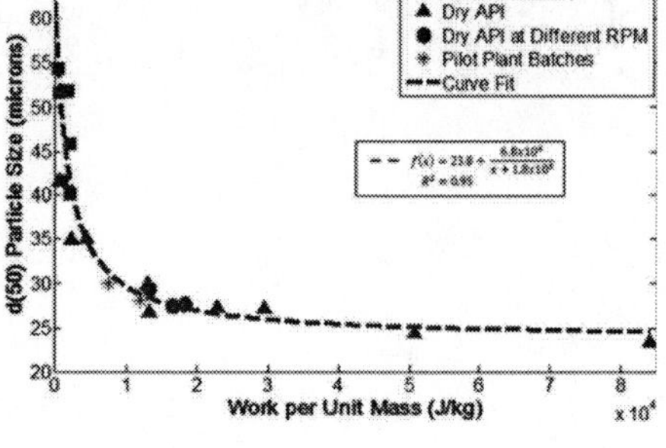

Remy et al. (2015)

Conclusions

23

- Bonded element, flexible fiber DEM a good method for fundamental studies
  - may be too computationally intensive for application-level studies
  - fragment size limited by element size
- Can make particles smoother by
  - including more bonded spheres: increases computational time, not ideal for large AR
  - using bonded sphero-cylinders: accuracy and time step are issues
- Can also include bond damage accumulation model

ITASCA
Consulting Group, Inc.

Bonded-Particle Modeling of Fracture and Flow

David Potyondy

Frontiers in Particle Science & Technology Conference:
Mitigation & Application of Particle Attrition
(Houston, TX, April 11-13, 2016)

Overview

Bonded-Particle Modeling
- Basic idea & essential features
- Microstructural physics of intact rock
- PFC model
- Microstructural models provided by BPM

Engineering Applications
- Rock Fracture
  - ➔ rock cutting (rock microstructure : cutter geometry)
- Material Flow
  - ➔ cave mining (physical & PFC models, theory ➔ REBOP)

Conclusions

Basic Idea

Rock Mass

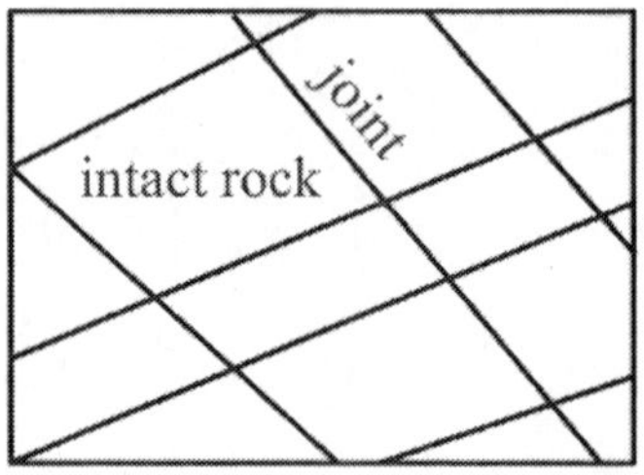

Mechanical behavior arises from interactions:

- microstructure (grain scale) of intact rock
- orientations & properties of joints

Basic Idea

Rock Mass

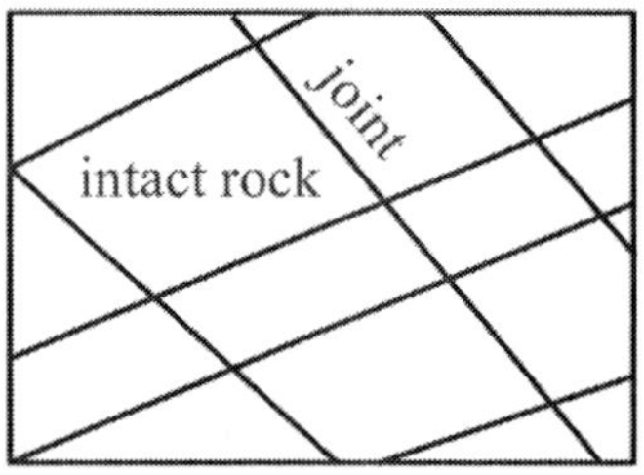

emergent phenomena

Mechanical behavior arises from interactions:

- microstructure (grain scale) of intact rock ➔ fracture & flow
- orientations & properties of joints ➔ rupture & sliding

Bonded-Particle Models mimic intact rock & joints

> Closer match to microstructural & structural features ➔
> Closer match to macroscopic behavior

Essential Features of a BPM

BPM consists of a base material (intact rock) to which larger-scale joints can be added.

- base material : bonded rigid grains
- joints : interfaces

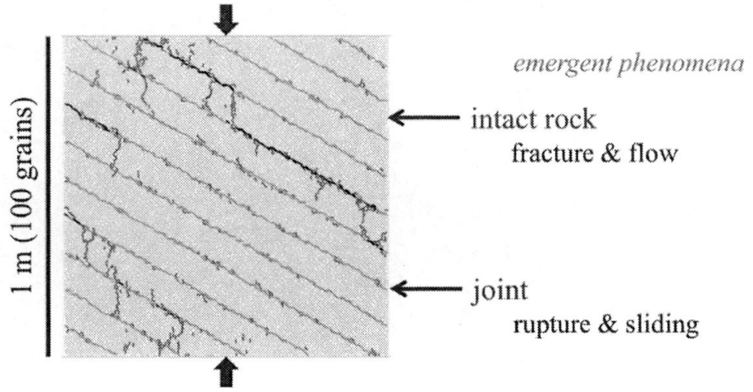

Damage consists of bond breakages.

ITASCA

Essential Features of a BPM

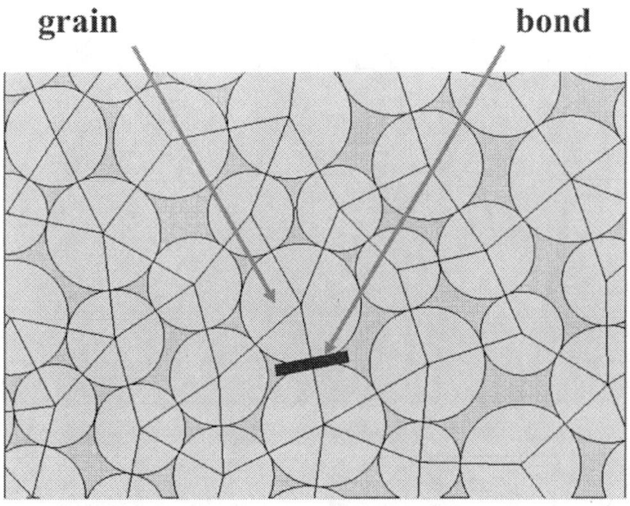

Damage consists of bond breakages.

ITASCA

Essential Features of a BPM

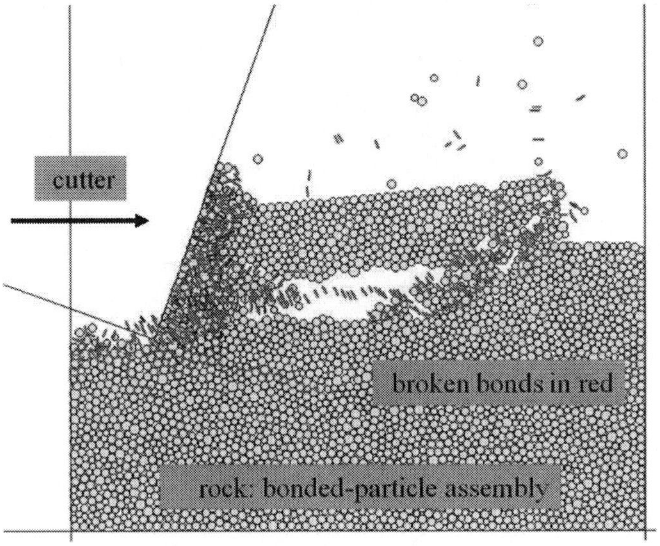

Synthetic material will fracture & flow in response to loading, and can evolve from a solid to a granular state.

Essential Features of a BPM

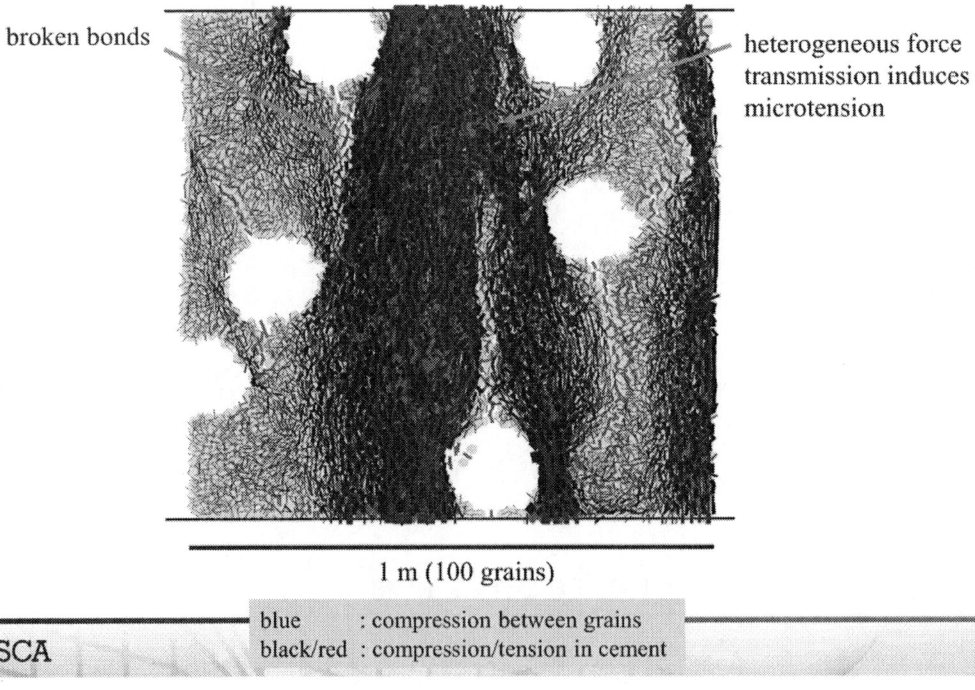

ITASCA

Essential Features of a BPM

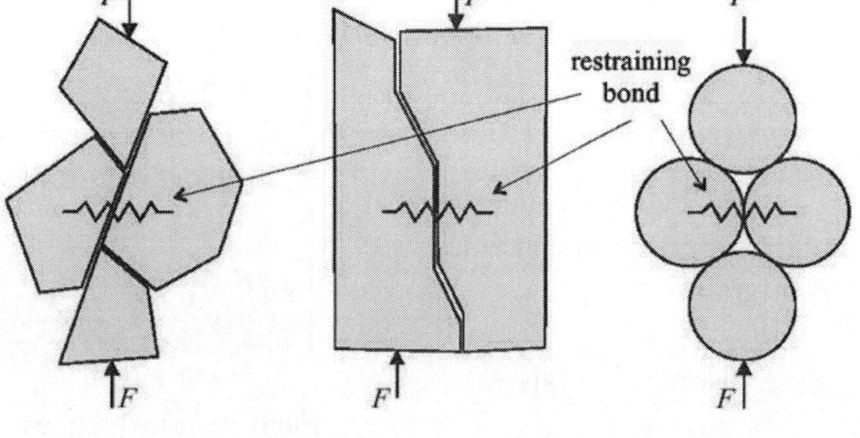

Microstructural mechanisms in cemented granular material to induce microtension and bond breakage.

ITASCA

Essential Features of a BPM

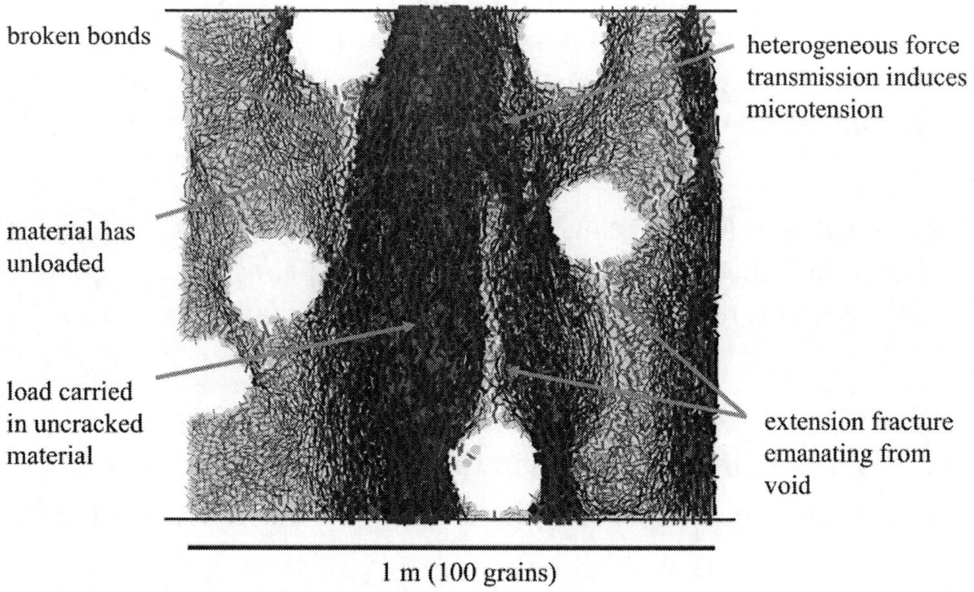

ITASCA

Microstructural Physics of Intact Rock

Intact rock can be viewed as an aggregate of crystals & amorphous particles joined by varying amounts of cementing materials.

Intact rock can be represented as…

Heterogeneous material comprised of cemented grains.

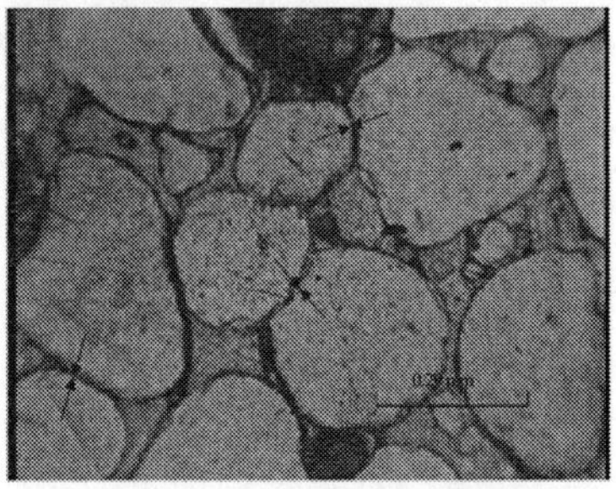

sandstone Dittes & Labuz (2002)

Much disorder in system:

grain size, shape & packing
grain & cement properties
degree of cementation
locked-in stresses

Each item influences mechanical behavior, and may evolve under load application.

ITASCA Dittes, M., and J.F. Labuz (2002) "Field and Laboratory Testing of St. Peter Sandstone," *J. Geotech. Engr.*, *128*(5), 372-380, May.

Microstructural Physics of Intact Rock

Compressive behavior is explained in terms of flaws & heterogeneities in the rock material on the microscopic scale in which brittle fracture is viewed as a localization in the proliferation of microcracking.

It is essential to have a realistic structural description of the rock as a basis for…theoretical developments, and to consider the evolution of the structure during the progression of the failure.

BPMs provide a micromechanical theory of brittle failure

- various physical models of the brittle-fracture process
- discern the effects of microstructure on macroscopic behavior

ITASCA

Historical Development (Peter Cundall & Itasca codes)

Peter Cundall

- Graduate student of Evert Hoek (late 1960's)

  Problem: Model a jointed rock mass ➔ Distinct-Element Method (Cundall, 1971)

ITASCA

Cundall, P.A. (1971) "A Computer Model for Simulating Progressive Large Scale Movements in Blocky Rock Systems," Proc. ISRM Symp. (Nancy, France, 1971), Vol. 1, Paper II-8.

Historical Development (Peter Cundall & Itasca codes)

Peter Cundall

- Graduate student of Evert Hoek (late 1960's)

  Problem: Model a jointed rock mass ➔ Distinct-Element Method (Cundall, 1971)

- University of Minnesota (1971 -)
  - BALL & TRUBALL (Cundall & Strack, 1979)

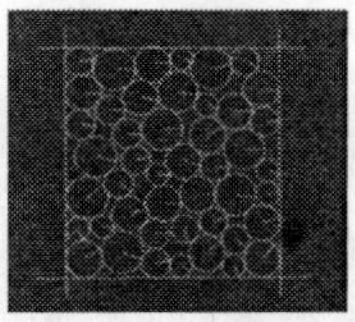

- Itasca (1981 -)
  - Itasca founded in 1981 to carry University research into practical application in rock engineering (2015: 185 personnel, 16 offices).
  - Universal Distinct Element Code (UDEC, 1984)
  - Fast Lagrangian Analysis of Continua (FLAC, 1986)
  - Particle Flow Codes (PFC2D/3D, 1994; PFC 5.0, 2014)

ITASCA

Cundall, P.A. (1971) "A Computer Model for Simulating Progressive Large Scale Movements in Blocky Rock Systems," Proc. ISRM Symp. (Nancy, France, 1971), Vol. 1, Paper II-8.
Cundall, P.A, and O.D.L. Strack (1979) "A Discrete Numerical Model for Granular Assemblies," *Géotechnique*, *29*, 47–65.

Bonded-Particle Modeling Methodology (PFC model)

PFC programs (PFC2D & PFC3D)* provide a general-purpose, distinct-element modeling framework that includes a computational engine and a graphical user interface. Itasca has sold ~1500 licenses since 1994.

Simulate movement & interaction of many finite-sized particles via distinct-element method, which provides an explicit dynamic solution to Newton's laws of motion.

Particles are rigid bodies with finite mass that move independently of one another and can both translate and rotate.

ITASCA

* Itasca Consulting Group, Inc. (2015) ***PFC — Particle Flow Code in 2 and 3 Dimensions***, Version 5.0. Minneapolis: Itasca.

Bonded-Particle Modeling Methodology (PFC model)

Particles may be balls or clumps

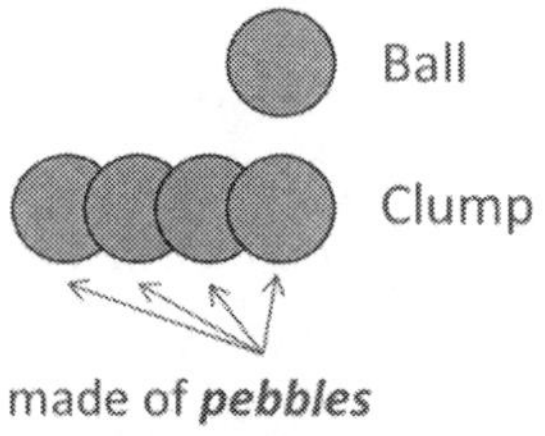

Behaves as rigid body with deformable boundary

ITASCA

Bonded-Particle Modeling Methodology (PFC model)

Particles interact at pair-wise contacts by means of internal force and moment.

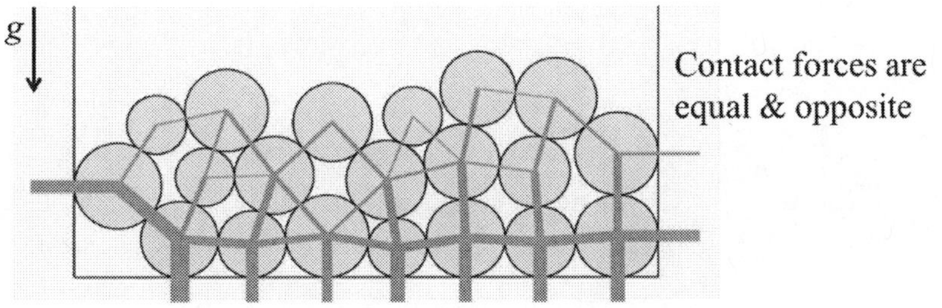

ITASCA

Bonded-Particle Modeling Methodology (PFC model)

Particles interact at pair-wise contacts by means of internal force and moment.

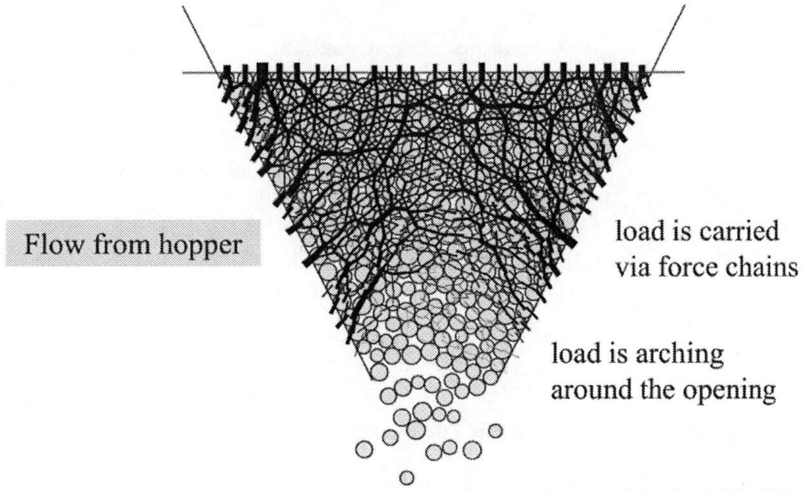

ITASCA

Bonded-Particle Modeling Methodology (PFC model)

Contact mechanics is embodied in particle-interaction laws that employ a soft-contact approach for which all deformation occurs at the contacts between the rigid bodies. The particle-interaction law (contact model) updates the internal force and moment.

ITASCA

soft-contact approach

Consider the system of me standing on two baseballs, pressed between steel plates. Top plate moves down by Δ .

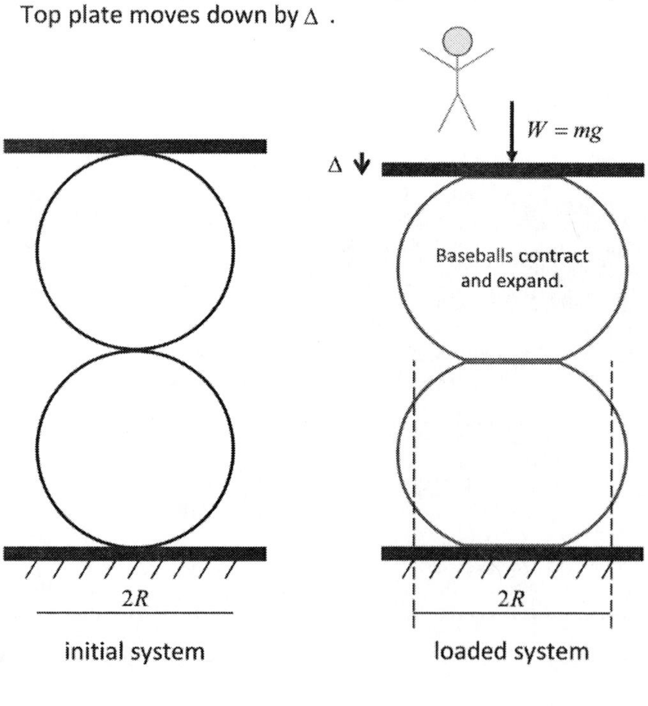

soft-contact approach

Consider the system of me standing on two baseballs, pressed between steel plates.
Top plate moves down by Δ .

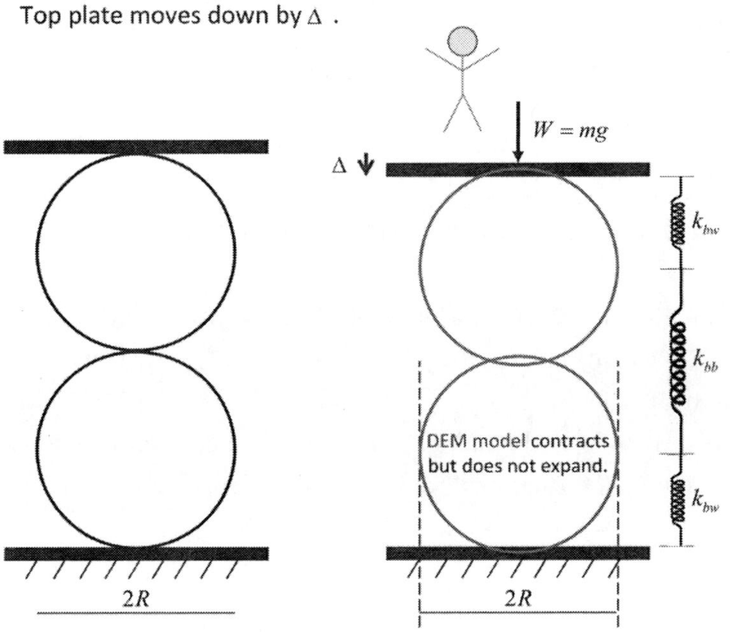

initial system　　loaded system

DEM model employs a "soft contact" approach:

all deformation occurs at the contacts between the rigid bodies.

The stiffnesses can be related to the effective modulus of an equivalent continuum. . .

soft-contact approach

Consider the system of me standing on two baseballs, pressed between steel plates.
Top plate moves down by Δ .

$$k_{bw} = \frac{AE^*}{R} = \pi R E^*$$

$$k_{bb} = \frac{AE^*}{2R} = \tfrac{1}{2} k_{bw}$$

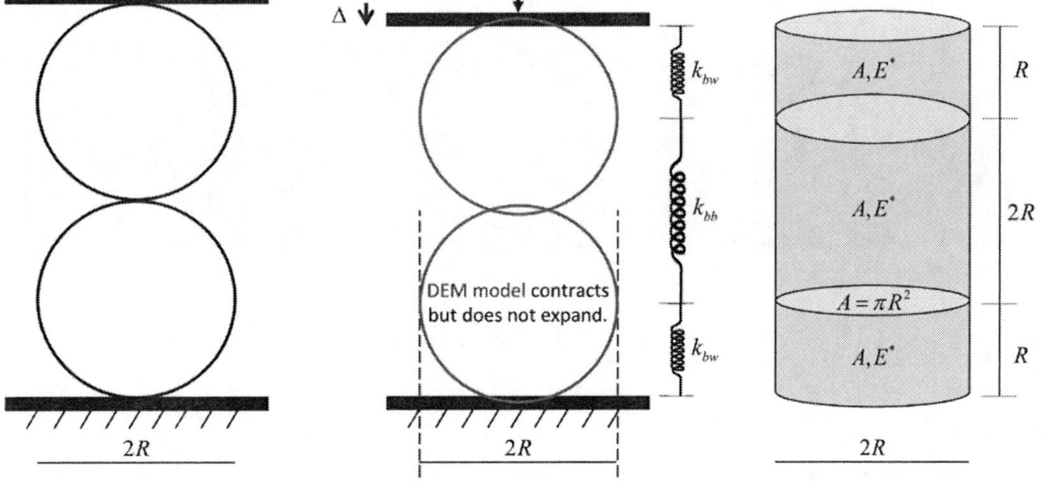

initial system　　loaded system　　equivalent continuum

Bonded-Particle Modeling Methodology (PFC model)

PFC model provides a synthetic material: rigid grains that interact at contacts, which encompasses a vast microstructural space --- only a small portion of this space has been explored.

PFC model includes both granular and bonded materials as well as an interface that can be inserted into the bonded materials.

The most up-to-date incarnation of the PFC model is provided in the form of the linear, contact-bonded, parallel-bonded, and flat-jointed materials to support:

- practical applications (via boundary-value models made from them)
- scientific inquiry (via further exploration of microstructural space)

ITASCA Potyondy, D. (2016) "Material-Modeling Support in PFC [fistPkg20]," Itasca Consulting Group, Inc., Technical Memorandum ICG7766-L (February 26, 2016), Minneapolis, Minnesota.

Microstructural Models Provided by BPM

Rich variety of models, described and classified

- base material itself (intact rock)
- overlay joints, voids & material regions

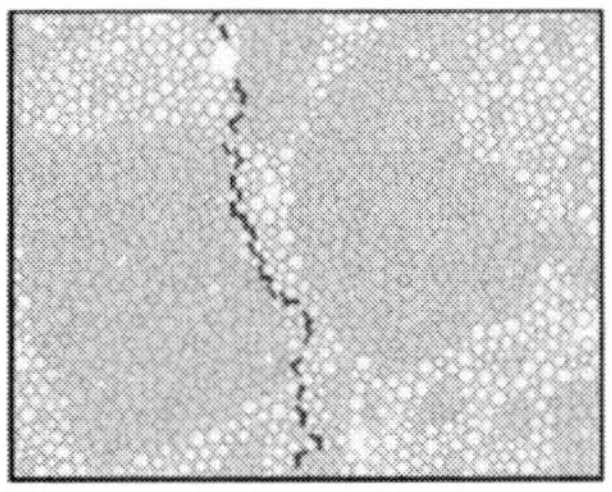

ITASCA Potyondy, D. O. (2015) "The Bonded-Particle Model as a Tool for Rock Mechanics Research and Application: Current Trends and Future Directions," *Geosystem Engineering*, *18*(1), 1–28.

Microstructural Models Provided by BPM

Base material itself can serve as model of intact rock

- rigid **grains** joined by deformable & breakable **cement**

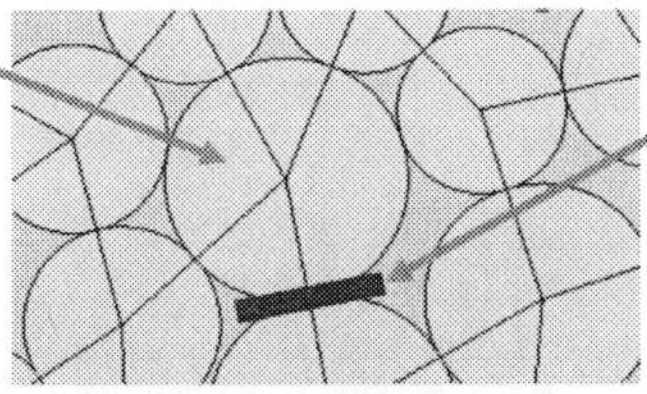

grains can be balls or clumps

cement can be

- contact-bonded **contact**
- parallel-bonded **contact**
- flat-joint **contact**

When bond breaks, behaves as linear contact.

ITASCA

PFC Model (linear contact, behavior summary)

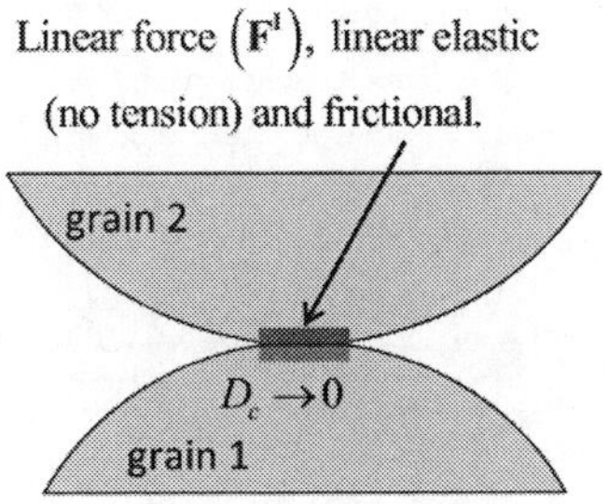

Behaves like glass beads.

PFC Model (parallel-bonded contact, behavior summary)

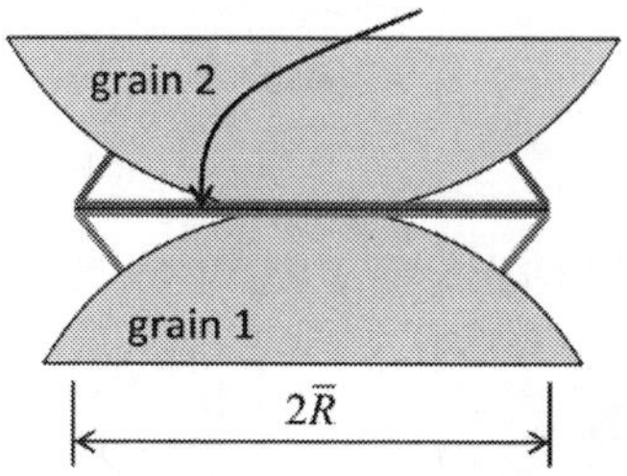

Behaves like
glass beads
cemented with epoxy.

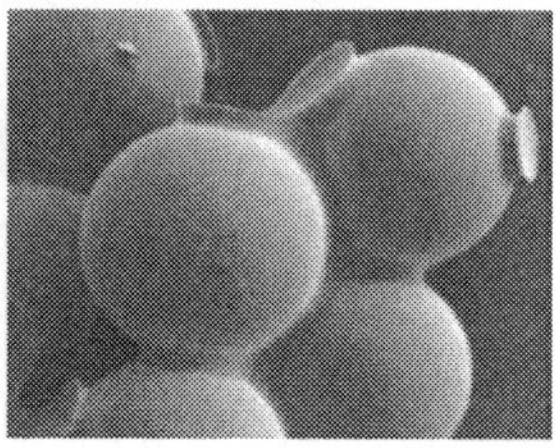

PFC Model (flat-jointed contact, behavior summary)

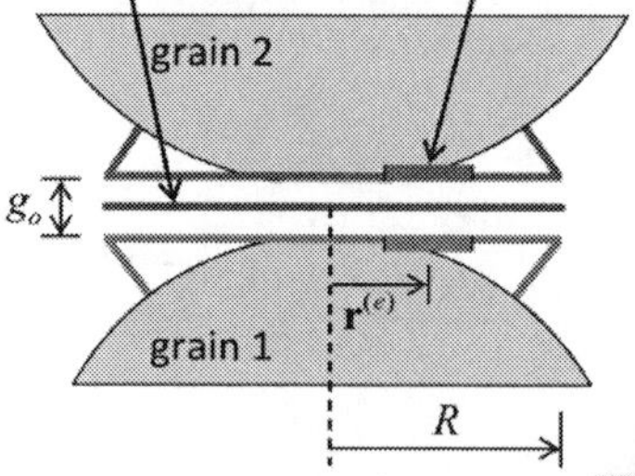

Behaves like
marble
with angular, interlocked grains.

PFC Model (flat-jointed material, behavior summary)

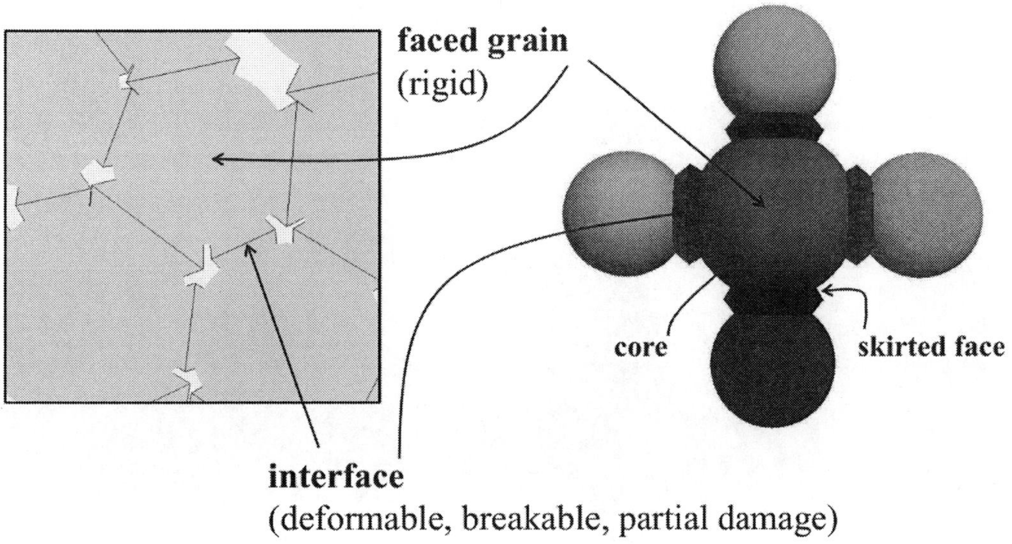

PFC Model (flat-jointed material, behavior summary)

Bending failure with. . .

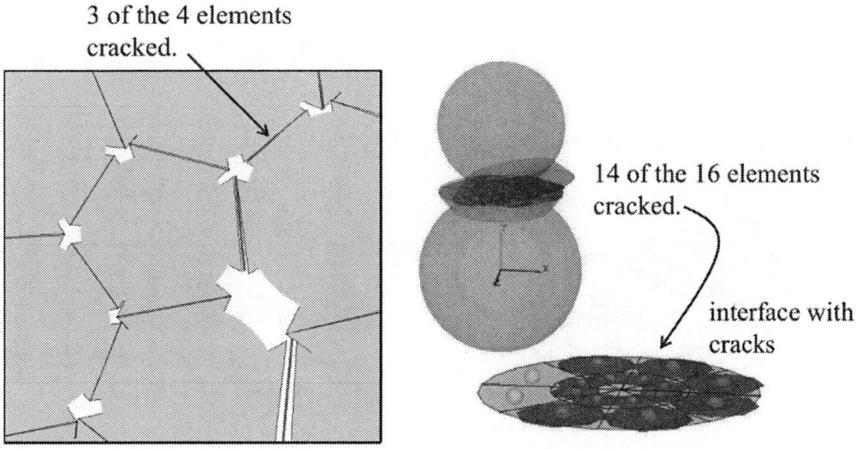

Crack thickness is proportional to gap.

The interface can sustain partial damage.

PFC Model (flat-jointed material, behavior summary)

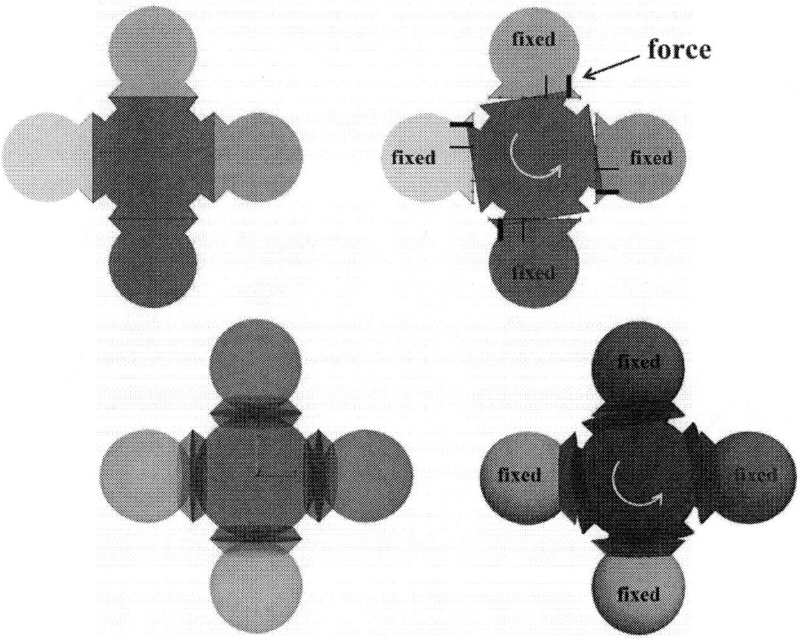

Even a fully broken interface continues to resist relative rotation.

Microstructural Models Provided by BPM

Joints can be overlaid on base material

- joint behaves as **interface**

interface is collection of smooth-joint contacts

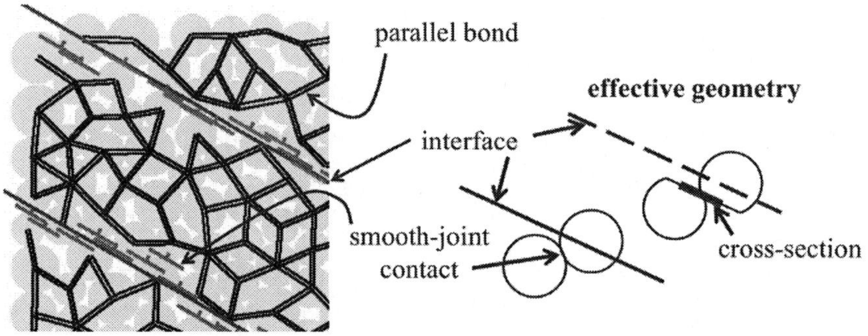

Smooth-joint contacts may be initially bonded. When bond breaks, they behave as linear contact.

ITASCA

Microstructural Models Provided by BPM

Void regions are identified…

- grains within these regions are removed

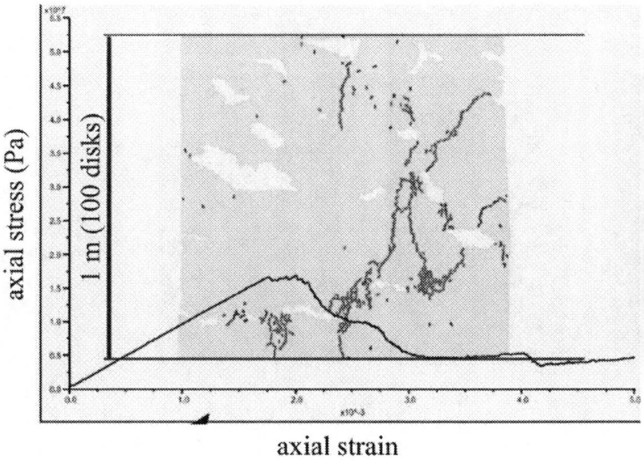

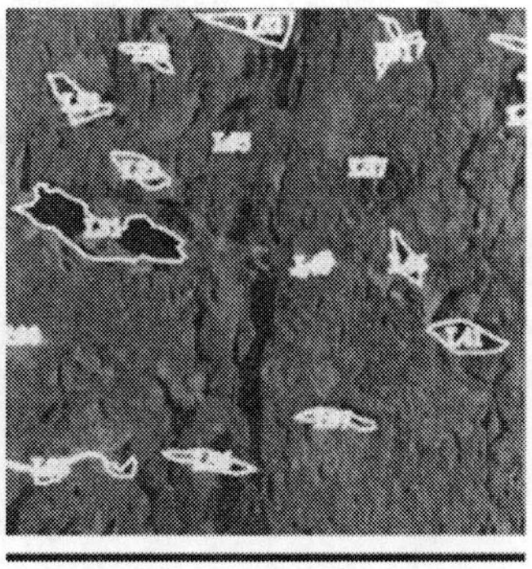

ITASCA Potyondy (2007) "The Effect of Voids on the Mechanical Properties of Rock," in Proc. 4th Int. Conf. on Discrete Element Methods, Brisbane, Australia, Cleary, Ed., MEI Conferences, CD.

Microstructural Models Provided by BPM

Material regions are identified…

- grains & contacts within these regions are assigned properties

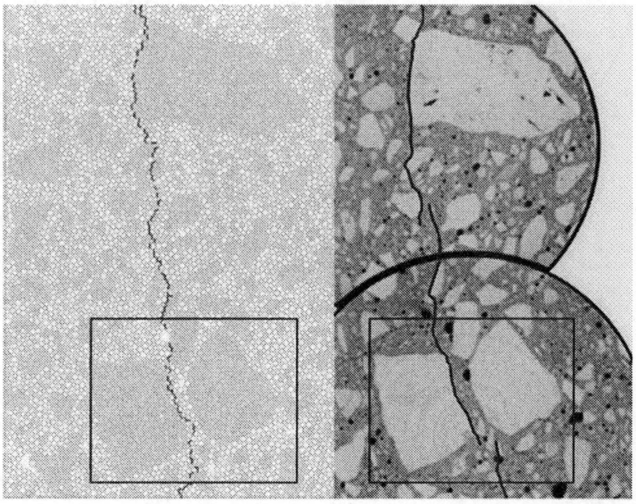

Katsaga (2010)

ITASCA Katsaga, T. (2010) "Geophysical Imaging and Numerical Modelling of Fractures in Concrete," Ph.D. Thesis, University of Toronto.

Microstructural Models Provided by BPM

Joints are identified…

- contacts that lie along these joints are assigned properties

Synthetic Rock Mass (SRM)

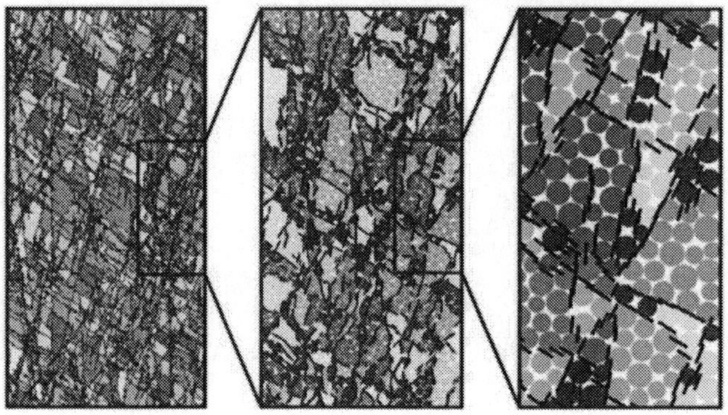

SRMs are used to represent a rock mass. The BPMs are remarkably complex, and used to study impact of joint fabric on rock mass strength, brittleness and fragmentation.

ITASCA Mas Ivars et al. (2011) "The Synthetic Rock Mass Approach for Jointed Rock Mass Modelling," *Int. J. Rock Mech. & Min. Sci.*, ***48***, 219–244.

Microstructural Models Provided by BPM

Material regions & joints are identified…

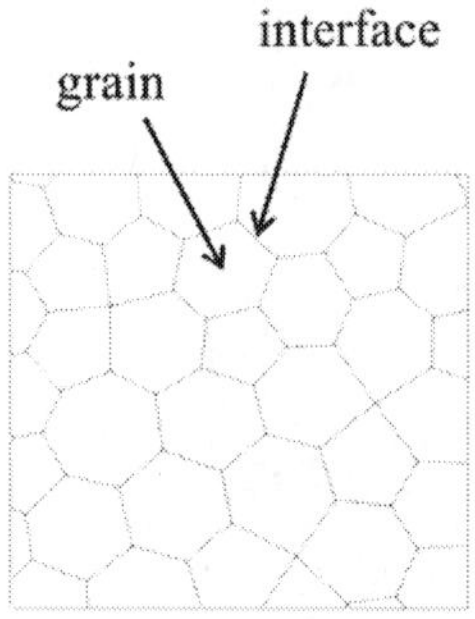

grain structure

Grain Based Model (GBM)

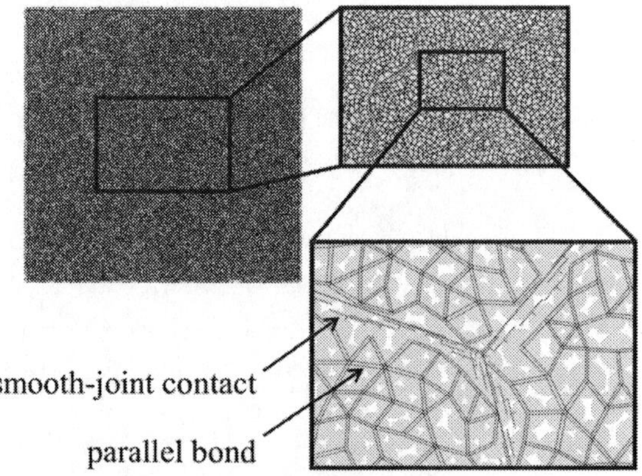

GBMs are used to represent intact compact rock, allow partial interface damage and grain breakage.

ITASCA Potyondy (2010) "A Grain-Based Model for Rock: Approaching the True Microstructure," in *Rock Mechanics in the Nordic Countries 2010*, pp. 225–234, C.C. Li, et al., Eds., ISBN: 978-82-8208-017-0, Kongsberg: Norwegian Group for Rock Mechanics.

Overview

Bonded-Particle Modeling

- Basic idea & essential features
- Microstructural physics of intact rock
- PFC model
- Microstructural models provided by BPM

Engineering Applications

- Rock Fracture
  - ➔ rock cutting (rock microstructure : cutter geometry)
- Material Flow
  - ➔ cave mining (physical & PFC models, theory ➔ REBOP)

Conclusions

Engineering Applications (Rock Fracture Example)

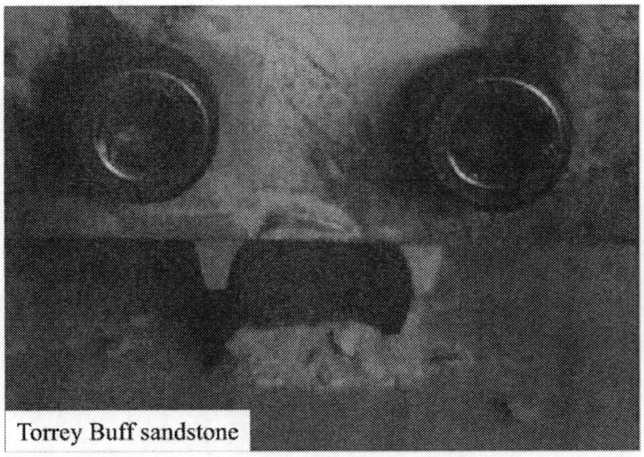

rock-cutting test (dry)

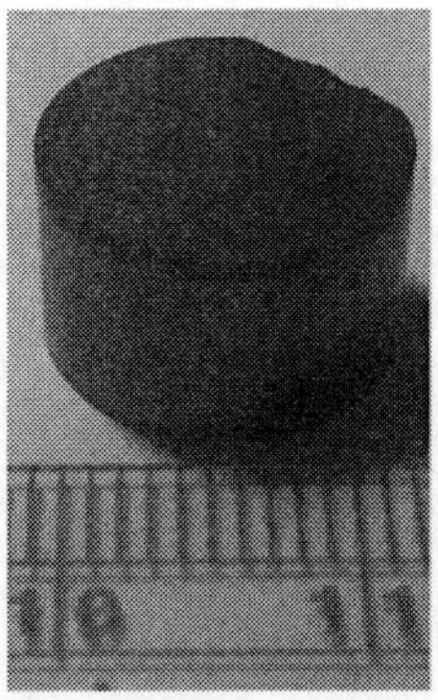

cutter (13-mm diameter)
depth of cut (1 mm)

Rock Cutting (system behavior)

Models of rock-cutting tests (dry & wet)

- Create synthetic rock (parallel-bonded material).
- Move cutter across surface of synthetic rock, while monitoring forces on the cutter & damage in the rock.
  - In a real borehole, drilling mud produces a pressure that acts on the rock surface and effectively strengthens the rock, thereby greatly increasing the energy requirements of drilling.
- Mimic the effect of drilling mud by identifying surface particles and applying pressure to those particles.
  - algorithms: chaining (2D) and shining-lamp (3D)
  - The pressure inhibits chip formation and leads to a more plastic-like zone of damage.

Rock Cutting (model, 2D)

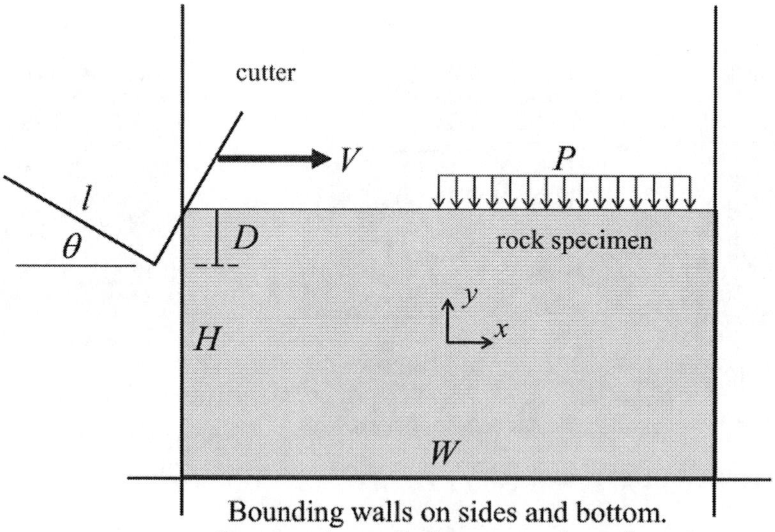

Bounding walls on sides and bottom.
Pressure is applied to top rock surface.

Rock Cutting (model, 2D)

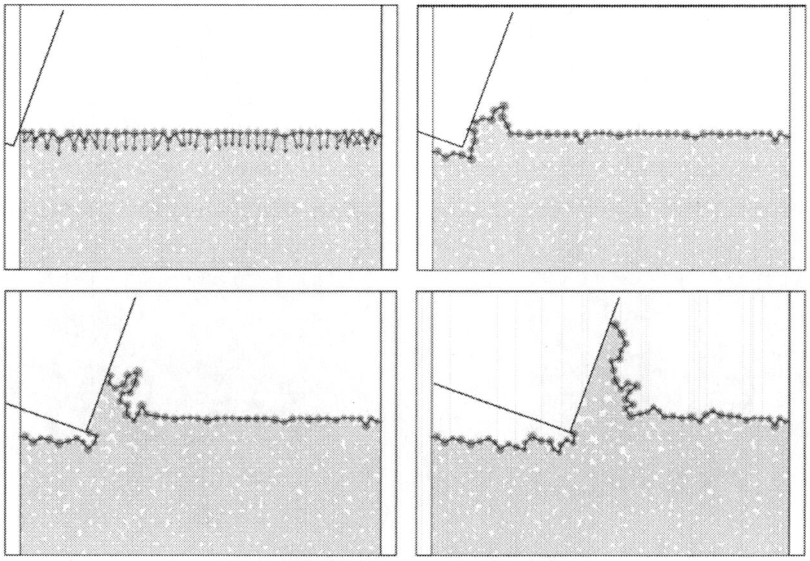

chaining algorithm

ITASCA

Rock Cutting (model, 2D)

initial period of damage formation

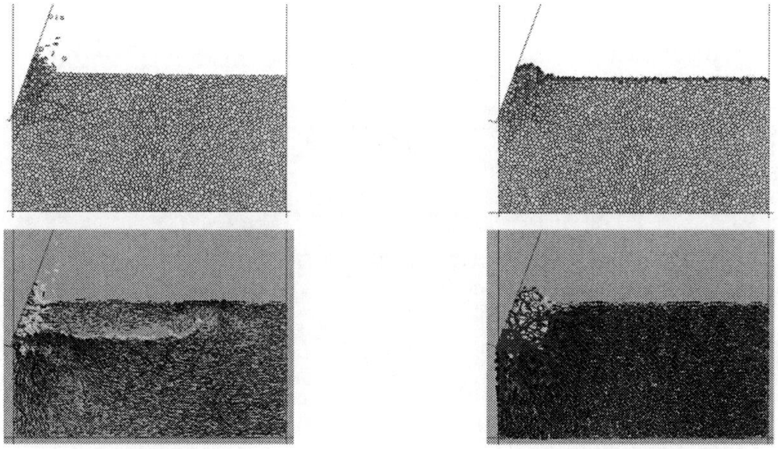

dry cutting

- damage
- forces

wet cutting
(5-MPa confining pressure)

- damage
- forces

ITASCA

Rock Cutting (model, 2D)

steady-state condition

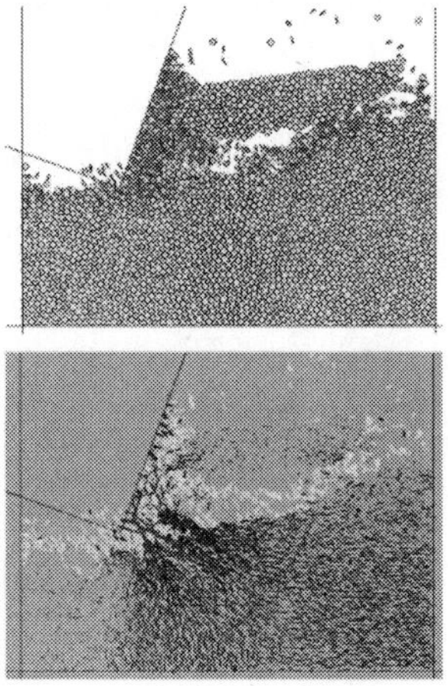

dry cutting

- damage
- forces

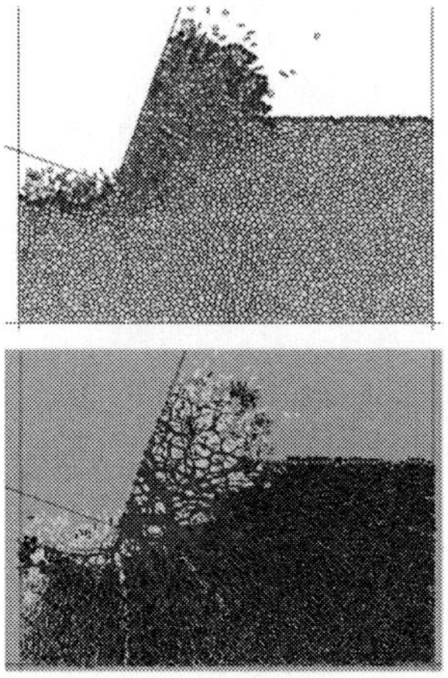

wet cutting
(5-MPa confining pressure)

- damage
- forces

Rock Cutting (model, 3D)

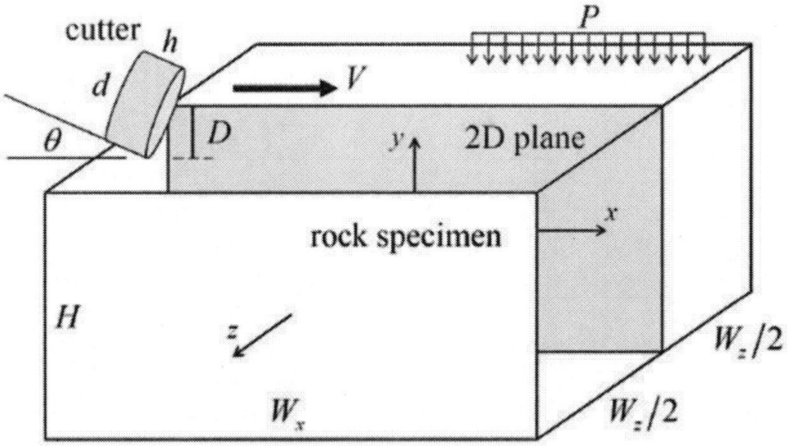

Bounding walls on sides and bottom.
Pressure is applied to entire rock surface.

Rock Cutting (model, 3D)

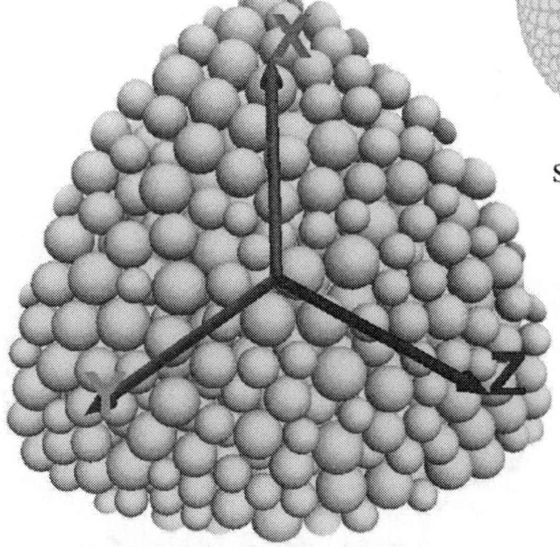

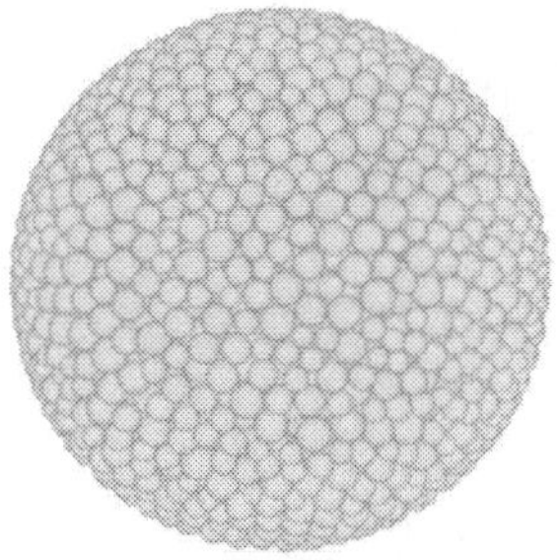

synthetic rock

shining-lamp algorithm

Potyondy, D. (2012) "PFC3D Pressure Boundary Condition," Itasca Consulting Group, Inc., Technical Memorandum ICG7233-L (January 30, 2012), Minneapolis, Minnesota.

Rock Cutting (model, 3D)

grid-cell intersections

(ratio of cell-size to average particle diameter: **1/2**)

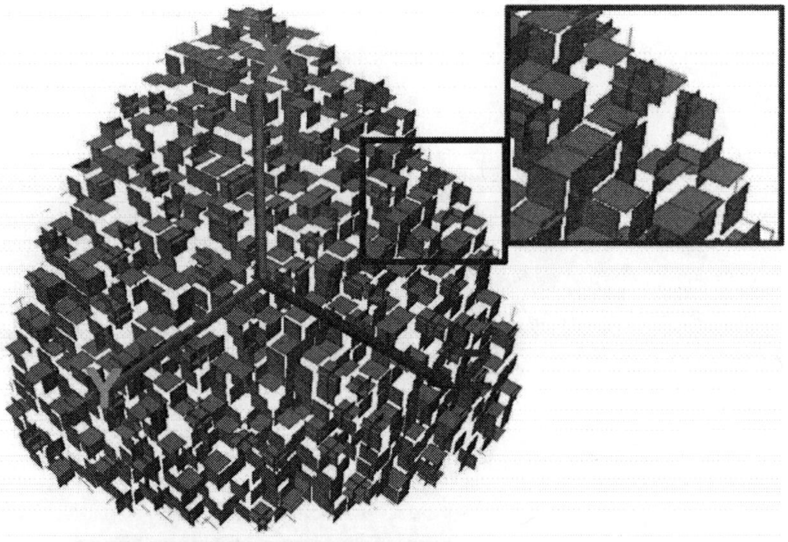

shining-lamp algorithm

Rock Cutting (model, 3D)

grid-cell intersections

(ratio of cell-size to average particle diameter: **1/4**)

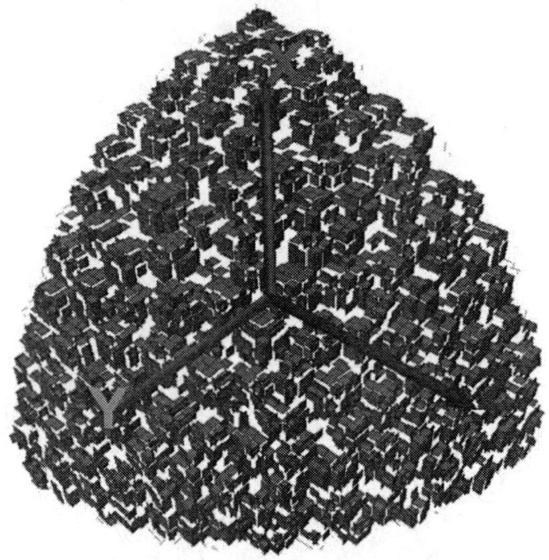

shining-lamp algorithm

ITASCA

Rock Cutting (model, 3D)

grid-cell intersections

(ratio of cell-size to average particle diameter: **1/8**)

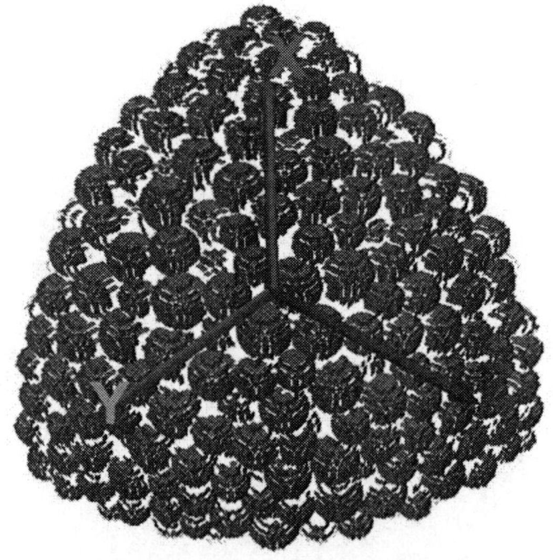

shining-lamp algorithm

ITASCA

Rock Cutting (model, 3D)

grid-cell intersections
(ratio of cell-size to average particle diameter: **1/16**)

shining-lamp algorithm

ITASCA

Rock Cutting (model, 3D)

wet cutting
(20-MPa confining pressure)

Blue particles have had all of their bonds broken.

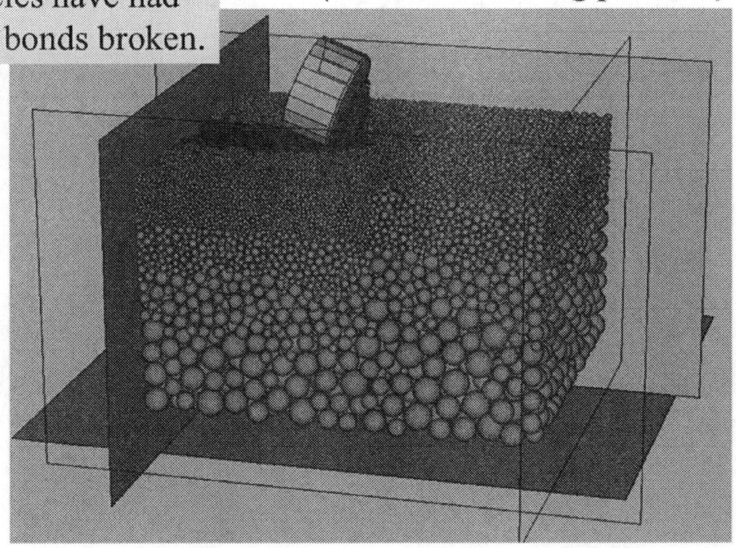

ITASCA

Ledgerwood, L.W. (2013) "PFC3D Modeling of Rock Cutting under Pressure Calibrated to the Inelastic Region of Triaxial Tests at High Strain," extended abstract in Proc. 6th Int. Conf. on Discrete Element Methods (DEM6), Colorado, USA, 5-6 August 2013.

Rock Cutting (mechanisms controlling system behavior)

After an initial period of damage formation, the cutter-rock system reaches a steady-state condition in which the cutter is no longer in contact with virgin rock, but instead is bearing against broken material.

The damaged region consists of broken material that extends in front of and slightly below the cutter, and the broken material flows beneath the cutter.

ITASCA

Rock Cutting (mechanisms controlling system behavior)

Damaged region differs for dry & wet cutting.

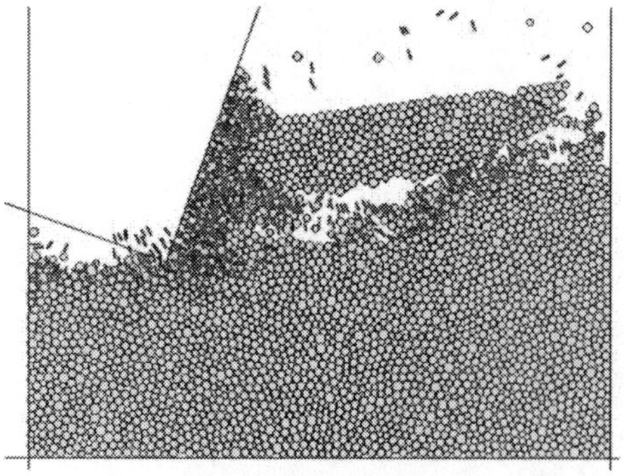

dry cutting
large fragments

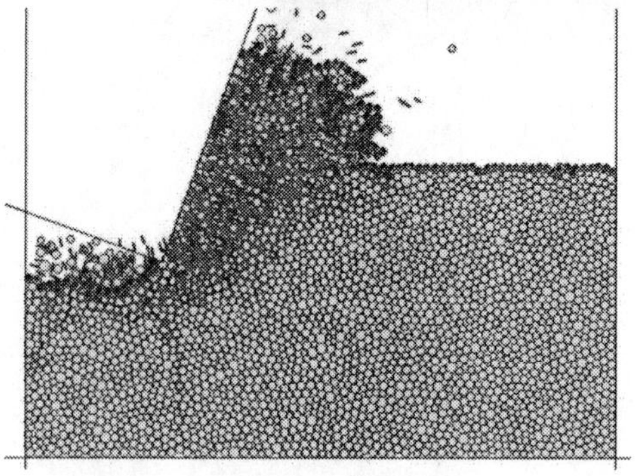

wet cutting
broken material extrudes up face

ITASCA

Rock Cutting (mechanisms controlling system behavior)

Forces flow through broken material into virgin rock.

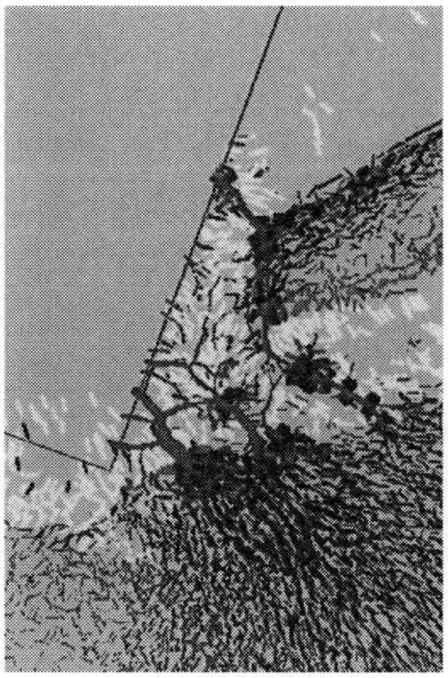

■ dry cut, forces

■ wet cut, forces

Rock Cutting (mechanisms controlling system behavior)

Mechanical response of cutter-rock system is controlled by:

- properties of rock on periphery of damaged region
- properties of broken material as it flows beneath the cutter

■ dry cut, forces

■ wet cut, forces

Rock Cutting (quantitative evaluation of 2D dry cutting)

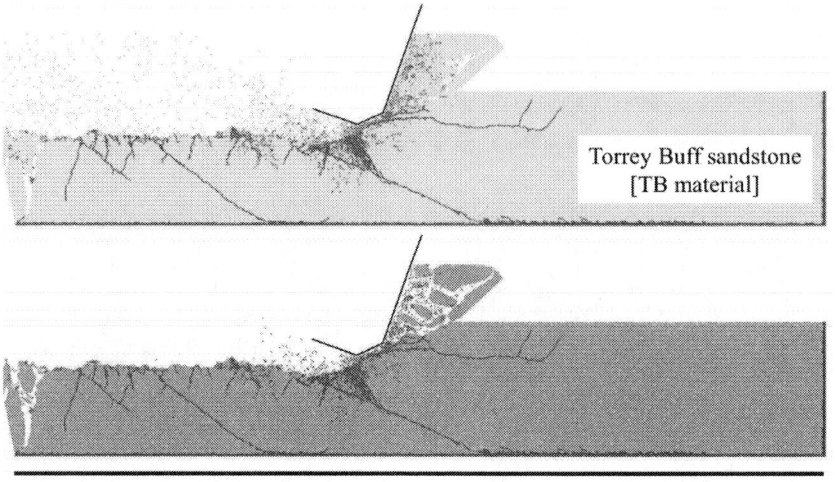

Rock Cutting (quantitative evaluation of 2D dry cutting)

The previous mechanisms occur at the grain scale. Bonded-particle model must have grain size similar to that of rock grains, or at least sufficient to resolve cutter chamfer.

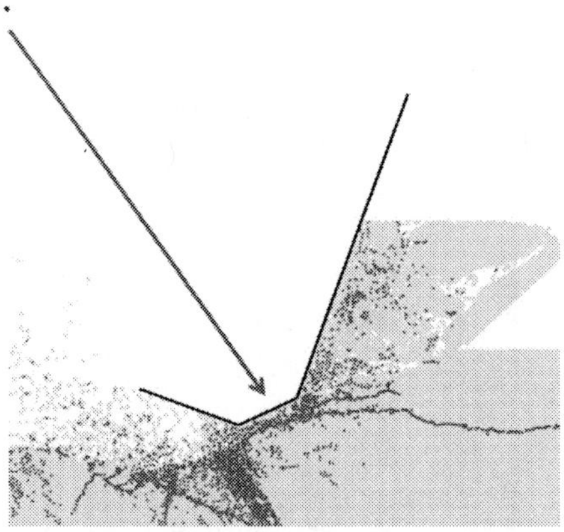

Rock Cutting (quantitative evaluation of 2D dry cutting)

The previous mechanisms occur at the grain scale. Bonded-particle model must have grain size similar to that of rock grains, or at least sufficient to resolve cutter chamfer.

What is grain size, and grain microstructure of Torrey Buff sandstone? This information will inform model creation.

ITASCA

Rock Cutting (quantitative evaluation of 2D dry cutting)

Microstructural features of TB material & Torrey Buff sandstone.

Grain size is 50 microns

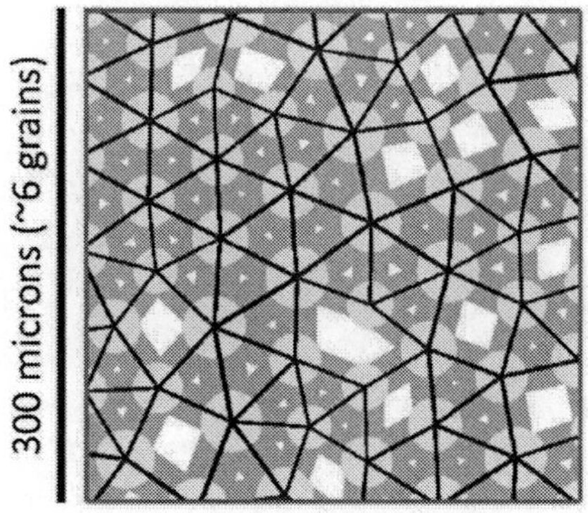

grains (yellow, 80% of true size)
cement (tan, 50% of true size)

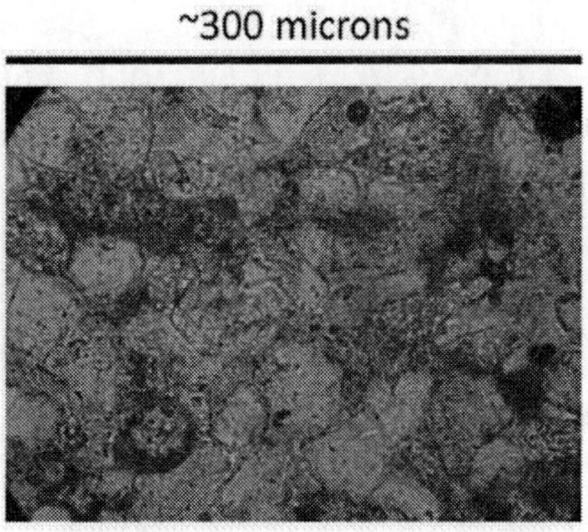

ITASCA

Rock Cutting (quantitative evaluation of 2D dry cutting)

Construct three TB materials that differ only in their grain size.

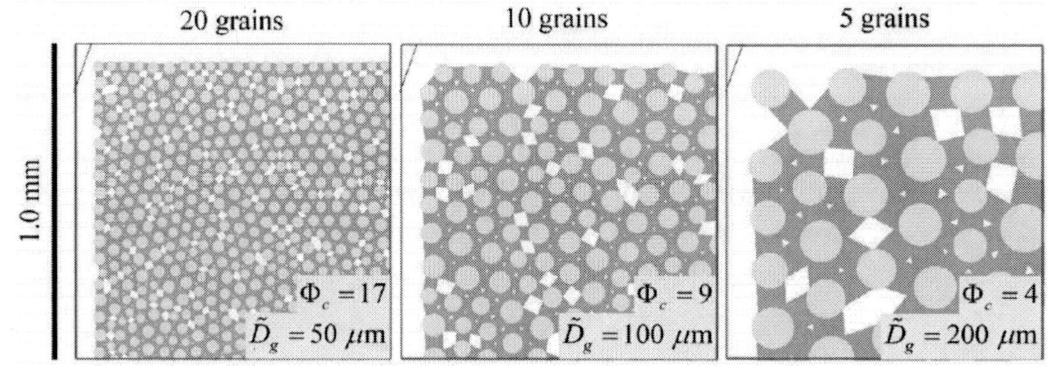

Φ_c is cutter resolution (number of grains across chamfer width)

Rock Cutting (quantitative evaluation of 2D dry cutting)

Microstructure of TB material is simplification of true microstructure; therefore, microproperties chosen via calibration process, attempting to match:

- Tensile strength
- Compressive strengths (at 0, 3 and 10 MPa)

Rock Cutting (quantitative evaluation of 2D dry cutting)

Microstructure of TB material is simplification of true microstructure; therefore, microproperties chosen via calibration process, attempting to match:

- Tensile strength
- Compressive strengths (at 0, 3 and 10 MPa)

TB materials (all grains sizes):

- **match** tensile strength, Young's modulus and Poisson's ratio of Torrey Buff sandstone
- **underestimate** the compressive strengths

We do not expect a quantitative match to forces in the 2D rock-cut test, because. . .

ITASCA

Rock Cutting (quantitative evaluation of 2D dry cutting)

The 2D synthetic cutter is not cutting a groove, but instead is cutting a slice completely across a unit-thickness rock specimen.

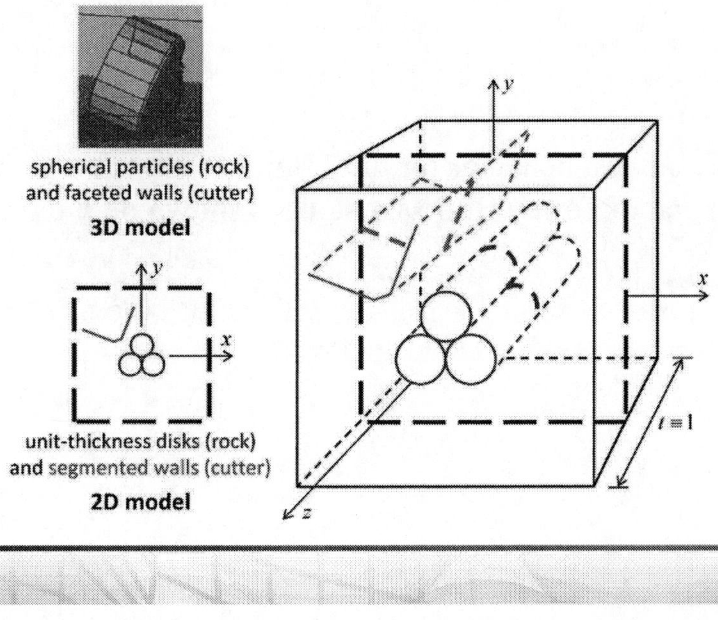

ITASCA

Rock Cutting (quantitative evaluation of 2D dry cutting)

We simulated the following 2D dry rock-cut test:

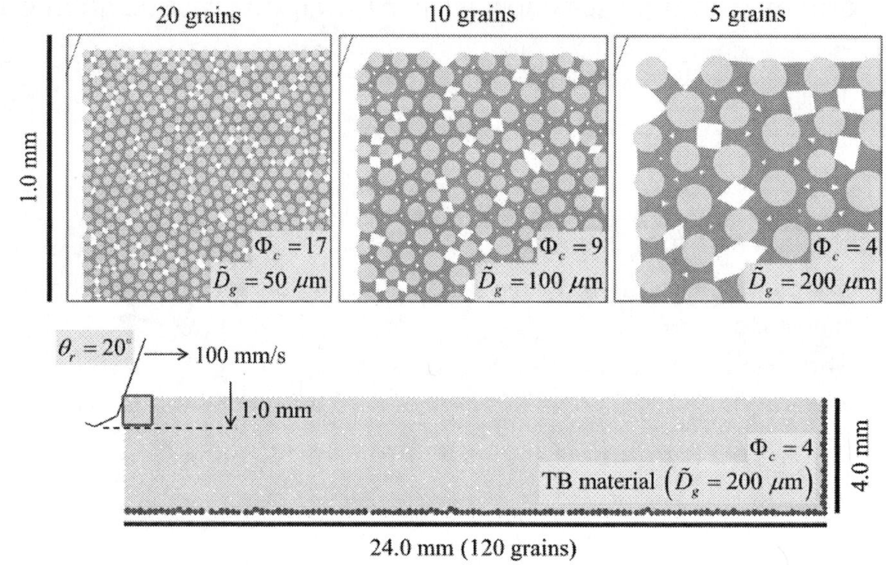

Rock Cutting (quantitative evaluation of 2D dry cutting)

The rock-cut test behavior for the TB material with a grain size equal to that of Torrey Buff sandstone is shown on the following 10 slides.

Rock Cutting (quantitative evaluation of 2D dry cutting)

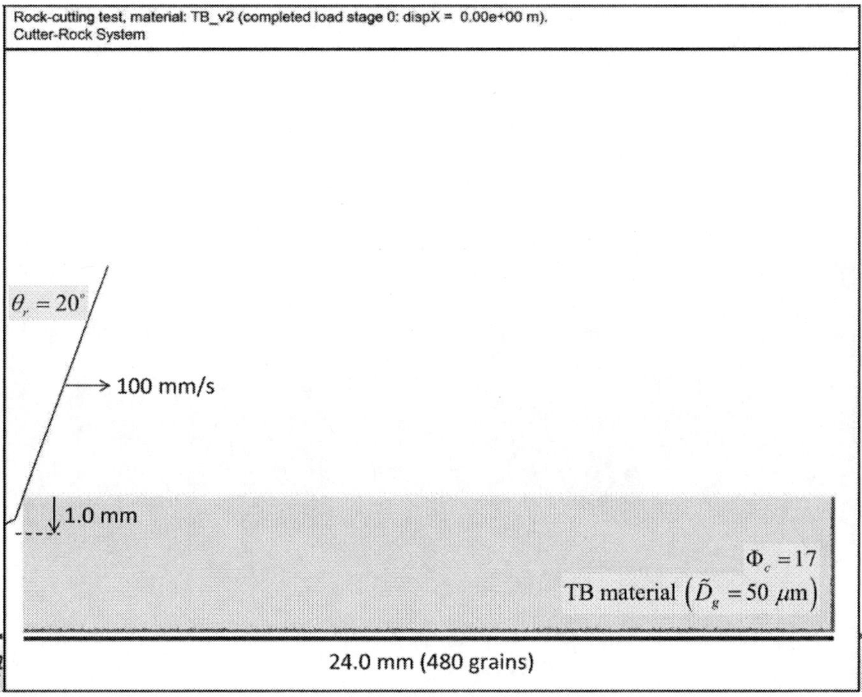

Rock Cutting (quantitative evaluation of 2D dry cutting)

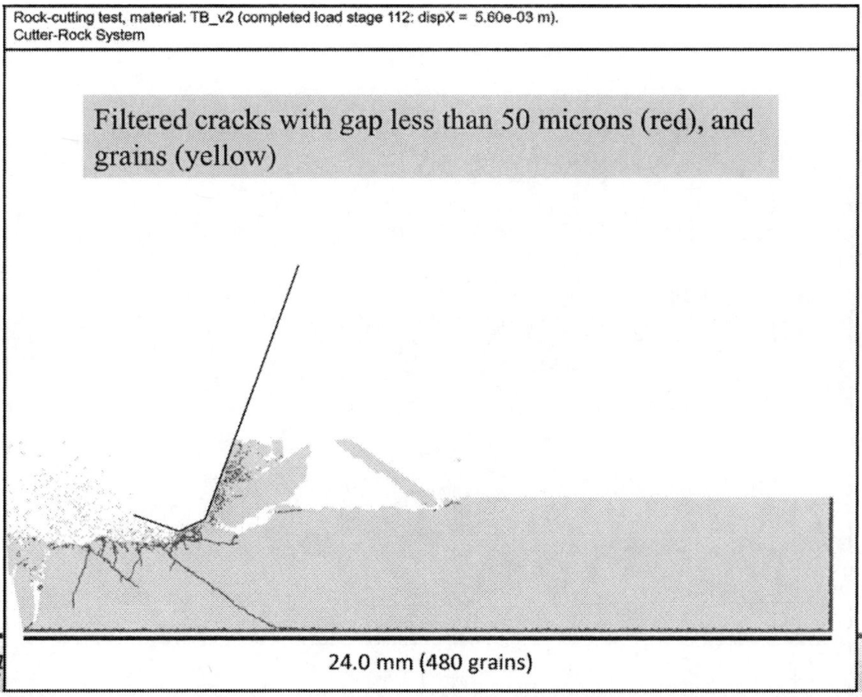

Rock Cutting (quantitative evaluation of 2D dry cutting)

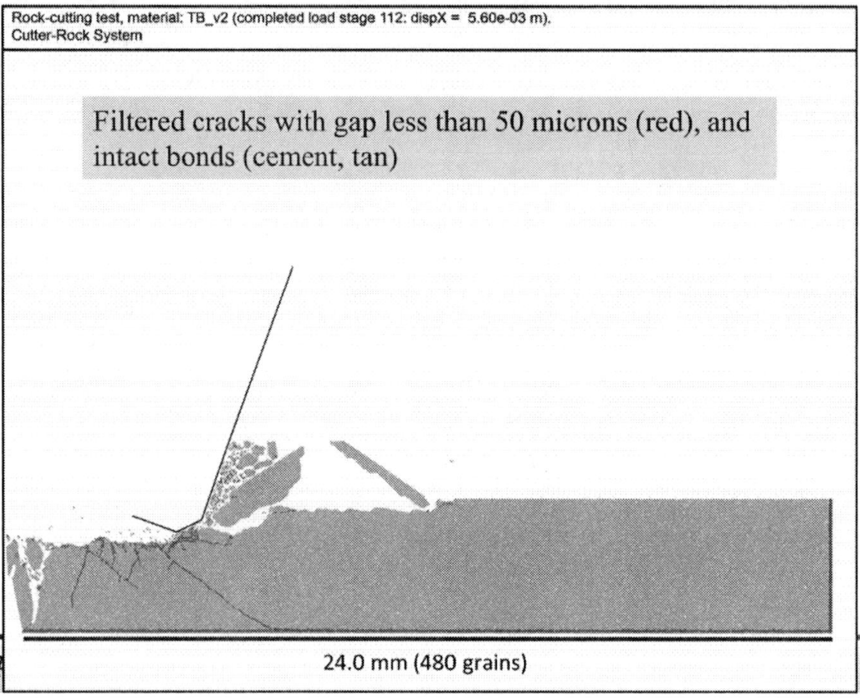

ITASCA

Rock Cutting (quantitative evaluation of 2D dry cutting)

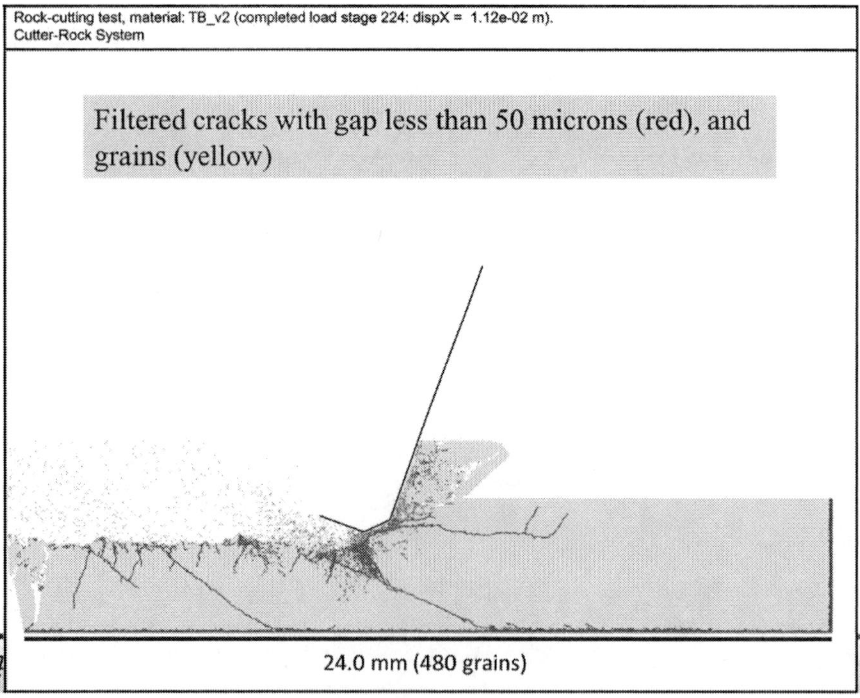

ITASCA

Rock Cutting (quantitative evaluation of 2D dry cutting)

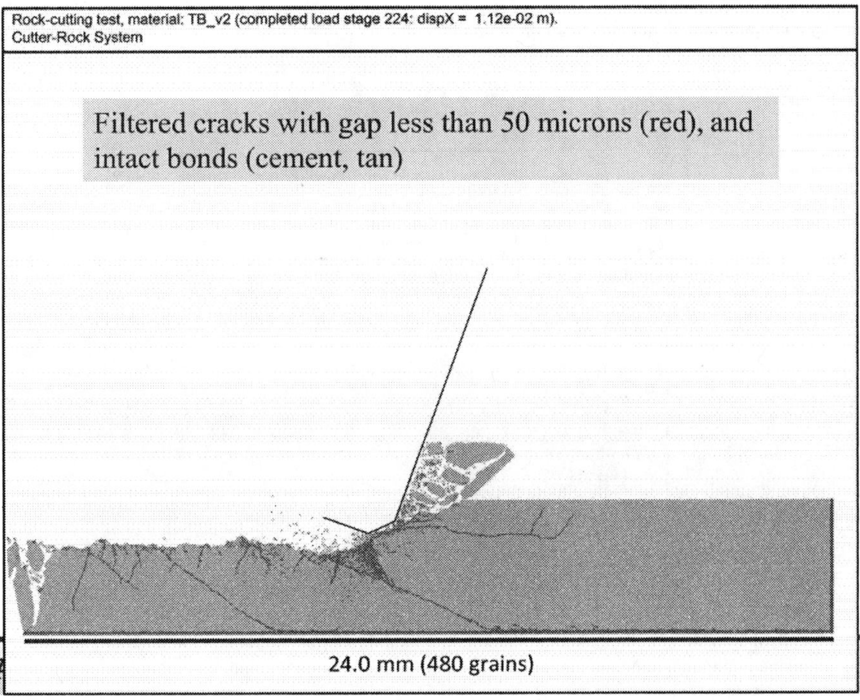

ITASCA

Rock Cutting (quantitative evaluation of 2D dry cutting)

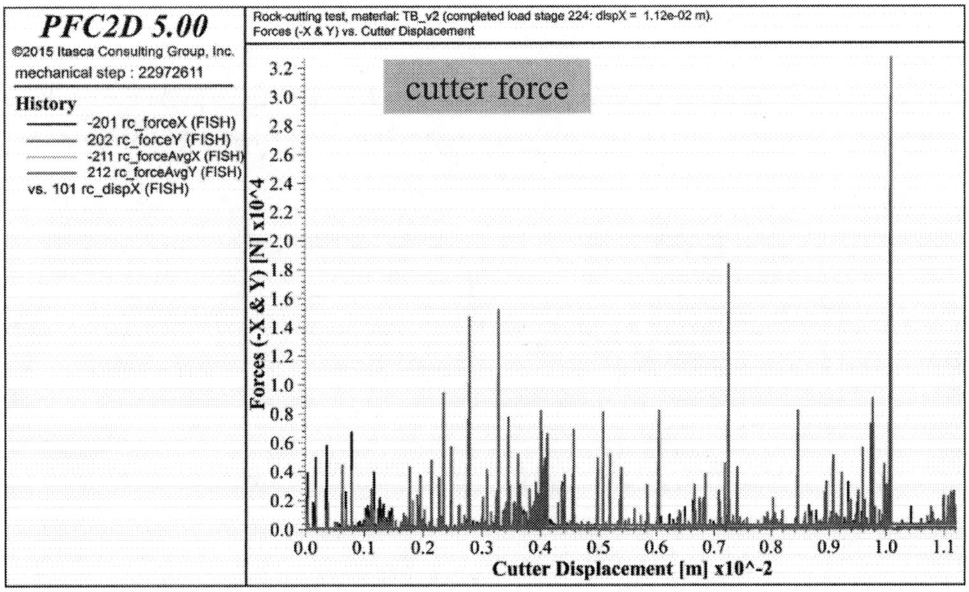

ITASCA

Rock Cutting (quantitative evaluation of 2D dry cutting)

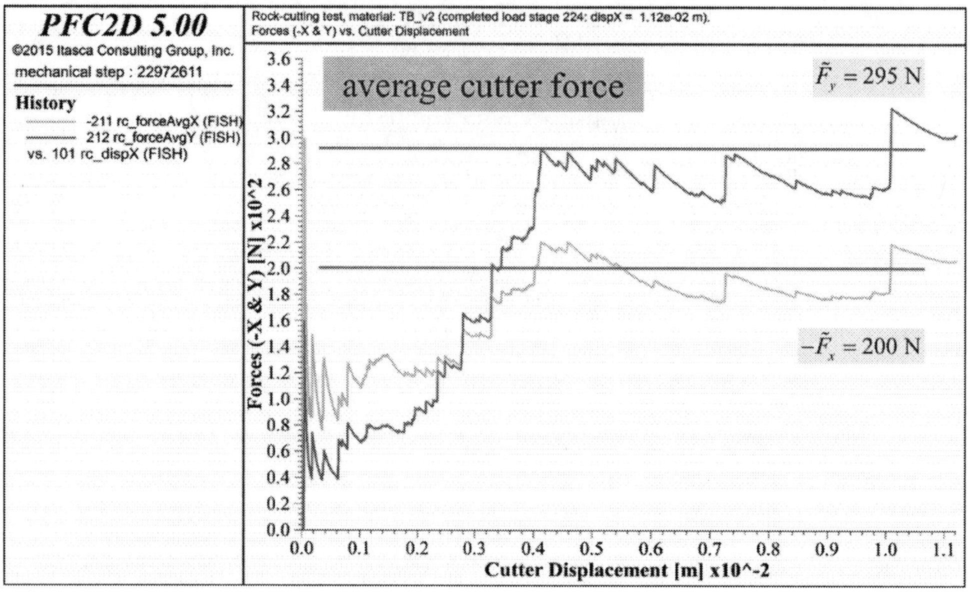

ITASCA

Rock Cutting (quantitative evaluation of 2D dry cutting)

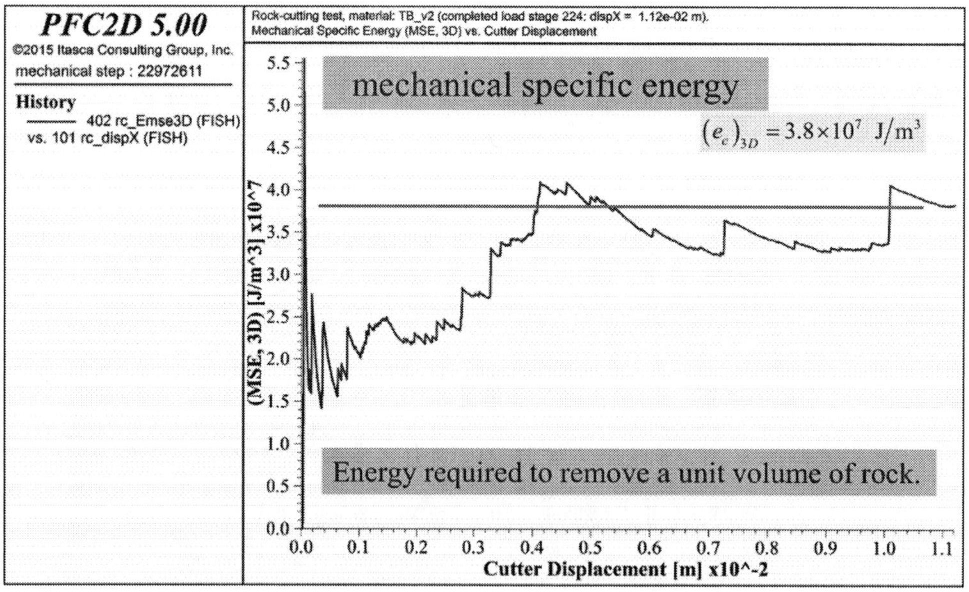

ITASCA

Rock Cutting (quantitative evaluation of 2D dry cutting)

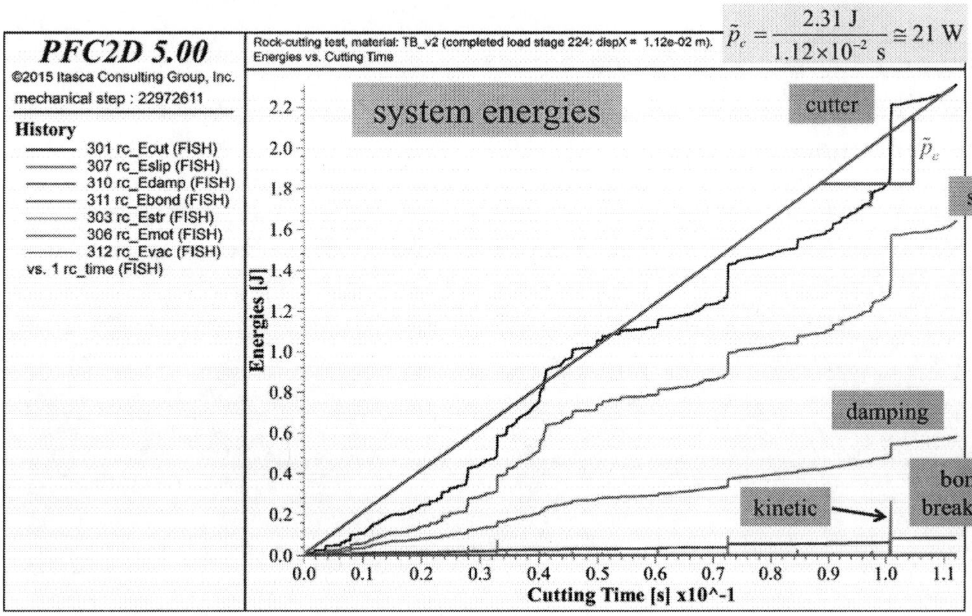

Rock Cutting (quantitative evaluation of 2D dry cutting)

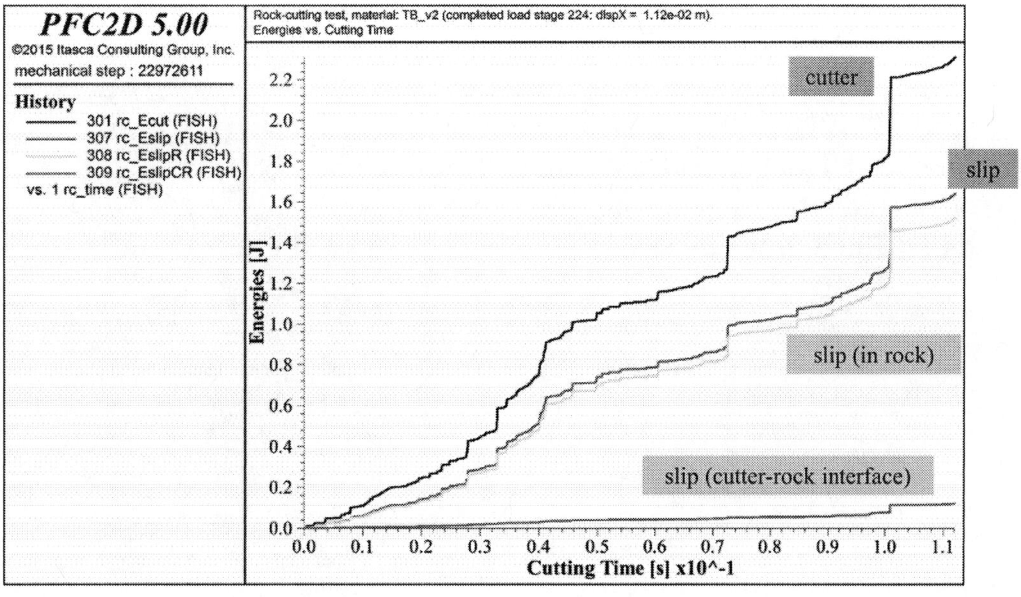

Rock Cutting (quantitative evaluation of 2D dry cutting)

The broken material in contact with the cutter consists of fully unbonded grains.

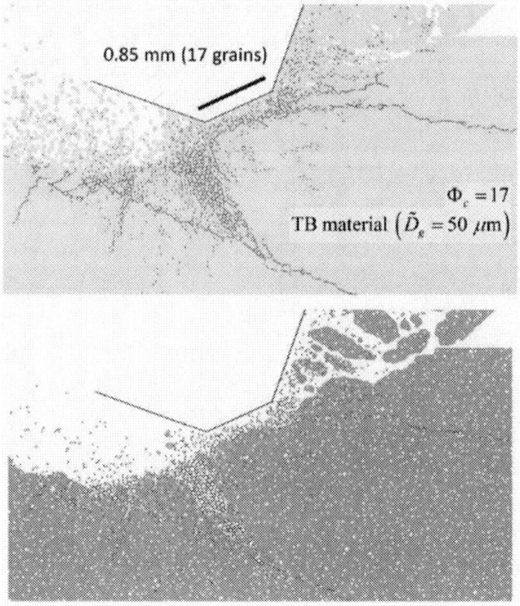

Near-cutter damage in TB material ($\tilde{D}_g = 50\ \mu$m) after 11.2 mm of cutter displacement.

ITASCA

Rock Cutting (quantitative evaluation of 2D dry cutting)

The behavior of the cutter-rock system consists of repeated episodes of cutter force build-up followed by localized damage near the cutter and stress redistribution during which the cutter force drops to zero.

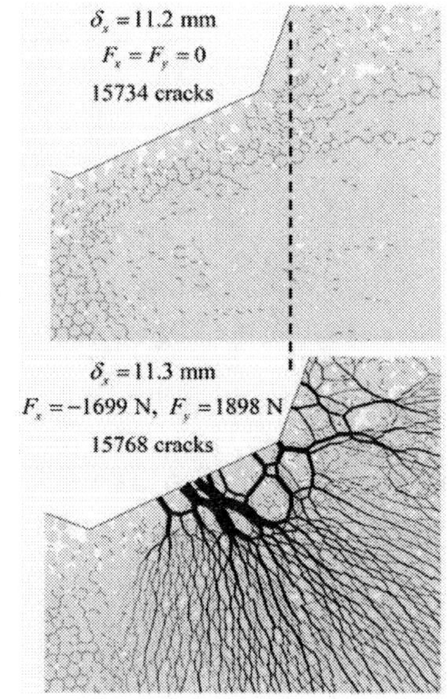

Cutter force build-up in TB material ($\tilde{D}_g = 50\ \mu$m) after 11.2 and 11.3 mm of cutter displacement showing force chains (scaled to same magnitude), filtered cracks with gap less than 50 microns (50% size), and grains.

ITASCA

Rock Cutting (quantitative evaluation of 2D dry cutting)

Similar rock-cutting properties are obtained for the TB material with a grain size equal to that of Torrey Buff sandstone, and for grain sizes that are two and four times larger.

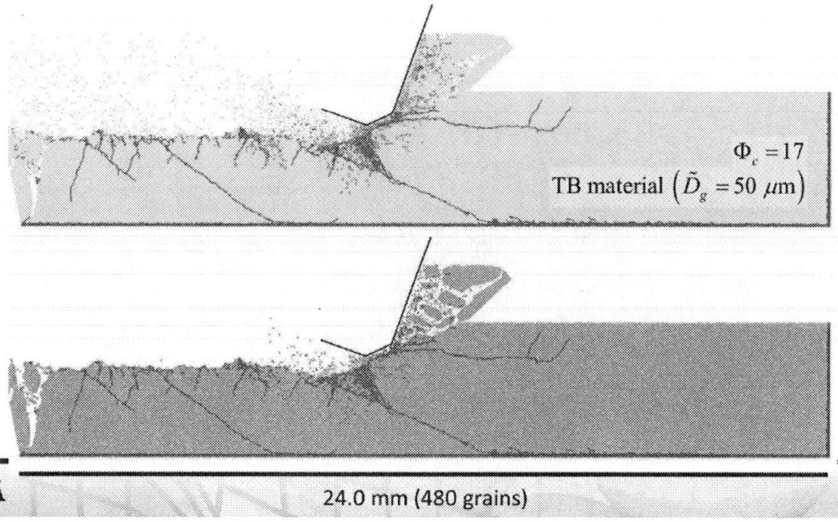

ITASCA

Rock Cutting (quantitative evaluation of 2D dry cutting)

Similar rock-cutting properties are obtained for the TB material with a grain size equal to that of Torrey Buff sandstone, and for grain sizes that are two and four times larger.

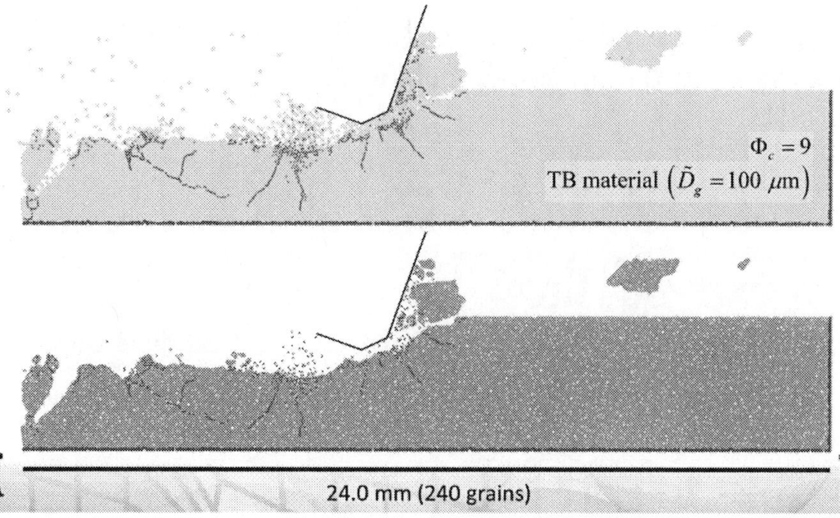

ITASCA

Rock Cutting (quantitative evaluation of 2D dry cutting)

Similar rock-cutting properties are obtained for the TB material with a grain size equal to that of Torrey Buff sandstone, and for grain sizes that are two and four times larger.

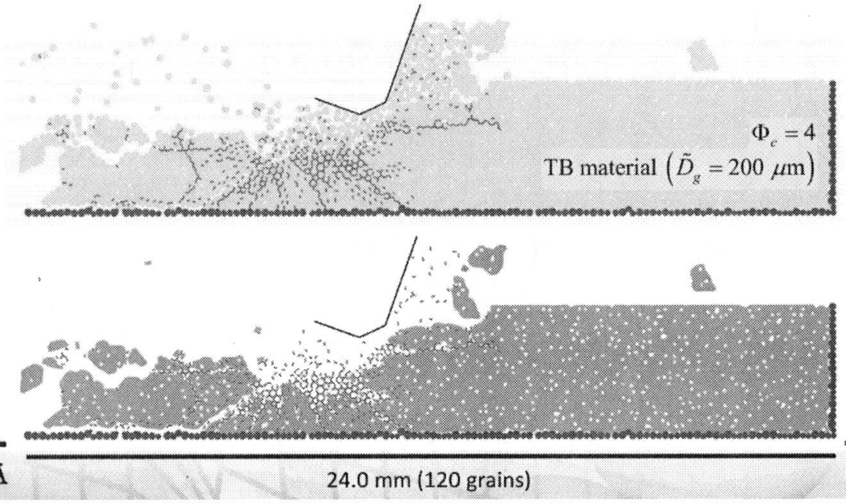

ITASCA

Rock Cutting (quantitative evaluation of 2D dry cutting)

The largest grain-size material has four grains across the chamfer width, and this appears to be sufficient to resolve the effect of this chamfer on the cutter force.

The largest grain-size material could be used to perform a parametric study of cutter-rock behavior.

	runtimes	
TB material $(\tilde{D}_g = 50\ \mu m)$:	11 days,	11-mm cutter displacement
TB material $(\tilde{D}_g = 100\ \mu m)$:	64 hours,	22-mm cutter displacement
TB material $(\tilde{D}_g = 200\ \mu m)$:	7 hours,	22-mm cutter displacement

ITASCA

Rock Cutting (conclusions)

- Relevant mechanisms are exhibited by the model.
  - Effect of dry & wet cutting (pressure inhibits chip formation, requires more energy to cut)
- Perform parametric studies of system behavior.
  - Different rock types can be created by varying the microstructural properties of the synthetic material. The model provides a link between rock microstructure and cutter geometry.
- The model produces qualitative agreement between synthetic and measured cutting forces; and thus, the model may be used to comparatively evaluate alternative rock types and cutter designs.

ITASCA

Conclusions & Future Directions

Bonded-particle modeling is in early phase of development

- have generic models for 2D & 3D intact compact rock

Future BPMs should be constructed to match particular types of rock

- granite, sandstone, limestone,…

All BPM modeling benefits from increased computational speed via

- multi-threading capability in PFC 5.0

ITASCA

Conclusions & Future Directions

With continuing increases in computational speed, BPM modeling will be increasingly applied to boundary-value problems

- How to incorporate scale effects that make in-situ strength less than intact strength?

Challenge is to keep models as simple as possible

- include features to allow relevant micromechanisms to occur

ITASCA

END OF DELIVERED TALK

ITASCA

EXTRA MATERIAL
(Gravity Flow of Fragmented Rock
in Block Caving Mines)

ITASCA

Overview

Bonded-Particle Modeling

- Basic idea & essential features
- Microstructural physics of intact rock
- PFC model
- Microstructural models provided by BPM

Engineering Applications

- Rock Fracture
  ➔ rock cutting (rock microstructure : cutter geometry)
- Material Flow
  ➔ cave mining (physical & PFC models, theory ➔ REBOP)

Conclusions

Gravity Flow of Fragmented Rock in Block Caving Mines

Matt Pierce

April 4, 2016

Pierce, M.E. (2010) ***A Model for Gravity Flow of Fragmented Rock in Block Caving Mines***. Ph.D. Thesis, University of Queensland, Brisbane, Australia.

Cave Mining

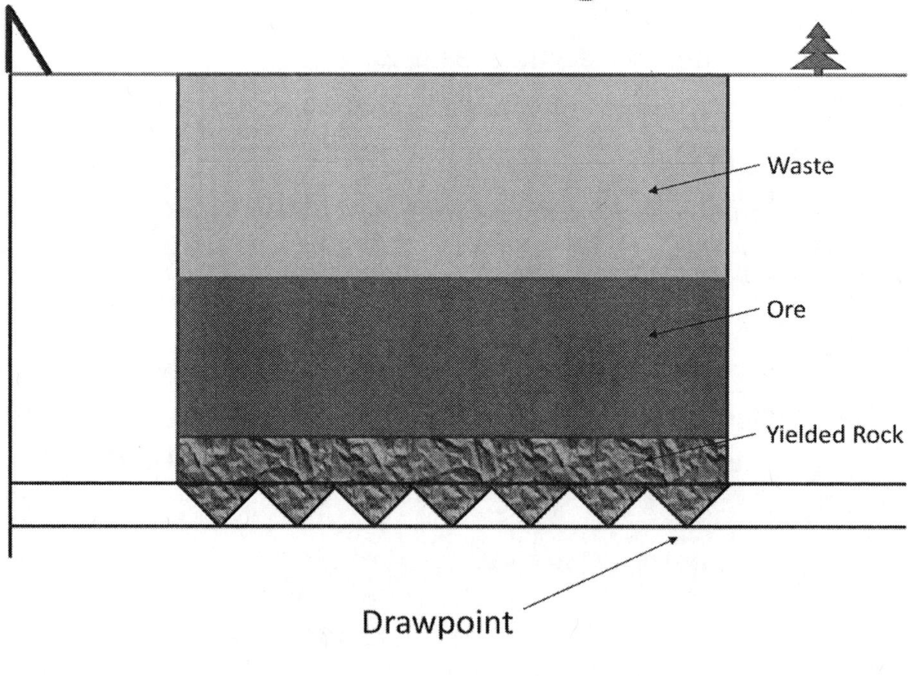

Cave Mining

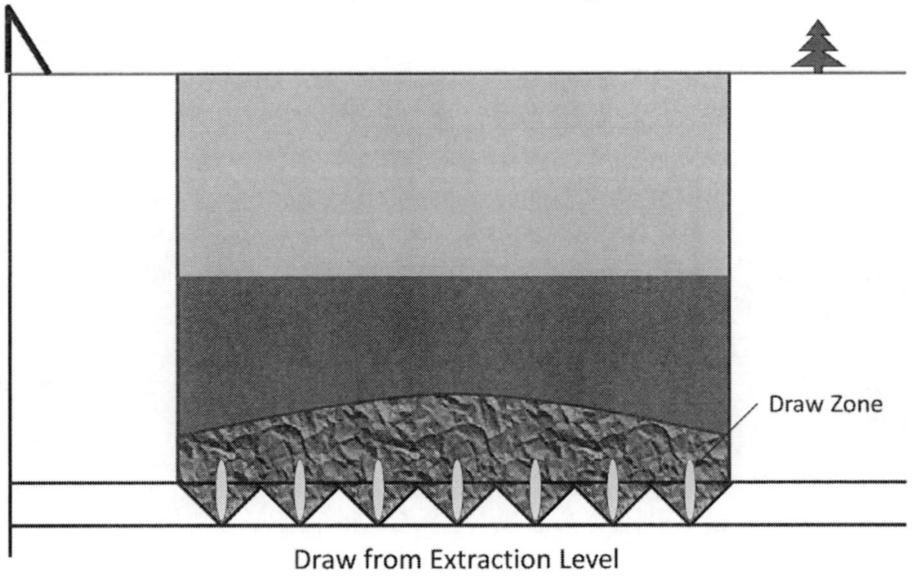

Cave Mining

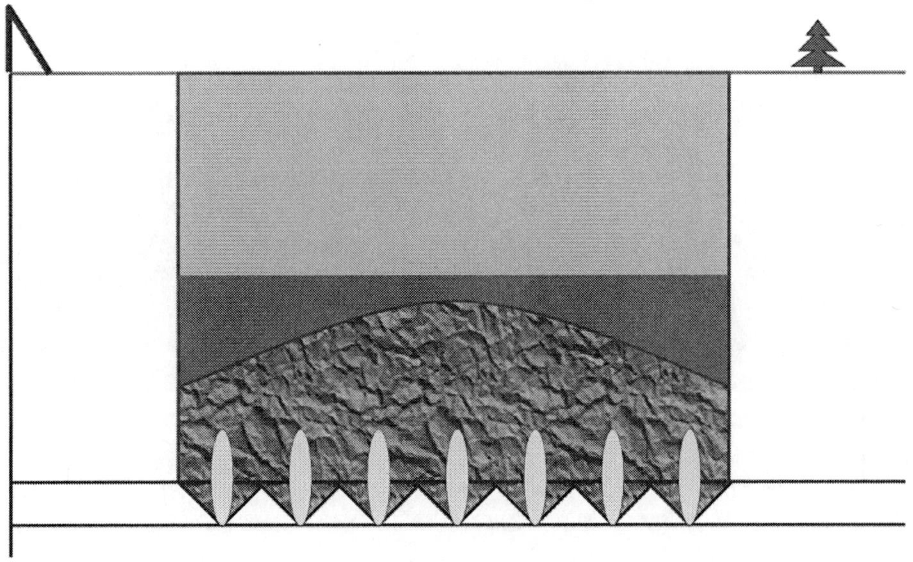

Caving

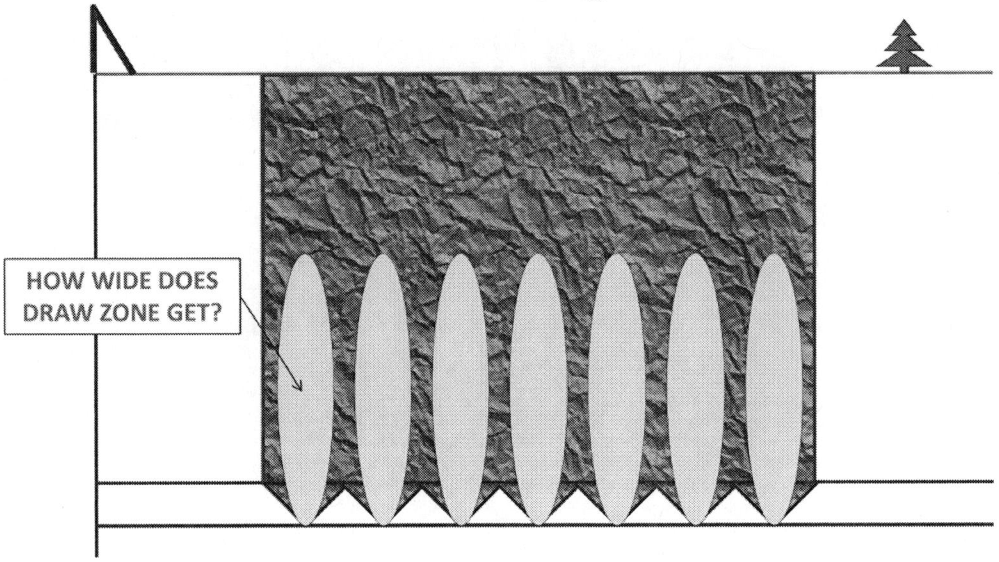

Caving

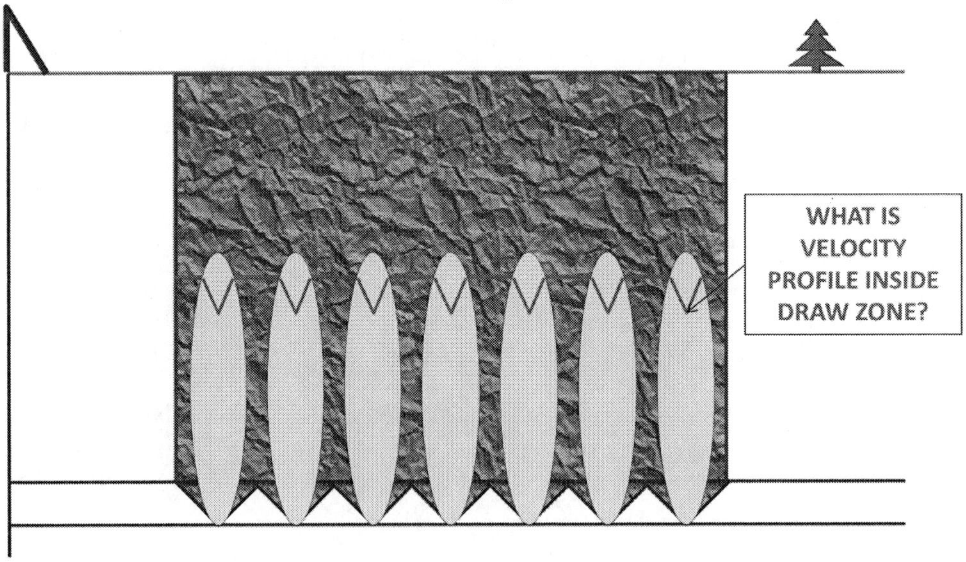

Caving

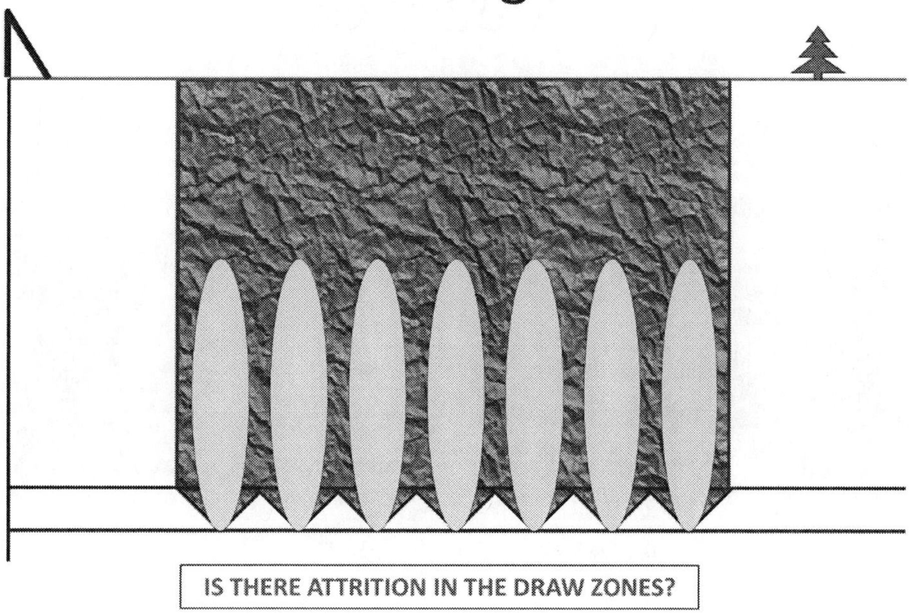

How Wide does Draw Zone get?

Draw Zone Width – 3D Physical Modelling

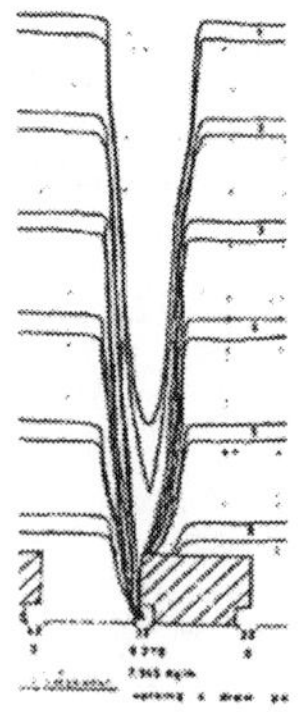

Marano (1980)
Sand

0.3mm

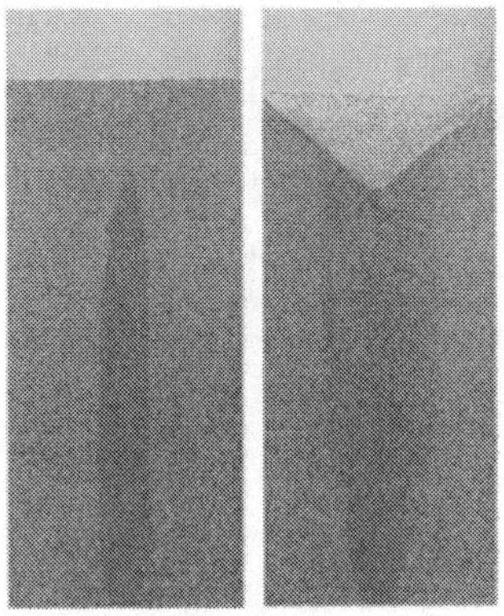

Ferjani (2003)
Crushed walnut shells

0.65mm

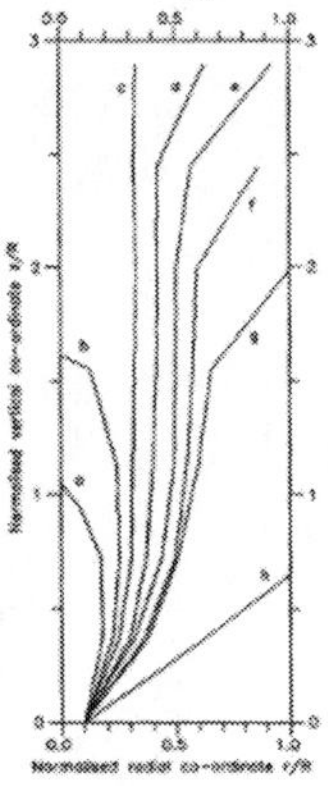

Watson (1993)
Polypropylene pellets
5mm

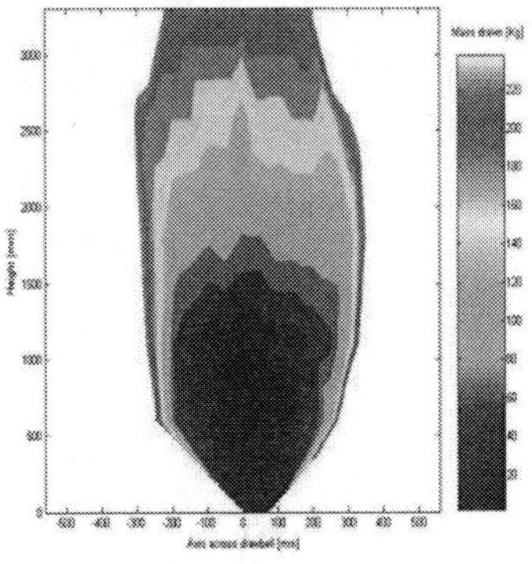

Castro (2006) - MMT
Crushed rock

8mm, 18mm

Draw Zone Width – PFC3D

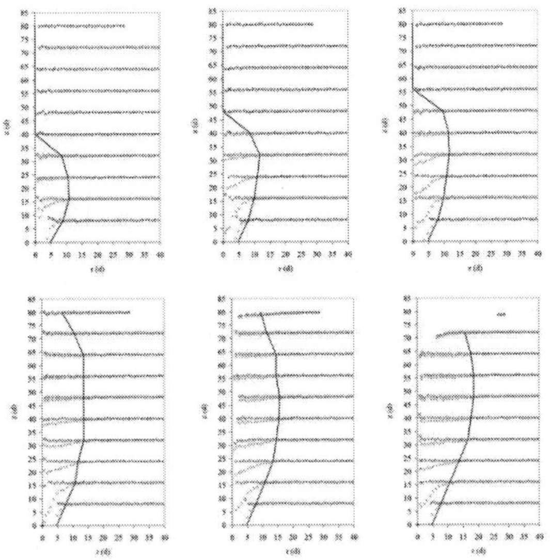

Allows systematic study of behavior: vary particle size, shape, friction, etc.

What is Velocity Profile Inside a Draw Zone?

Velocity Profile - 2D Physical Modeling

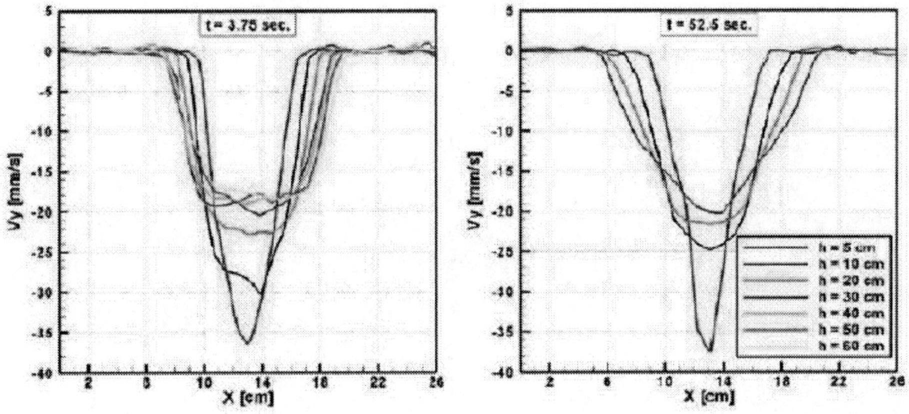

Figure 2-17 Profiles of vertical velocity (Vy) observed by Sielamowicz et al. (2005) during 2D plane hopper experiments employing amarantus seed (outlet width = 1 cm). Lines represent velocity profiles at different elapsed times , *t*, and elevations above the outlet, *h*.

Velocity Profile – PFC3D

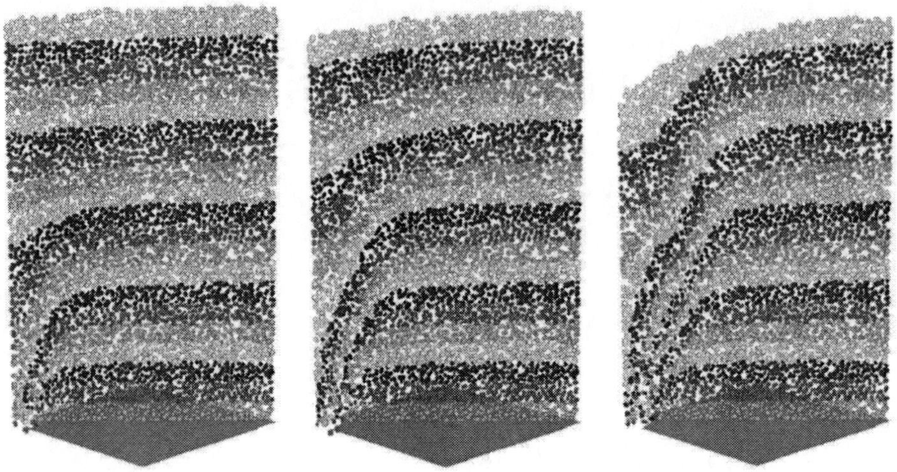

Velocity profile examined in PFC3D simulations of draw employing quarter symmetry

Velocity Profile - PFC3D

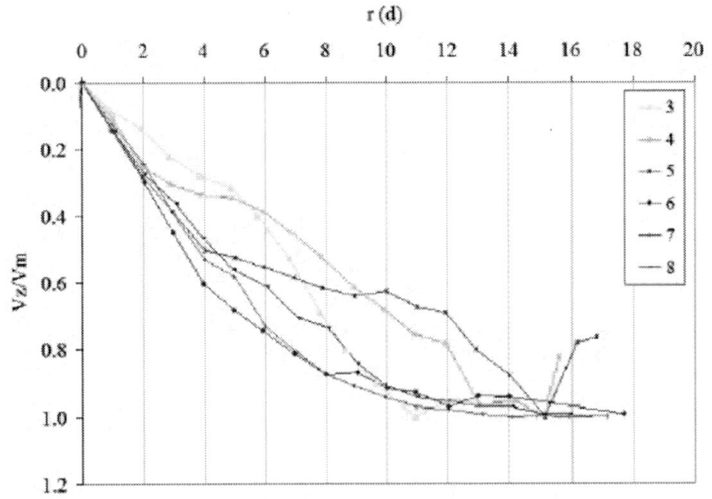

Velocity profiles from PFC3D model suggest that shear band also exists in 3D, and that it is approximately 10 mean particle diameters wide.

REBOP

- Understanding of draw zone width and velocity profile from PFC (and physical models, theory) embedded within **REBOP**
- **R**apid **E**mulator **B**ased **O**n **P**FC
- Developed by Peter Cundall in 1999 and extended by M Pierce as a means to rapidly simulate the flow of broken rock in caving mines
- REBOP has now been applied at approximately 20 different caving panels around the world, including successful back analysis of drawn grades
- Used routinely for drawpoint layout design and forecasting of recovery and dilution based on expected fragmentation

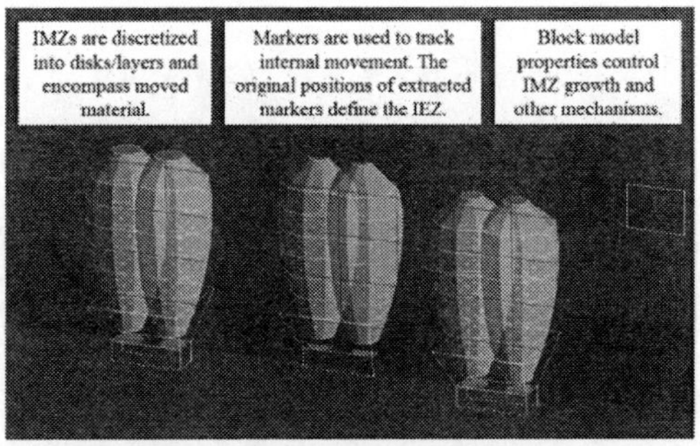

IMZ: Isolated Movement Zone

IEZ: Isolated Extraction Zone

Is There Attrition in the Draw Zones?

Professor John Bridgwater

Position	Emeritus professor
College	St Catharine's
Qualifications/honours	MA, PhD, ScD
Address	Department of Chemical Engineering and Biotechnology University of Cambridge New Museums Site Pembroke Street Cambridge CB2 3RA, UK
Email	jb231@cam.ac.uk
Research group	

Research description

1. Mechanics and Modelling for the Mixing of Powders and Granular Materials
2. Attrition of granular materials
3. Paste Flow and Extrusion

Bridgwater Model for Attrition

- Bridgwater et al. (2003) developed an empirical model for attrition based on annular shear cell experiments on manufactured materials
- Model suggests that under shear, the mass produced by splitting and rounding is strongly controlled by the product of stress and shear strain, normalized to tensile strength:

$$W = K_N \left(\frac{\sigma_N \Gamma^{\phi}}{\sigma_{scs}} \right)^{\beta}$$

Normal stress (σ_N); Shear strain (Γ); Tensile strength of fragments (σ_{scs})

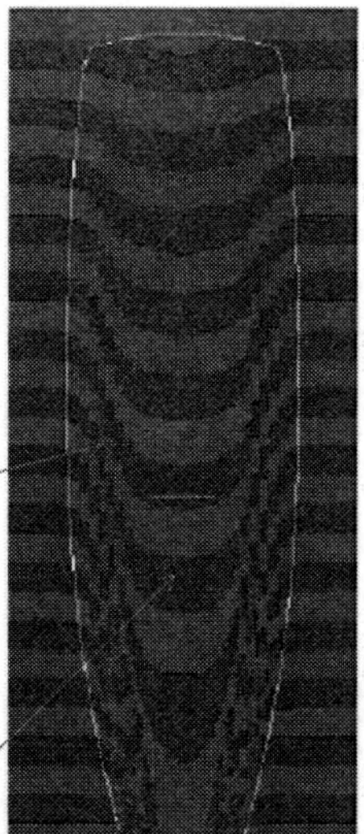

Outer annulus is in shear. This is where most of attrition would be expected to occur. Shear strain is a function of the velocity profile and mass drawn.

Stress inside IMZ can be estimated using Janssen's equation for steady state stresses in a tall silo.

Attrition Model: Henderson Mine

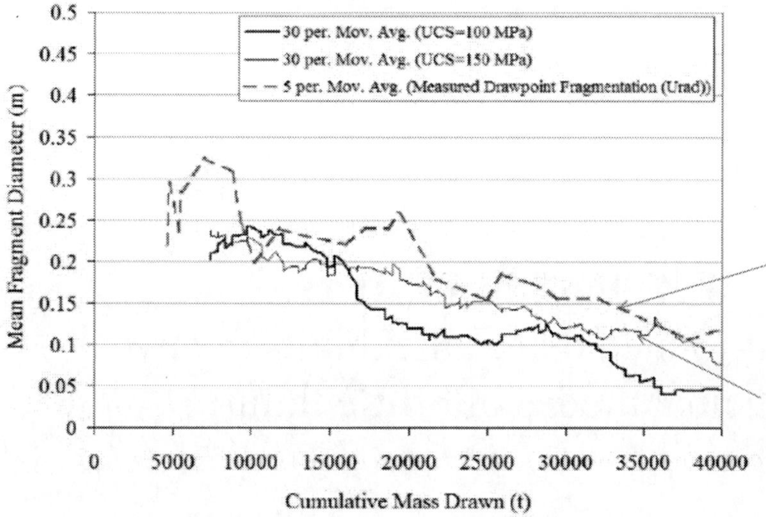

Actual mean fragment size extracted

REBOP prediction of fragmentation extracted USING Bridgwater model

- Reasonable match to change in fragment size measured over time

Extending Birdgwater Attrition Model - PFC3D

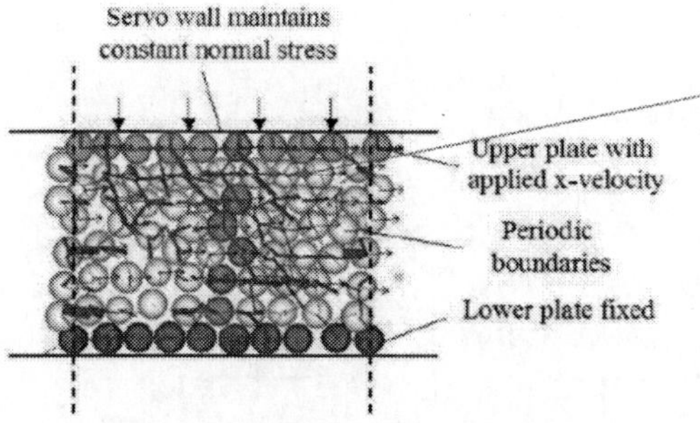

Figure 4-24 Illustration of PFC3D annular shear-cell model

Microphysical law at contact level to relate slip work to attrition/fines production. Can be tuned to match Bridgwater relations.

...then extend work to study attrition for a range of fragment sizes and shapes (more like caved rock)....ONGOING

First demonstrate that one can reproduce Bridgwater attrition experiments on uniformly sized spherical particles with PFC3D.....

EXTRA MATERIAL
(2D flat-jointed material exhibits the two principal modes of brittle failure)

Microstructural Physics of Intact Rock

Brittle mechanical properties measured:
- uniaxial & triaxial compression tests
- tension tests

Two principal modes of brittle failure:
- shear fracture
- extension fracture

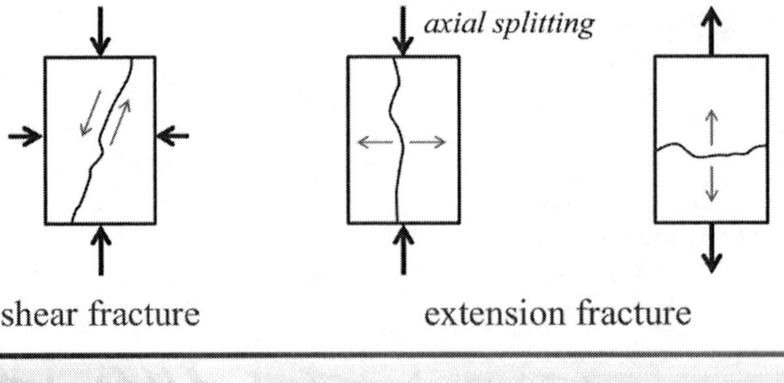

Example (flat-jointed material)

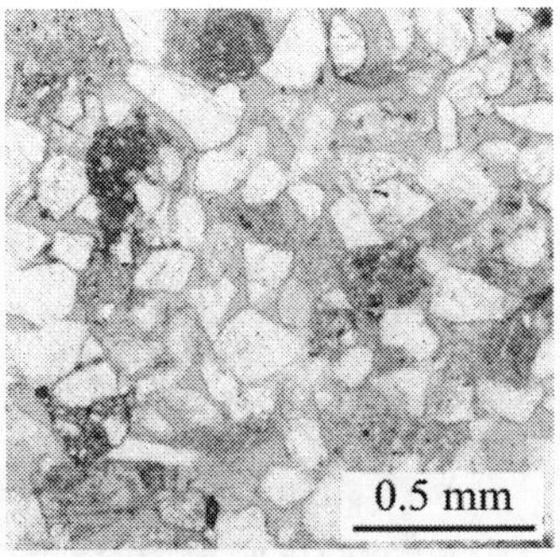

Castlegate sandstone

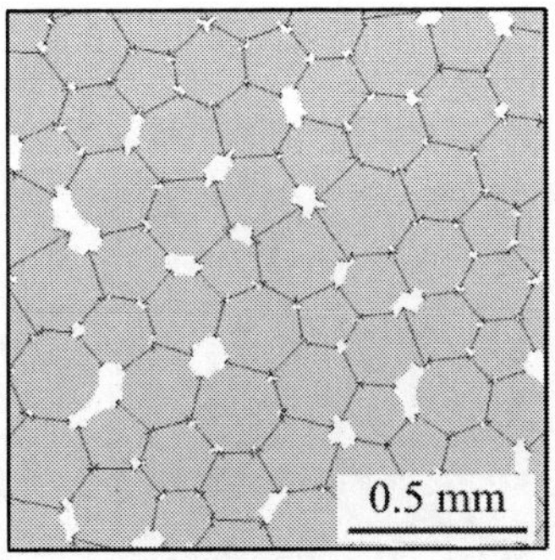

CG material

ITASCA Potyondy, D.O., and Y. Xiao. (2014) "Simulating Perforation Damage with a Flat-Jointed Bonded-Particle Material," to be submitted to *Int. J. Rock Mech. & Min. Sci.* pending approval by BP, manuscript dated September 4, 2014.

Example (flat-jointed material)

Direct-tension test

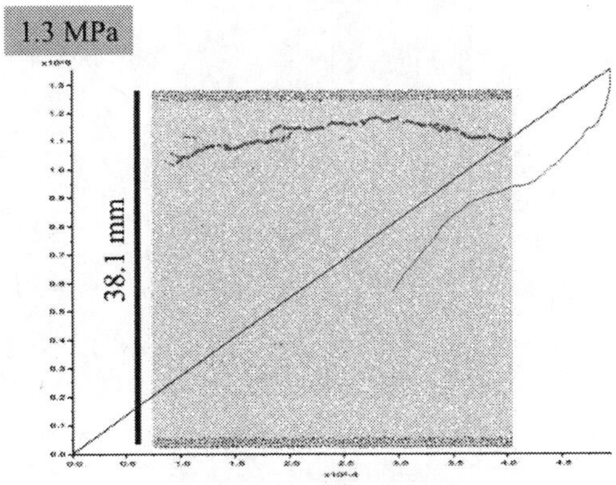

Fracture-toughness test

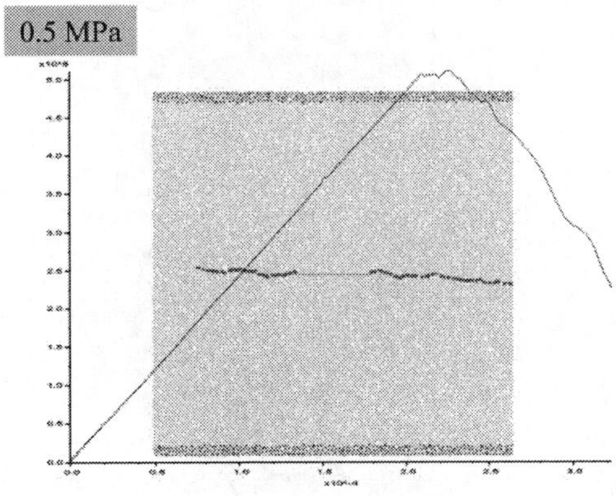

$K_{Ic} \cong 5.7 \times 10^4 \ \text{Pa}\sqrt{\text{m}}$

ITASCA

Example (flat-jointed material)

Unconfined-compression test

18 MPa

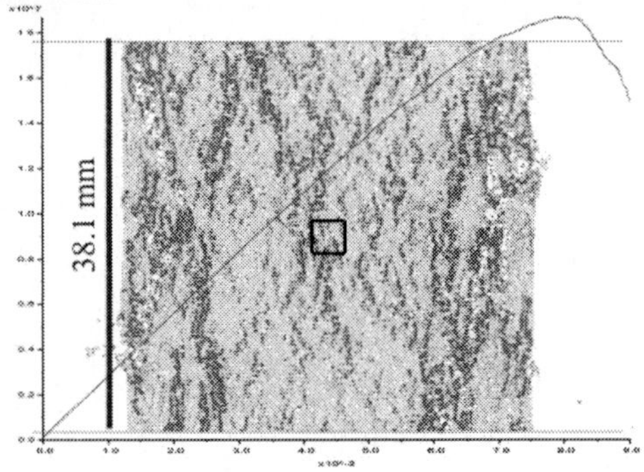

Confined-compression test

38 MPa

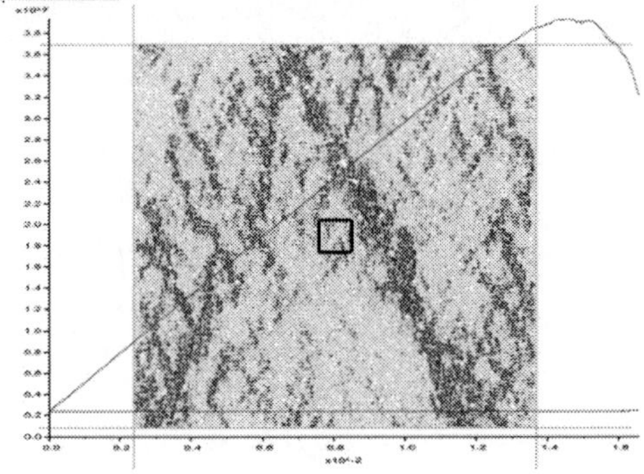

post-peak states

Example (flat-jointed material)

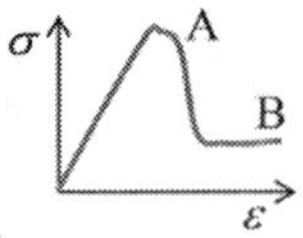

Compression tests

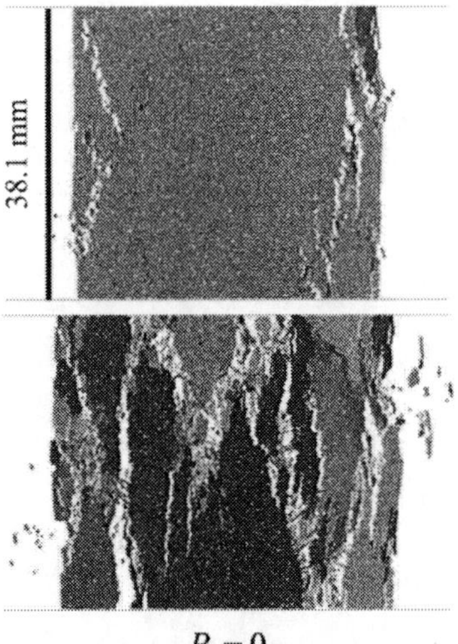

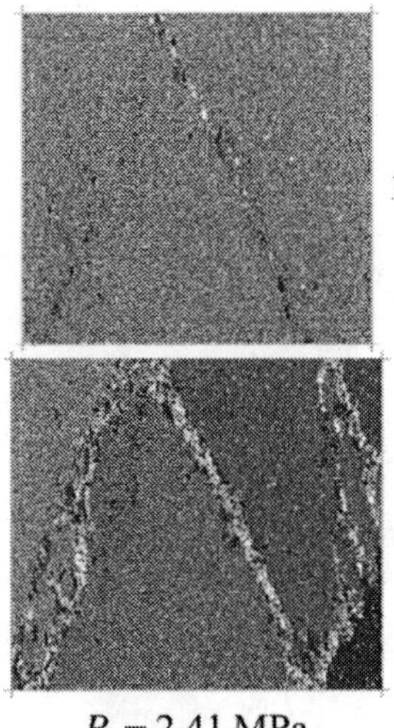

post-peak (A)

residual (B)

$P_c = 0$

unconfined

axial splitting

$P_c = 2.41$ MPa

confined

shear fracture

Approaches for Accurate Modeling of Particle Attrition in DEM Simulations

Potapov, A., Rocky DEM Inc.

Originally presented on: 4/12/2016 14:00:00 - 14:30:00

The accurate modeling of particle attrition and breakage in a Discrete Element Model (DEM) are critical in the performance prediction of processes such as high-pressure grinding rolls, cone crusher and mills. These simulations need to consider the reduction in the particle size and subdivision of particles into smaller fragments, which provide significant numerical and programing challenges. The presentation focuses on two approaches for these type of DEM breakage simulations – instantaneous and discrete grain breakage. The instantaneous breakage model treats every particles as a single entity and breaks it into fragments instantaneously based on force or energy breakage criterion. On the other hand, the discrete grain breakage approach uses elementary particles to be "glued" together by a virtual "glue" that can withstand tensile forces up to a threshold limit before breakage occurs. Advantages and disadvantages of both these methods will be presented. In addition, approaches for particles attrition simulations and simulations of boundary attrition caused by particles interactions will also be considered.

CD-adapco
A Siemens Business

Frontiers in Particle Science & Technology

FPST

Investigating Jet Cup Attrition: Experiments and CFD-DEM Simulations

Getnet Agegnehu, Ravindra Aglave, Oleh Baran
(CD-adapco)
&
Ray Cocco (PSRI)
April 12, 2016
Houston

Outline

CD-adapco
A Siemens Business

- Motivation
- Description of the problem
- Necessary models to solve
  - DEM particle types
  - CFD and DEM coupling
  - Coarse grain approach
- Model Set-up
  - Boundary and operating conditions
- Results

Motivation

- To understand the phenomenon of attrition
- Investigate
  - the attrition characteristics of materials
  - the breaking energy supplied by the system
  - the distribution of attrition rates among the particles.
- Unknown how the components of attrition relate to each other in a jet cup
  - abrasion
  - fragmentation,
  - if that relationship is preserved in a cyclone.
- Fluidized Bed (Air Jet) Test
  - ASTM-D-5757-95
- Jet Cup Test
  - Uses less amount of particles
  - Considered as adequate
  - mimic attrition in cyclones which is largest contributed source of attrition in a well-designed unit.

Gangwal et. al., Comparison of Attrition Test Methods: ASTM Standard Fluidized Bed vs Jet Cup, Ind. Eng. Chem. Res. 2000, 39, 1155-1158

Problem Description

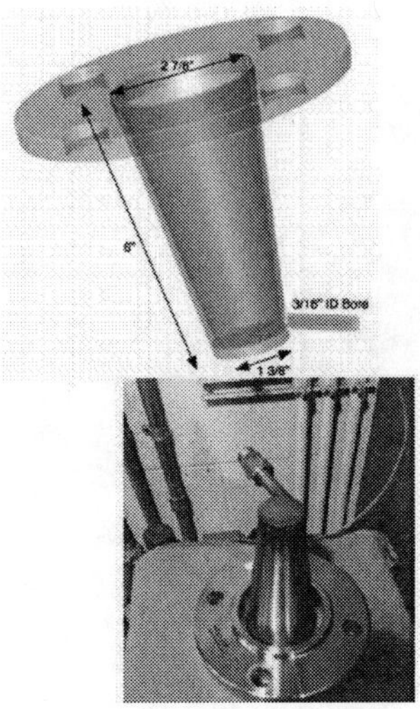

- Particles were injected and settled in the lower part of the Jet Cup,
- Injected volume fraction =~0.15
- A CDF curve was provided using a an input table
- The mass of the particles was set to 100g,
- A random injector was efficient enough to add particles in short period of time,
- The actual number of particles could reach to ~280 million
  - Coarse Grain (CG) Model was used instead.

FV Flow

STAR-CCM+

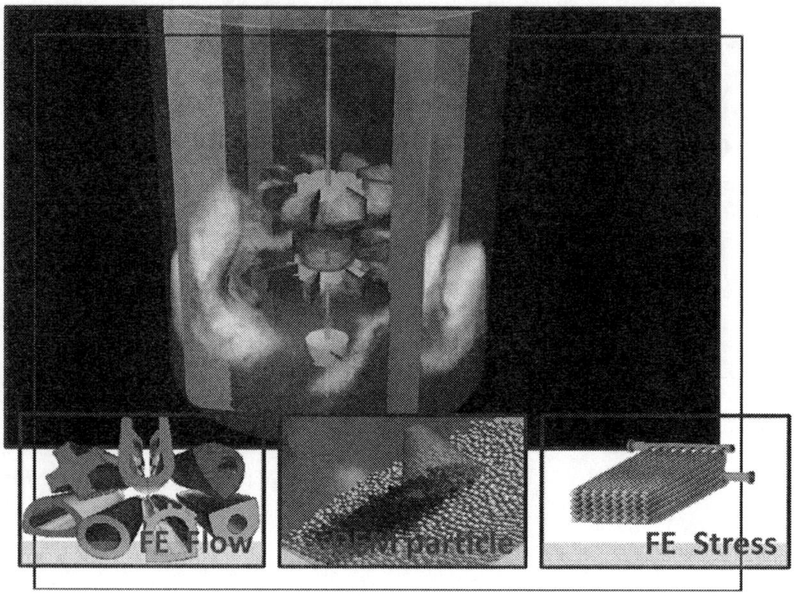

DEM integration into STAR-CCM+

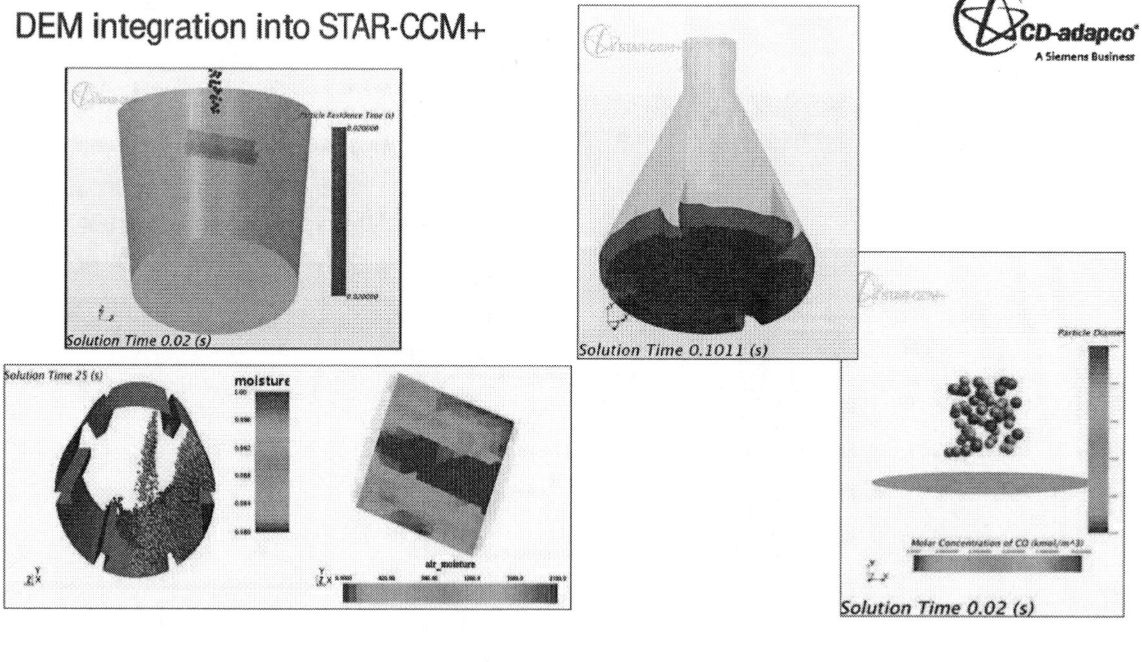

Particle Types in STAR-CCM+

Spherical	Most basic DEM particle	
Coarse Grain	Sphere represents many particles	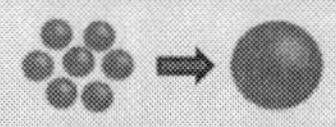
Composite	Rigid, unbreakable	
Clumps	Flexible, breakable	
Cylinders	True shapes, choice of radius and lengths	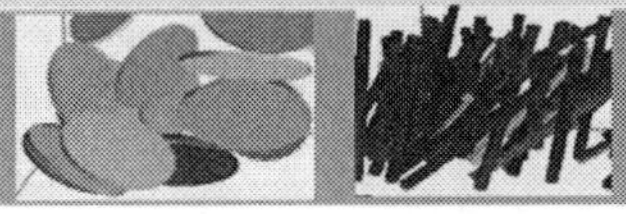

DEM Frxsdqj to Fluid Flow

Drag Force	Di Felice
	Schiller-Naumann
	Haider and Levenspiel
	Gidaspow (for 2-way coupling only)
	User defined field function
Drag Torque	Sommerfeld Rotational Drag or user defined rotational drag coefficient
Lift Force	Shear Lift: Choice of Saffman, Sommerfeld, user-set coefficients
	Spin Lift: Choice of Sommerfeld or user-set coefficients
Pressure Gradient Force	Buoyancy force
Two-way coupling	Fluid is affected by particles: Momentum source is applied to continuous phase
Other interactions	Gravity force, User-Defined Body Force, Particle Radiation, Energy model, Passive Scalar, Chemical reactions

Coarse Grain Particle

- DEM parcel represents some number of identical unresolved particles

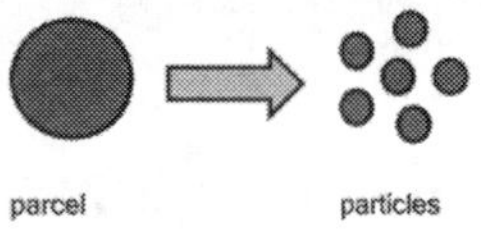

$$d_{parcel} = m\, d$$
$$m \geq 1$$

m^3 -number of particles in parcel

- Fluid-particle interaction (drag, lift etc.) are calculated for a representative particle and applied to the entire parcel
  - while the contact dynamics are calculated on the parcel scale

- Example Fluidized bed using Coarse Grain
- Each parcel represent 47 fine particles
- Total 1.7 billion particles simulated

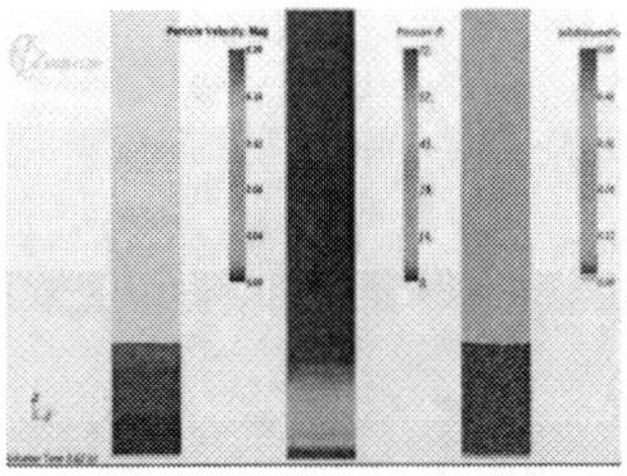

➡ **Faster DEM computing time**

Coarse Grain Details

- Example for Gidaspow drag force calculation

$$C_d = \begin{cases} \frac{4}{3}\left(\frac{150(1-\alpha_p)}{\alpha_p Re_p} + 1.75\right); & \text{if } \alpha_p < \alpha_{min}; \text{ Ergun equation based} \\ \frac{24 + 3.6 Re_p^{0.687}}{Re_p} \alpha_p^{-3.65}; & \text{if } \alpha_p \geq \alpha_{min}; \text{ Wen Yu model} \end{cases}$$

– Here α_p is solid void fraction, α_{min} is the cutoff void fraction (=0.8), Re_p is the particle Reynolds number, d is particle diameter

$F_{drag}(parcel) = m^3 \frac{1}{2} \rho_f v^2 C_d A_d$ - this drag force is applied to parcel containing m^3 particles, as a result:

Acceleration, velocity and displacement of parcels due to scaled drag is similar to acceleration, velocity and displacement and of fine particles due to original unscaled drag force

Coarse Grain Model and Length Scales

Length Scales	Notation, Field Function	Relation to other scales	Setting in STAR-CCM+
Smallest important geometry feature	l_{geom}	Diameter of thinnest pipe, baffle thickness, outlet/inlet dimension, etc	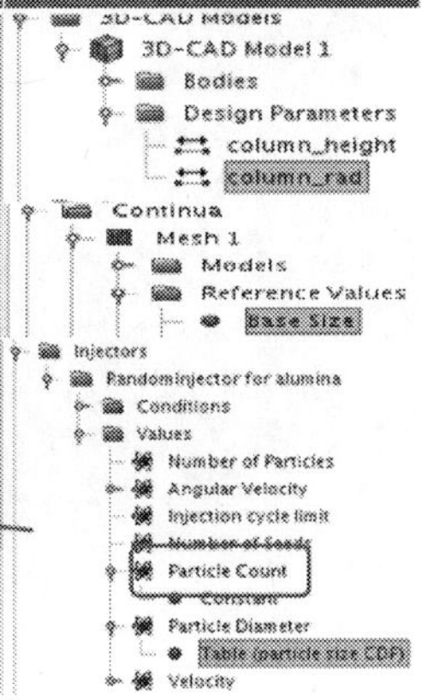
Minimum Cell Size	l_{cell}	$l_{cell} \ll l_{geom}$	
Maximum Parcel Diameter	$d_{parcel} = d\, m$ Here $m = (\$ParticleCount)^{1/3}$	$d_{parcel} \ll l_{cell}$	
Maximum Particle Diameter	d	$d < d_{parcel}$	

Limits in Coarse Grain Model

- Parcel diameter must be smaller than mesh size for 2-way coupled simulations
- Increasing $ParticleCount
  - Decreases total parcel count -> improves DEM performance
  - Reduces accuracy of contact physics
    - Negligible for 'fluidized bed' applications
  - Does not affect the DEM timestep
    - It is still the same as for fine particles, proportional to $d\sqrt{\frac{\rho}{E}}$

- For fluidized bed applications, the Young's modulus E of particles can be reduced up to $\frac{100Mpa}{(\$ParticleCount)^{\frac{2}{3}}}$
  - Larger DEM timestep but reduced contact physics accuracy
  - Restitution coefficients > 0.8

Geometry & Mesh

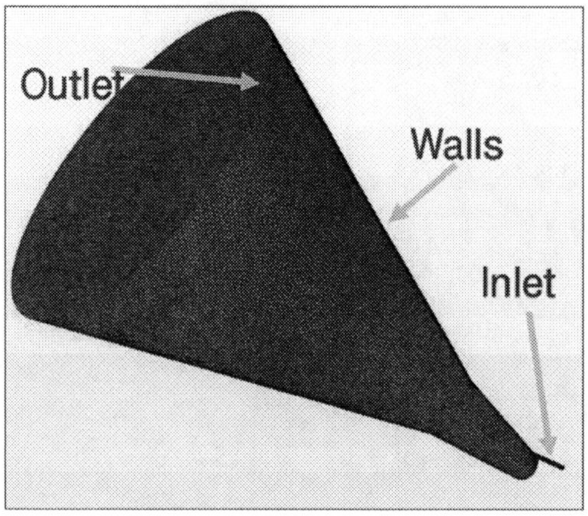

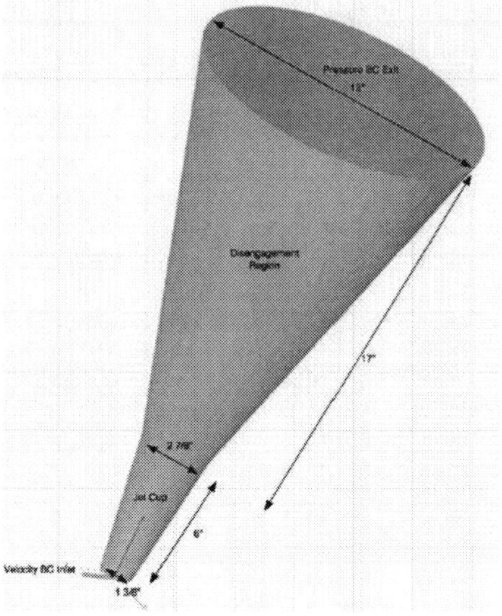

Physics selection

CD-adapco
A Siemens Business

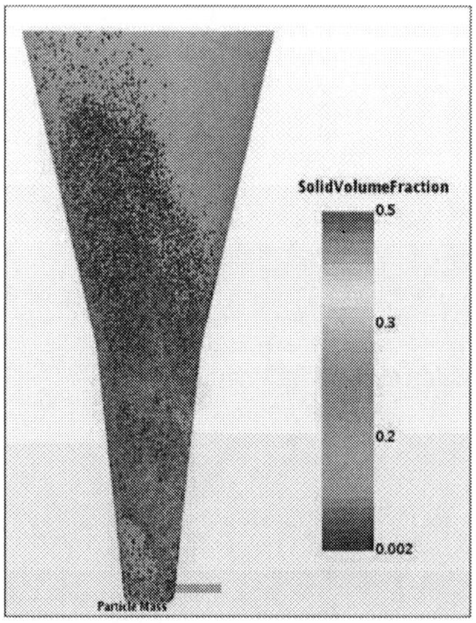

- Coupled flow solver,
- Air as a fluid,
- Implicit unsteady,
- Particles with density of ~1491 kg/m3,
- Two Way Coupled DEM simulation,
- Drag Force: Gidaspow
- Contact force: Hertz-Mindlin

Particles

- CG Parcel diameter = 13 x particle size,
- Total CG Parcels = 69400,
- Median diameter of experiment = 78 microns,
- Median diameter of DEM particle = 83 microns,

Contacts	Friction	Restitution
Particle-Particle	0.61	0.1
Particle-Wall	0.4	0.2

	Young's (MPa)	Poisson	Density (kg/m³)
Particle	200	0.3	1491
Wall	68000	0.33	2702

Performance: 16 hours to simulate 1 second on 56 cores

Indications of Particle Attrition

- In STAR-CCM+, attrition prone areas can be identified via different approaches such as:
- Through Contact Forces: STAR-CCM+ gives the Contact Elastic Forces between particles and particles with a wall,
- Through impact Wear: STAR-CCM+'s DEM model predicts the impact wear rate due to collision of particles with a wall,
- The DEM model also estimates an abrasive wear potential when particles dragged along walls.

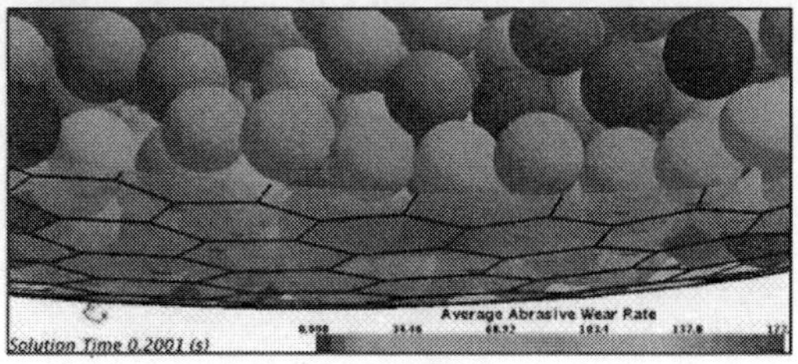

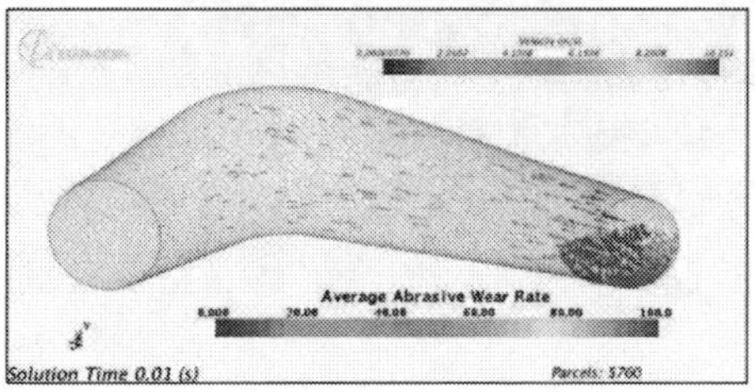

Particle Velocity (U_{in} = 137 m/s)

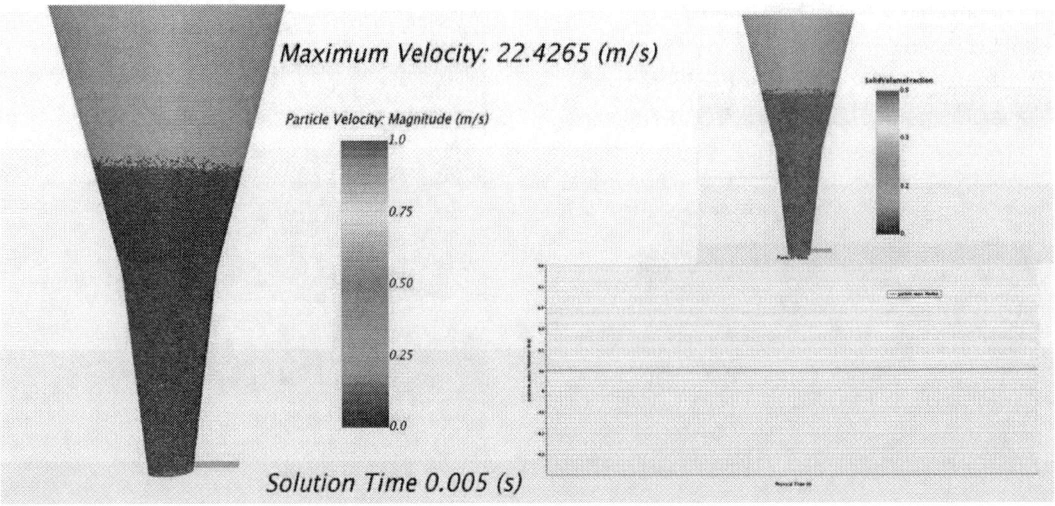

Visual Comparison with Experiment (U_{in} = 137 m/s)

Comparison of average particle velocity

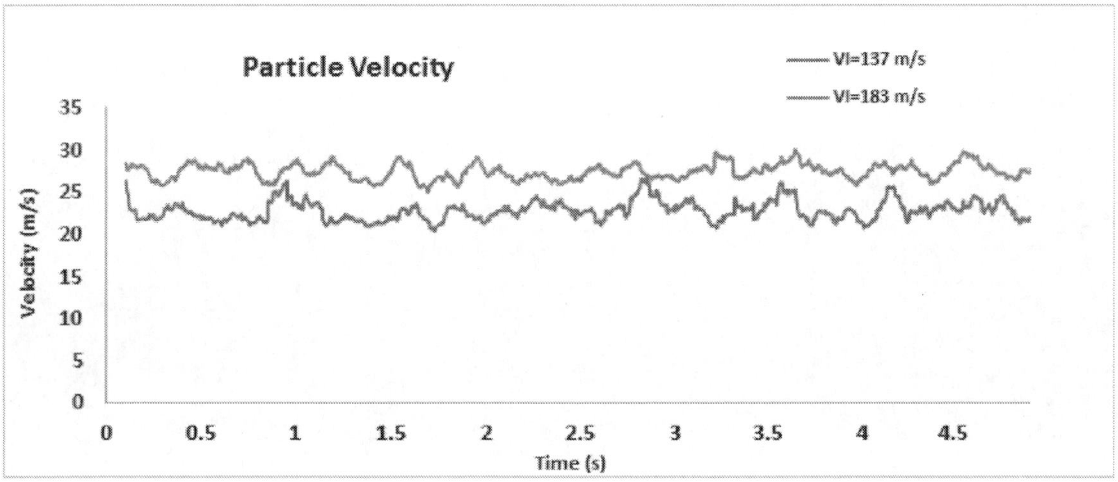

Comparison of Contact Force

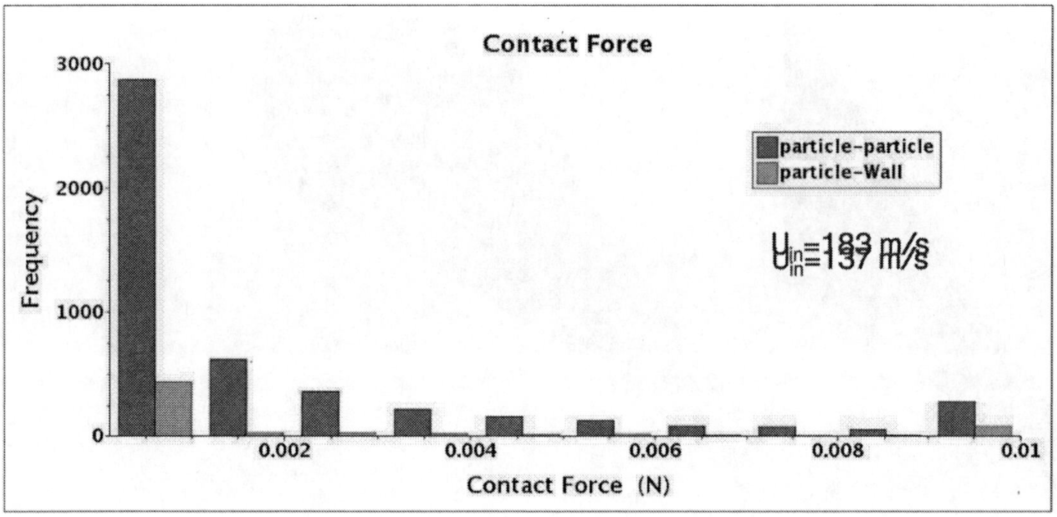

Comparison of P-P and P-W Contact %age

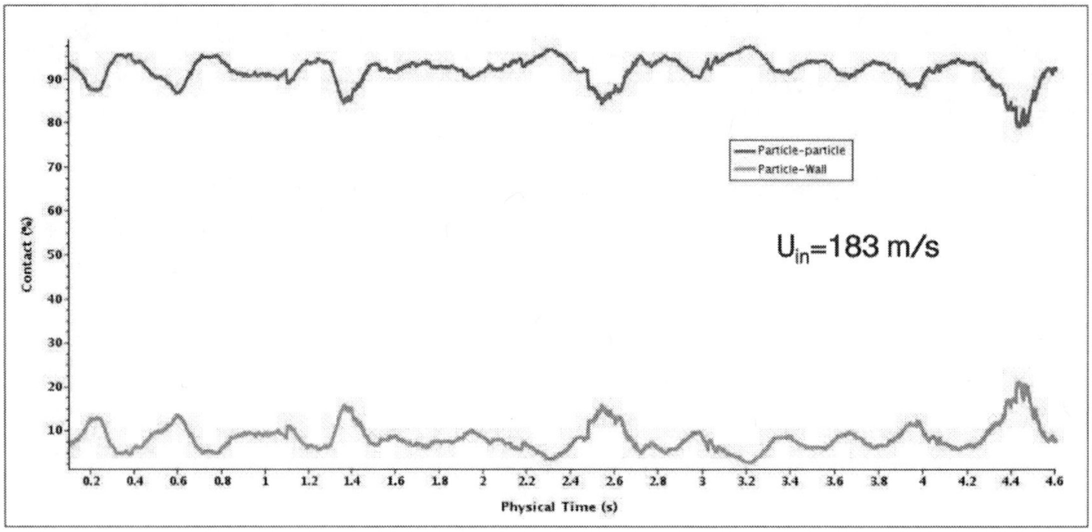

Abrasive Wear Comparison (Indicative)

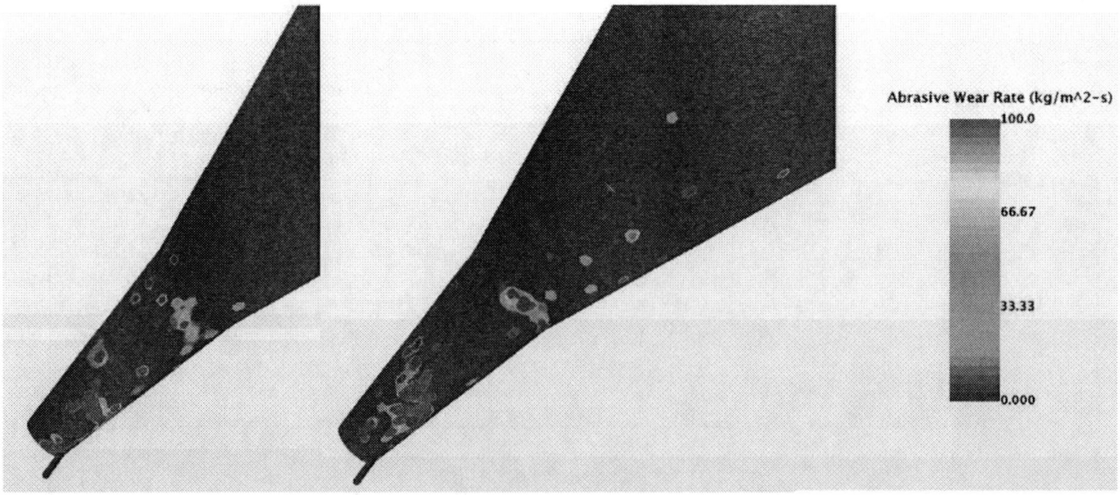

Summary of the work

- A method to investigate jet-cup attrition with CFD + DEM simulation established
  - Reasonable set-up and simulation times
- Some insight in to the distribution of contact forces
  - Kinetic energy of particle can tell the breakage energy
- Lot more analysis (quantitative) to be under taken

Population Balance Models of Attrition

Bert Diemer
University of Delaware
Professor of Practice &
Co-Director, MEPT Program

University of Delaware MEPT Department of Chemical & Biomolecular Engineering

New MEPT Program

Master of Engineering in Particle Technology

1 Year Professional Masters Degree Program
In its 2nd year!

Fall Semester (12 credits)
- Particle Kinetics
- Particle Transport
- Particle Properties
- Math for Particle Systems

Spring Semester (12 credits)
- Particle Unit Ops
- Particle Product Engineering
- Elective 1
- Elective 2

Summer Semester (6 credits) – Industrial Internship

University of Delaware MEPT Department of Chemical & Biomolecular Engineering

What is Attrition?

Original abstract and submitted slides presume the dictionary definition

- Dictionary:
  - A gradual wearing down, weakening
    - of materials by friction; abrasion/erosion
    - of a military enemy by constant harassment, or attack (a war of attrition)
    - of a workforce by not replacing losses to retirement, resignation or death
- My particle technologist colleagues:
  - *Any* unwanted size reduction
    - Can be by any mechanism, not just erosion
- This conference?
  - Not just mitigation but *application*
    - Application implies it may be desirable
  - Perhaps *occurring naturally* rather than by design?
    - *Adventitious* size reduction?

University of Delaware MEPT Department of Chemical & Biomolecular Engineering

Breakage Population Balance

Can model any breakage process that is linear in the number or mass of a given size or size class

- Continuous spectrum mass form:

$$\frac{\partial m(V)}{\partial t} = \underbrace{\int_V^\infty S_{br}(\varphi) b_W(V;\varphi) m(\varphi) d\varphi}_{\text{birth term}} - \underbrace{S_{br}(V) m(V)}_{\text{death term}}$$

- Sectional mass model:

$$\frac{dm_k}{dt} = \underbrace{\sum_{j=k+1}^{K} S_{br,j} b_{k,j}^W m_j}_{\text{birth term}} - \underbrace{S_{br,k} m_k}_{\text{death term}}$$

- Rate of breakage as a function of size known as the "selection" function: $S_{br}(V)$ or $S_{br,j}$
- Distribution of fragments is conditional on the parent size: $\underbrace{b_W(V;\varphi)}_{\text{mass}} = \frac{V}{\varphi} \underbrace{b(V;\varphi)}_{\text{number}}$ or $b_{k,j}^W$

University of Delaware MEPT Department of Chemical & Biomolecular Engineering

Sectional Model In Matrix Form

$$\boxed{\dot{\mathbf{m}} = \mathbf{A}\mathbf{m} \quad ; \quad \mathbf{m}(0) = \mathbf{m}_o}$$

$$\mathbf{A} = \begin{bmatrix} -S_{br,1} & S_{br,2}b^W_{1,2} & \cdots & S_{br,K-2}b^W_{1,K-2} & S_{br,K-1}b^W_{1,K-1} & S_{br,K}b^W_{1,K} \\ 0 & -S_{br,2} & \cdots & S_{br,K-2}b^W_{2,K-2} & S_{br,K-1}b^W_{2,K-1} & S_{br,K}b^W_{2,K} \\ \vdots & \vdots & \ddots & \vdots & \vdots & \vdots \\ 0 & 0 & \cdots & -S_{br,K-2} & S_{br,K-1}b^W_{K-2,K-1} & S_{br,K}b^W_{K-2,K} \\ 0 & 0 & \cdots & 0 & -S_{br,K-1} & S_{br,K}b^W_{K-1,K} \\ 0 & 0 & \cdots & 0 & 0 & -S_{br,K} \end{bmatrix} ; \mathbf{m} = \begin{bmatrix} m_1 \\ m_2 \\ \vdots \\ m_{K-2} \\ m_{K-1} \\ m_K \end{bmatrix}$$

- Since $\mathbf{A}$ is upper triangular, eigenvalues are its diagonal elements

$$\lambda = \left\{-S_{br,1}, -S_{br,2}, \ldots, -S_{br,K-2}, -S_{br,K-1}, -S_{br,K}\right\}$$

University of Delaware MEPT Department of Chemical & Biomolecular Engineering

Analytical Solution

- Herbst, J. A. and Fuerstenau, D. W., "Scale-Up Procedure for Continuous Grinding Mill Design Using Population Balance Models," *Intl. J. Min. Proc.*, **7** (1980) 1-31.

$$\boxed{\dot{\mathbf{m}} = \mathbf{A}\mathbf{m} \quad ; \quad \mathbf{m}(0) = \mathbf{m}_o \quad \Rightarrow \quad \mathbf{m}(t) = \mathbf{E}\,\mathbf{J}(t)\,\mathbf{E}^{-1}\mathbf{m}_o}$$

$$\mathbf{J}(t) = \begin{bmatrix} e^{\lambda_1 t} & 0 & \cdots & 0 \\ 0 & e^{\lambda_2 t} & \cdots & 0 \\ \vdots & \vdots & \ddots & \vdots \\ 0 & 0 & \cdots & e^{\lambda_K t} \end{bmatrix} ; \quad \mathbf{E} = \begin{bmatrix} 1 & e_{12} & \cdots & e_{1,K-2} & e_{1,K-1} & e_{1K} \\ 0 & 1 & \cdots & e_{2,K-2} & e_{2,K-1} & e_{2K} \\ \vdots & \vdots & \ddots & \vdots & \vdots & \vdots \\ 0 & 0 & \cdots & 1 & e_{K-2,K-1} & e_{K-2,K} \\ 0 & 0 & \cdots & 0 & 1 & e_{K-1,K} \\ 0 & 0 & \cdots & 0 & 0 & 1 \end{bmatrix} ; \quad \mathbf{E}^{-1} = \begin{bmatrix} 1 & \tilde{e}_{12} & \cdots & \tilde{e}_{1,K-2} & \tilde{e}_{1,K-1} & \tilde{e}_{1K} \\ 0 & 1 & \cdots & \tilde{e}_{2,K-2} & \tilde{e}_{2,K-1} & \tilde{e}_{2K} \\ \vdots & \vdots & \ddots & \vdots & \vdots & \vdots \\ 0 & 0 & \cdots & 1 & \tilde{e}_{K-2,K-1} & \tilde{e}_{K-2,K} \\ 0 & 0 & \cdots & 0 & 1 & \tilde{e}_{K-1,K} \\ 0 & 0 & \cdots & 0 & 0 & 1 \end{bmatrix}$$

- $\mathbf{J}(t)$ is time-dependent diagonal matrix
  - Elements are exponentials of □$_k t$
- $\mathbf{E}$ is matrix of eigenvectors, $\mathbf{E}^{-1}$ is its inverse
  - Both will be upper triangular

University of Delaware MEPT Department of Chemical & Biomolecular Engineering

Eigenvector Matrix & Inverse

Elements of E

$$e_{im} = \begin{cases} 0, m < i \\ 1, m = i \\ \sum_{j=i+1}^{m} \frac{S_{br,j} b_{i,j}^{W}}{S_{br,i} - S_{br,m}} e_{jm}, m > i \end{cases}$$

Elements of E^{-1}

$$\tilde{e}_{im} = \begin{cases} 0, m < i \\ 1, m = i \\ \sum_{j=i+1}^{m} -e_{ij} \tilde{e}_{jm}, m > i \end{cases}$$

University of Delaware MEPT Department of Chemical & Biomolecular Engineering

Required Number of Parameters

- For completely general model with n size classes
  - Selection function values $= n$
  - Fragment distribution values $= (n^2 - n)/2$
  - Total $= (n^2 + n)/2$
    - 20 size classes requires 210 parameters
    - 100 size classes requires 5050 parameters
- Number of parameters can be greatly reduced by models for:
  - Selection function
  - Fragment distribution

University of Delaware MEPT Department of Chemical & Biomolecular Engineering

Power-Law Selection Function Model

- Requires just 2 parameters instead of n
  - Parameters are
    - Size-independent constant $S^o{}_{br}$
    - Order in size □ (typically $\frac{1}{4} \le \eta \le 1$)
- Basis in 1st-principles & single-particle measurement
  - Vogel & Peukert, *Powder Tech.*, **129** (2003) 101-110.
  - Ghadiri & Zhang, *Chem. Eng. Sci*, **57**, (2002), 3659-3669, 3671-3686.
  - Moreno-Atanasio & Ghadiri, *Chem. Eng. Sci.*, **61** (2006) 2476-2481.

$$S_{br}(V) = S^o_{br} V^{\eta}$$

continuous spectrum

or

$$S_{br,k} = S^o_{br} \overline{V}_k^{\ \eta}$$

sectional

UNIVERSITY OF DELAWARE MEPT DEPARTMENT OF CHEMICAL & BIOMOLECULAR ENGINEERING

Fragment Distribution Models

- Look at 2 types:
  - Fixed small fragment distribution models
    - Fragment size independent of parent size
  - Self-similar models
    - Fragments get smaller as parent gets smaller
- Reduces $(n^2 - n)/2$ parameters to a handful describing fragment distribution function

UNIVERSITY OF DELAWARE MEPT DEPARTMENT OF CHEMICAL & BIOMOLECULAR ENGINEERING

Fixed Small Fragment Distribution Model

- Attrition leaves one larger fragment and a distribution of smaller fragments
- Small fragment distribution is fixed
  - Example:
    - Log-normal fragments with specified mass median and geometric standard deviation (2 parameters)
- Mass allocated to large fragment is mass of next smaller size class
- Remaining mass allocated to small fragments according to small fragment distribution

Fixed Small Fragment Distribution Model In Matrix Form for Geometric Spacing

small fragments

$$
\mathbf{A} = \begin{bmatrix}
-S_{br,1} & S_{br,2}b_1^W & S_{br,3}b_1^W & \cdots & S_{br,K-3}b_1^W & S_{br,K-2}b_1^W & S_{br,K-1}b_1^W & S_{br,K}b_1^W \\
0 & -S_{br,2} & S_{br,3}b_2^W & \cdots & S_{br,K-3}b_2^W & S_{br,K-2}b_2^W & S_{br,K-1}b_2^W & S_{br,K}b_2^W \\
0 & 0 & -S_{br,3} & \cdots & S_{br,K-3}b_3^W & S_{br,K-2}b_3^W & S_{br,K-1}b_3^W & S_{br,K}b_3^W \\
\vdots & \vdots & \vdots & \ddots & \vdots & \vdots & \vdots & \vdots \\
0 & 0 & 0 & \cdots & -S_{br,K-3} & S_{br,K-2}b_L^W & 0 & 0 \\
0 & 0 & 0 & \cdots & 0 & -S_{br,K-2} & S_{br,K-1}b_L^W & 0 \\
0 & 0 & 0 & \cdots & 0 & 0 & -S_{br,K-1} & S_{br,K}b_L^W \\
0 & 0 & 0 & \cdots & 0 & 0 & 0 & -S_{br,K}
\end{bmatrix}
$$

large fragments

Example Trajectory

$$S_{br}^{o},\ \text{cm}^{-3/2}\text{min}^{-1} = \begin{cases} 0.1, & d > 33\mu\text{m} \\ 0.0001, & d < 33\mu\text{m} \end{cases}$$

$$\eta = \frac{1}{2}$$

$$d_{50} = 6\mu\text{m}$$

$$\sigma_g = 1.48$$

Requires 5 parameters for 100 sections

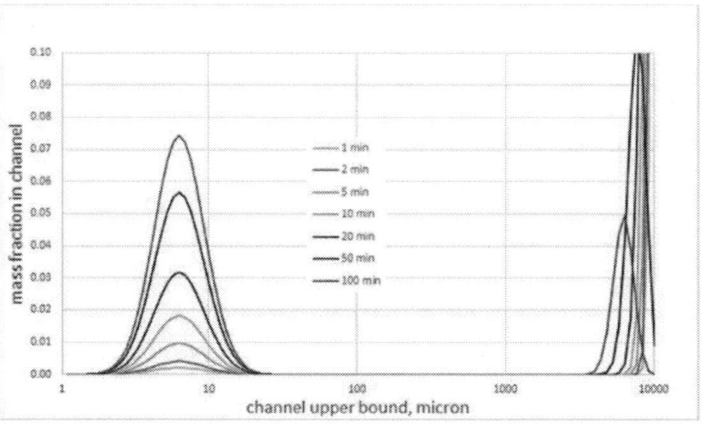

University of Delaware MEPT Department of Chemical & Biomolecular Engineering

Self-Similar Fragment Model

- Assumes fragment distribution depends on ratio of fragment to "parent" size
  - Continuous spectrum form:

$$\underbrace{b_W(V;\varphi)}_{\text{mass}}dV = \frac{V}{\varphi}\underbrace{b(V;\varphi)}_{\text{number}}dV = \frac{V}{\varphi}\times\frac{\overbrace{\hat{b}(V/\varphi)}^{\text{self-similar fragments}}}{\varphi}dV = z\hat{b}(z)dz$$

  - Sectional form:

$$b_{k,j}^{W} = \int_{V_k/V_j}^{V_{k+1}/V_j} z\,\hat{b}(z)\,dz$$

University of Delaware MEPT Department of Chemical & Biomolecular Engineering

Diemer-Olson-Hill-Ng Fragment Distributions

Hill & Ng, *AIChEJ*, 42 (6), 1600-1611 (1996).

Diemer & Olson, *Chem. Eng. Sci.*, 57 (19), 4187-4198 (2002).

Derived from Joint Multivariate Distributions… Example: Power-law Product for p daughters

$$C\prod_{i=1}^{p} z_i^{q-1}$$

Eliminate one daughter via conservation of mass and integrate over $p-2$ daughters to get marginal distribution for single daughter - gives *beta* distribution

q parameter has mechanistic meaning

$q<1$ favors small z: erosion/attrition mechanisms
$q\sim 1$ favors no z: uniform breakage
$q>1$ favors large z: fracture mechanism

$$\hat{b}(z) = p\frac{z^{q-1}(1-z)^{r-1}}{\mathrm{B}(q,r)}$$

$$p = \frac{q+r}{q} \quad \text{or} \quad r = q(p-1)$$

Has been shown to describe most literature self-similar daughter distributions

UNIVERSITY OF DELAWARE MEPT DEPARTMENT OF CHEMICAL & BIOMOLECULAR ENGINEERING

DOHN Fragments in Sectional Form

- DOHN continuous self-similar number distribution $$\hat{b}(z) = \frac{q+r}{q} \times \frac{z^{q-1}(1-z)^{r-1}}{\mathrm{B}(q,r)}$$
- DOHN self-similar mass-based sectional fragments

$$b_{k,j}^{W} = \frac{q+r}{q}\left[\frac{\mathrm{B}_{inc}\left(q+1,r;V_{k+1}/V_j\right)-\mathrm{B}_{inc}\left(q+1,r;V_k/V_j\right)}{\mathrm{B}(q,r)}\right]$$

UNIVERSITY OF DELAWARE MEPT DEPARTMENT OF CHEMICAL & BIOMOLECULAR ENGINEERING

Binary Breakage (p = 2, r = q)

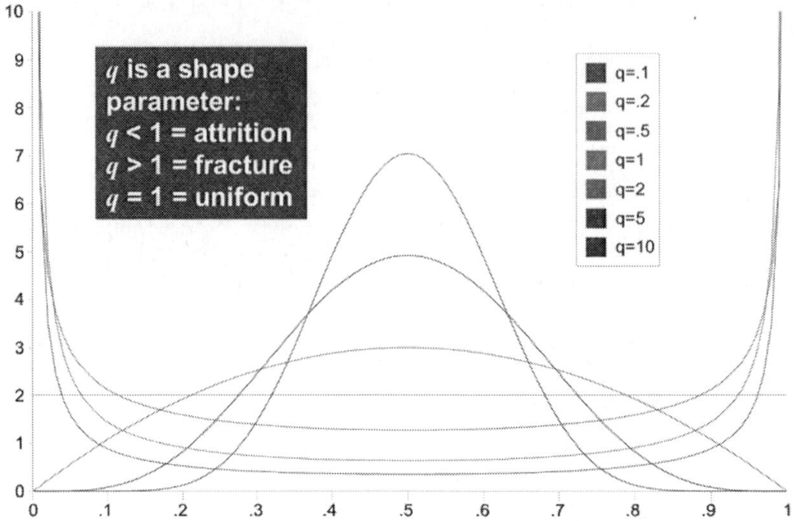

q is a shape parameter:
q < 1 = attrition
q > 1 = fracture
q = 1 = uniform

UNIVERSITY OF DELAWARE MEPT DEPARTMENT OF CHEMICAL & BIOMOLECULAR ENGINEERING

Example Trajectory

S^o_{br} = 5.44 cm$^{-3/2}$min^{-1}, □ = ½ , q = 0.1, r = 1, p = 11

Reaches a similarity solution

Small fragments are in fine tail

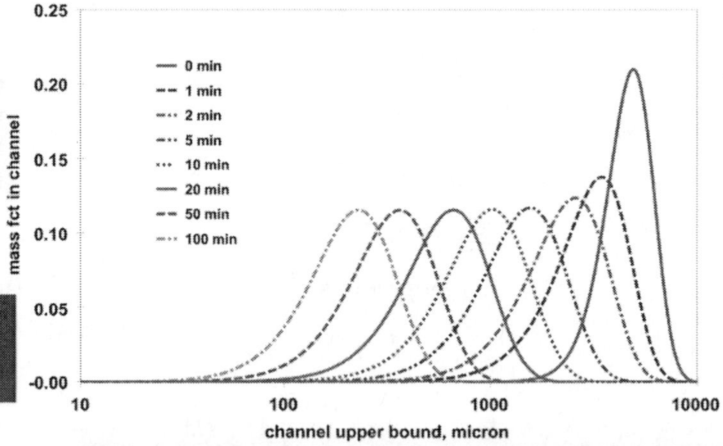

Requires 4 parameters for 100 sections

UNIVERSITY OF DELAWARE MEPT DEPARTMENT OF CHEMICAL & BIOMOLECULAR ENGINEERING

Peterson Similarity Solution

Peterson, T. W., *Aerosol Sci. & Tech.*, 5, 93-101 (1986).

- Special case with r = 1, p = (q +1)/ q
  - Example is a member of this class
- Yields analytical similarity solution
  - Modified Gamma Distribution

$$x = \frac{V}{V_n(t)} \quad ; \quad y = \frac{\Gamma\left((q+1)/\eta\right)}{\Gamma\left(q/\eta\right)} x$$

$$\underset{\text{number probability density}}{f(y)dy} = \frac{\eta}{\Gamma\left(q/\eta\right)} y^{q-1} e^{-y^\eta} dy$$

Note that when r = 1, $p \geq 2$ implies $q \leq 1$, i.e., the Peterson fragment distribution is an *attrition* distribution

UNIVERSITY OF DELAWARE MEPT DEPARTMENT OF CHEMICAL & BIOMOLECULAR ENGINEERING

Sectional Model Results Approach Analytical Solution

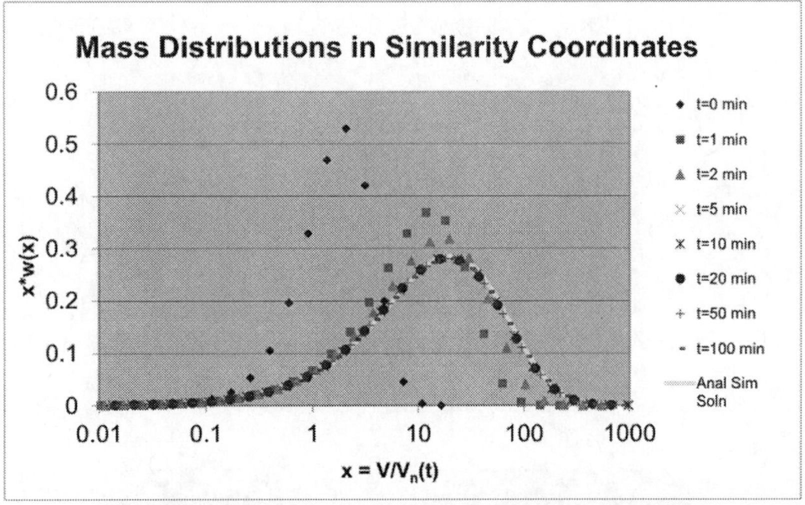

UNIVERSITY OF DELAWARE MEPT DEPARTMENT OF CHEMICAL & BIOMOLECULAR ENGINEERING

Sensitivity to Fragment Distribution

□ = ½ , r = 1 (Peterson)… vary q

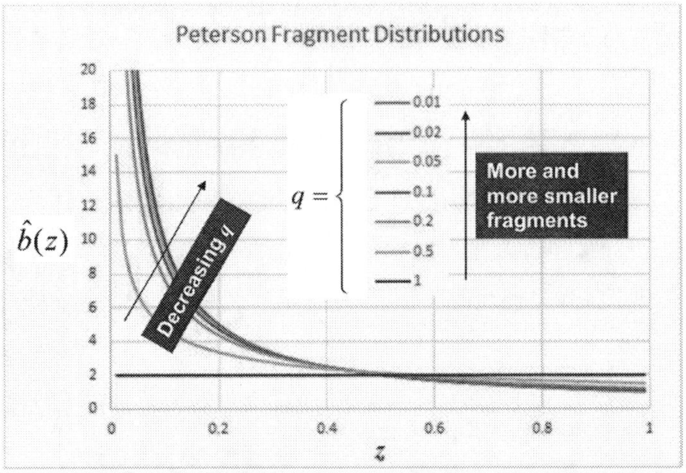

Sensitivity to Fragment Distribution

Similarity Solutions for: □ = ½ , r = 1 (Peterson)… vary q

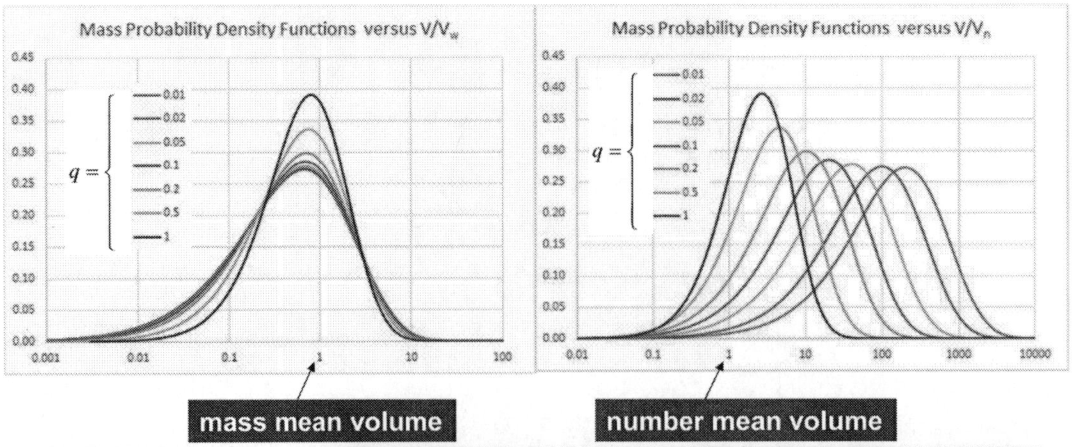

Sensitivity to Breakage Order

Similarity Solutions for: q = 0.01 , r = 1 (Peterson)… vary □

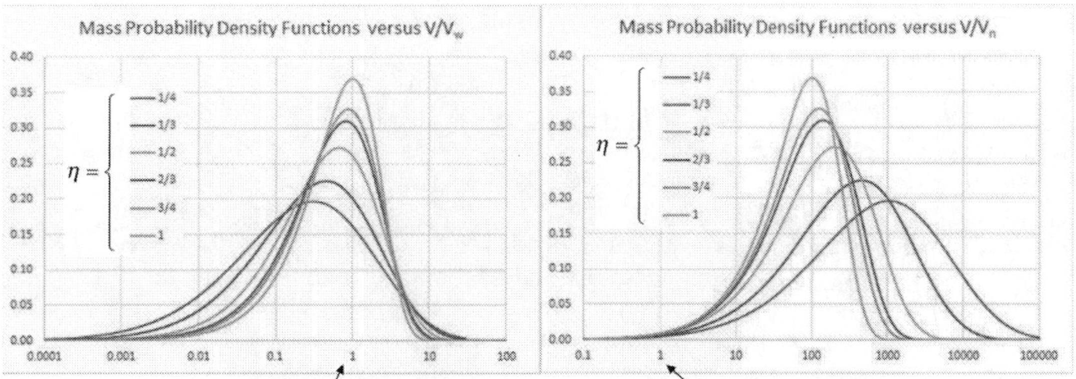

mass mean volume

number mean volume

Summary Remarks

- Standard breakage formalism can model linear attrition processes however "attrition" may be defined
  - General sectional form requires lots of parameters
  - Models for the selection function and fragment distribution greatly reduce parameters
- There are multiple ways to specify fragment distributions that conform to the dictionary definition of "attrition"
  - Self-similar fragment distributions give unimodal self-preserving product distributions
  - Fixed fragment distributions can describe bimodal product distribution
- Questions:
  - Do we actually see both fixed and self-similar type behaviors?
  - How do we exploit these tools in attrition "applications"?

ExxonMobil

April 11-13, 2016

MODELING CATALYST EXTRUDATE BREAKAGE BY IMPULSIVE FORCES

Energy lives here

Jean W. L. Beeckman, Natalie A. Fassbender, Theodore E. Datz

ExxonMobil Research and Engineering Company,
Process Technology Department, Catalyst Technology Division
1545 Route 22 East, Annandale, New Jersey 08801

2016 Frontiers in Particle Science and Technology Conference, Houston, TX

Background

- ❑ Fuels/Lubes/Chemicals catalysts are typically extruded bodies of constant cross-section but with variable lengths.
- ❑ "Natural or forced" catalyst sizing by breakage is important in
  - ➢ Catalyst manufacturing plants
  - ➢ Reactor loading/unloading practices
  - ➢ Catalyst regeneration facilities
- ❑ Catalyst geometry is characterized by aspect ratio, shape and diameter.
  - ➢ Aspect ratio is the average length to diameter ratio (L/D)
- ❑ Why aspect ratio specifications ?
  - ➢ Small aspect ratios lead to excessive pressure drop
  - ➢ Large aspect ratios can lead to low reactor weight fills
- ❑ Remedies for controlling the aspect ratio are often "ad hoc"
  - ➢ Engineering analysis often lacking
  - ➢ Literature data scarce

ExxonMobil 2

Goals of the study

- ❑ Develop a simple drop test that allows to gather the relevant data on the reduction of the aspect ratio due to collision
- ❑ Develop an adequate parametric model to predict the reduction in aspect ratio
- ❑ Finally, link the parameters to independently measurable properties of the materials and the conditions of the test.

ExxonMobil 3

Goal 1: Experimental set-up

- The aspect ratio is called Φ and stands for the number average L/D ratio of a sample of typically 100-300 extrudates
- 25 catalysts used here cover a wide range of characteristics
  active phase - binder - extrusion – thermal conditions

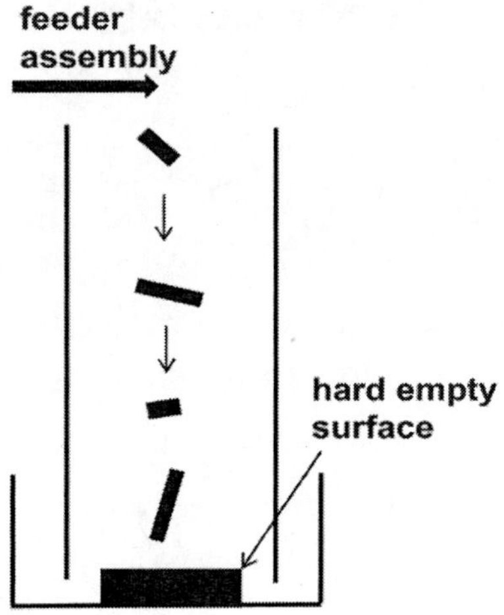

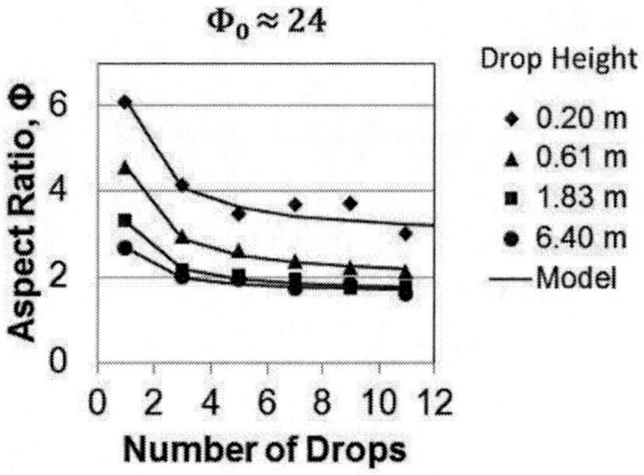

ExxonMobil 4

Extrudate drop and impact via high speed photography at 10,000 frames/sec

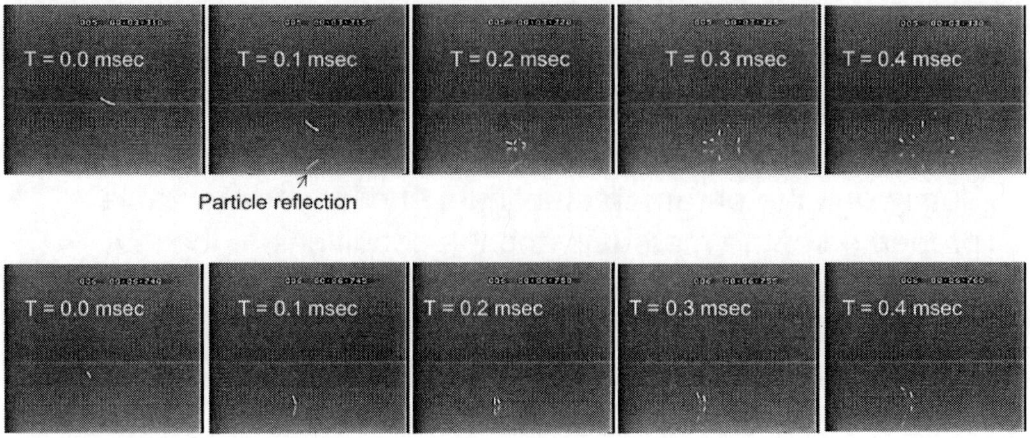

ExxonMobil

5

Data reveal two asymptotic parameters Φ_∞ and Φ_α

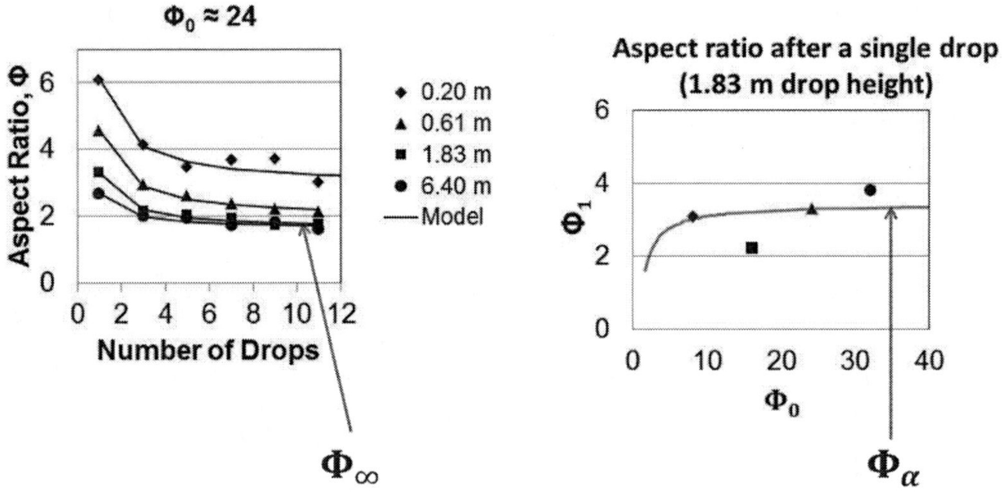

ExxonMobil

6

Goal 2 : Formal model equations

The general two parameter model is obtained via a 1st order Padé approximation:

$$\Phi_j = (\gamma\Phi_o + j\Phi_\infty)/(\gamma + j) \;;\; j = 0,1,2,\ldots$$

$$\gamma = (\Phi_\alpha - \Phi_\infty)/(\Phi_o - \Phi_\infty)$$

The above model satisfies 4 conditions

1. Any Φ_0 allowed
2. $\Phi_j \rightarrow \Phi_\infty$ for large j
3. $\Phi_1 \rightarrow \Phi_\alpha$ for large Φ_0
4. any Φ_j as starting point

Presented at NAM24 2015, Pittsburgh PA
Published in AICHE Journal, DOI 10.1002/aic.15046 (wileyonlinelibrary.com)

ExxonMobil

7

Use the model to optimize plant severity

Severity Sequencing

	HS → LS	LS → HS
Φ_0	24.00	24.00
drop (m)	1.83	0.61
Φ_1	3.30	4.57
drop (m)	0.61	1.83
Φ_2	2.87	2.74

A 0.1 L/D advantage for HS → LS confirmed in the laboratory

Severity Conditioning

	1 HS drop	3 LS drops
Φ_0	24.00	24.00
drop (m)	1.83	0.61
Φ_1	3.30	4.57
drop (m)		0.61
Φ_2		3.33
drop (m)		0.61
Φ_3		2.88

ExxonMobil

8

Goal 3: Modeling the collision

- The force of the impact is found from Newton's second law applied just prior to breakage

$$F_i = 2mv/\Delta\tau$$

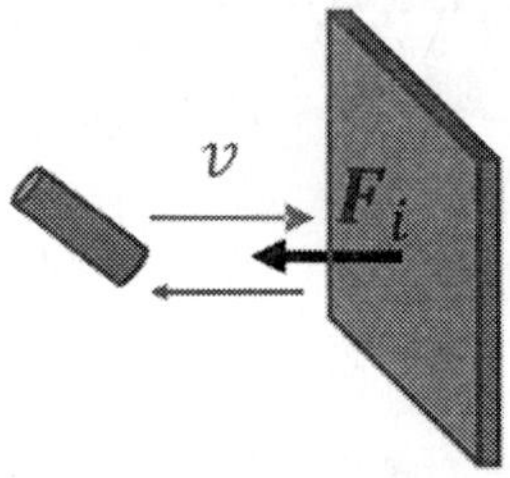

- The strength of the extrudate is characterized by the Euler-Bernoulli modulus of rupture σ. The rupture force is:

$$F_r = \sigma D^3/\varphi w$$

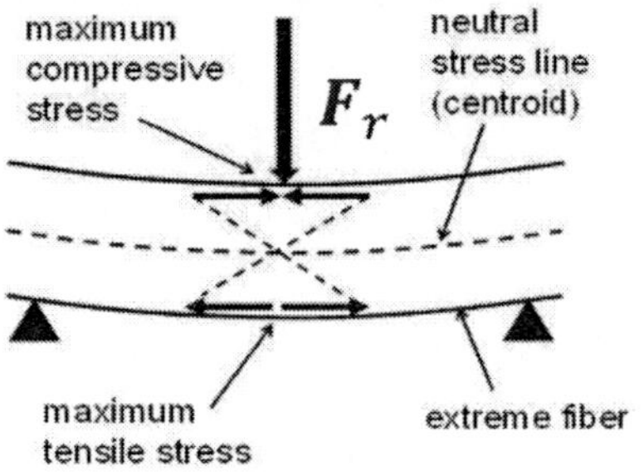

AIChE Journal, 2016,
DOI# 10.1002/aic.15231

ExxonMobil

9

Impact force & extrudate strength balance at twice the asymptotic aspect ratio Φ_∞

- The force balance leads to a dimensionless group Be

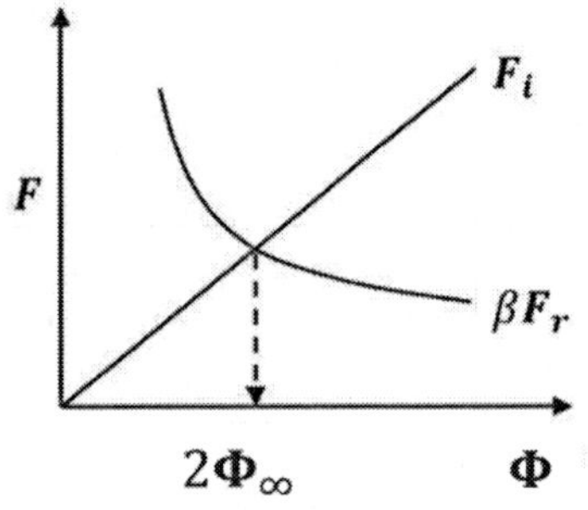

$$\Phi_\infty = C\sqrt{Be} \quad ; \quad Be = \sigma/\psi\rho Da$$

ψ : shape factor
ρ : density
D : diameter
a : deceleration

- Normalizing the deceleration "a" to the gravitational acceleration "g" allows to introduce the concept of a dimensionless severity S_∞ for the impact:

$$\Phi_\infty = \frac{1}{S_\infty}\sqrt{Be_g} \; ; \; Be_g = \sigma/\psi\rho Dg \quad S_\infty = C'\sqrt{v/g\Delta\tau}$$

ExxonMobil

10

The slope of Φ_∞ versus $\sqrt{\sigma/\psi\rho D v}$ is a measure of the contact time of the collision

$$slope = \sqrt{\frac{\beta\Delta\tau}{8}} \rightarrow \Delta\tau \cong 0.6\, msec$$

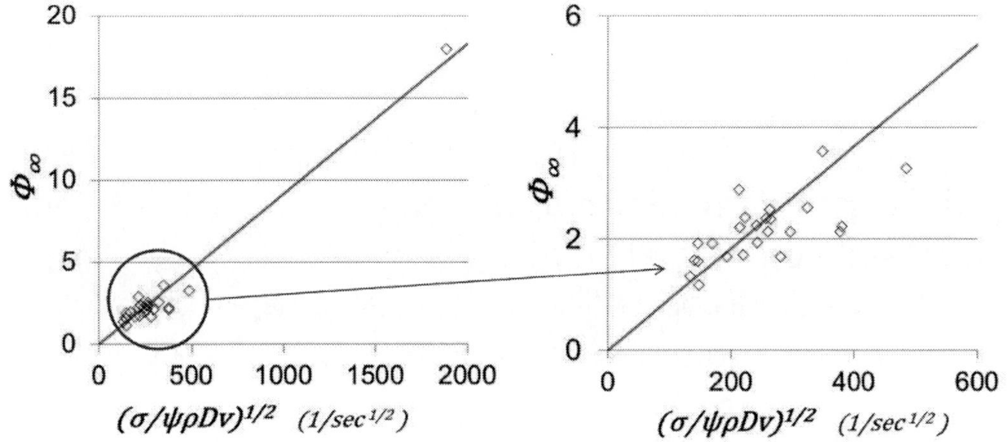

ExxonMobil

11

Data demonstrate the proposed functionality for impact velocity and bending strength

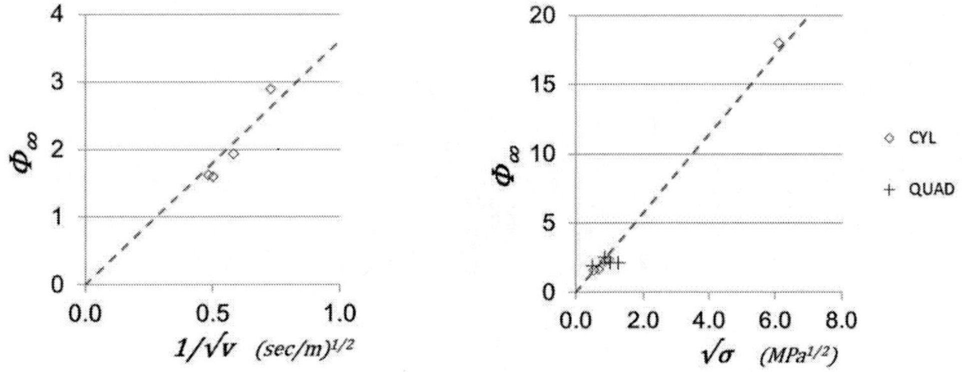

ExxonMobil

12

Conclusion

- A two parameter model successfully predicts the extrudate impulsive breakage phenomenon.
- Both parameters have a physical – mechanical meaning.
- Newton's 2nd law and the Euler-Bernoulli modulus of rupture adequately describe the collision.
- The approach allows to quantitatively describe the severity of the collision.
- The methodology allows for a direct application to individual equipment and plants.

ExxonMobil 13

QUESTIONS

ExxonMobil 14

Laboratory Evaluation of the Attrition Resistance of FCC Catalysts

BASF
We create chemistry

Melissa Clough, Ph.D. – Technology Specialist
Frontiers in Particle Science & Technology – AIChE
Houston TX – April 12, 2016

Overview: Attrition in FCCUs

BASF
We create chemistry

Attrition Sources in an FCCU

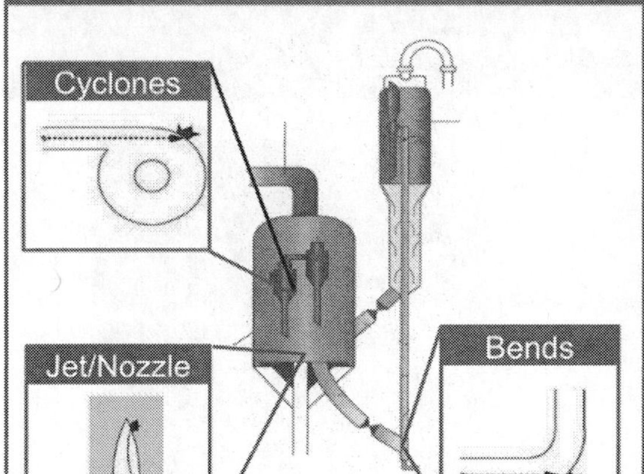

Attrition Mechanisms in an FCCU

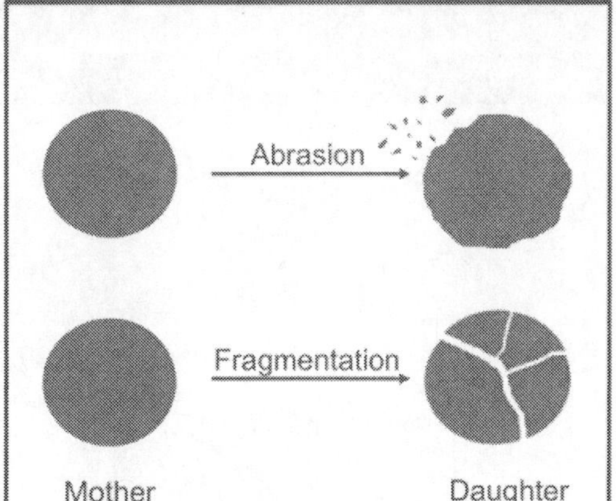

Ted Knowlton (2008), Particle Solid Research Institute
Amblard B., et al. (2015), Development of an attrition evaluation method using a Jet Cup rig, Powder Technology, 274, 455-465

2

Controlling Particulates is More Important Than Ever

BASF We create chemistry

Worldwide Regulatory Pressures

- Tightening PM regulations worldwide
- Standards for reconstructed FCCUs
  - < 1 lb / 1000 lb coke burned or < 0.04 grains / dry scf
- National Ambient Air Quality Standard (40 CFR part 50)
  - PM2.5 ≤ 35 mg / m^3 per 24 hour average
  - Pressure on non-attainment areas to comply

FCC Operational Pressures

- Opacity constraints
- Expander blade vibrations
- Cyclone dipleg deposits
- Air grid plugging
- Waste heat boiler fouling
- Fuel oil quality specifications

3

Defining a "Good" Test Method

BASF We create chemistry

Goal: Mimic refinery results

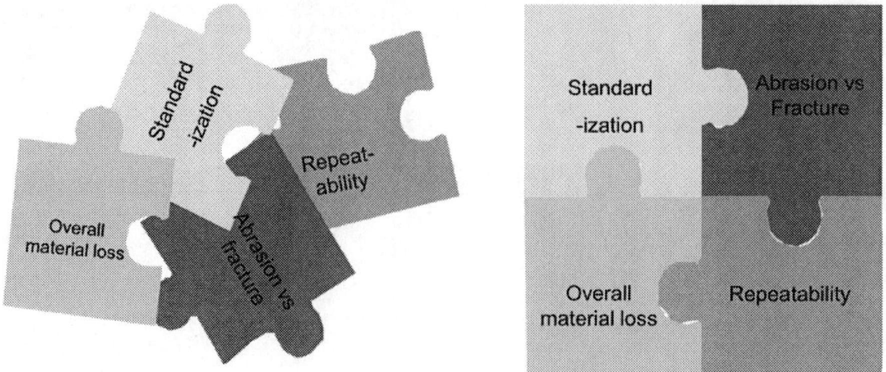

4

Major test methods: Air-jet & Jet cup

BASF We create chemistry

ASTM Air Jet (RT or hot)

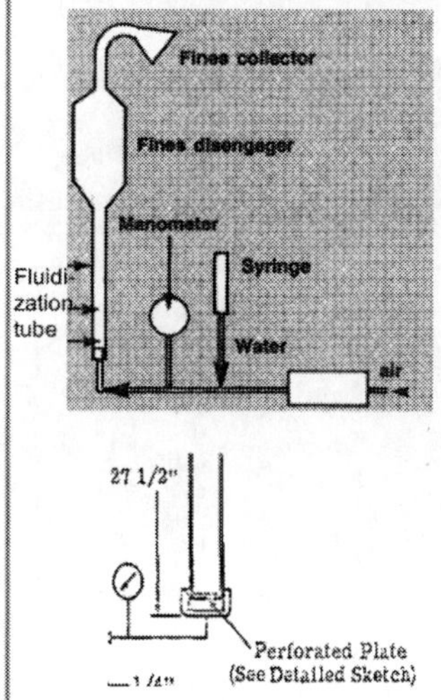

Grace Davison Jet Cup

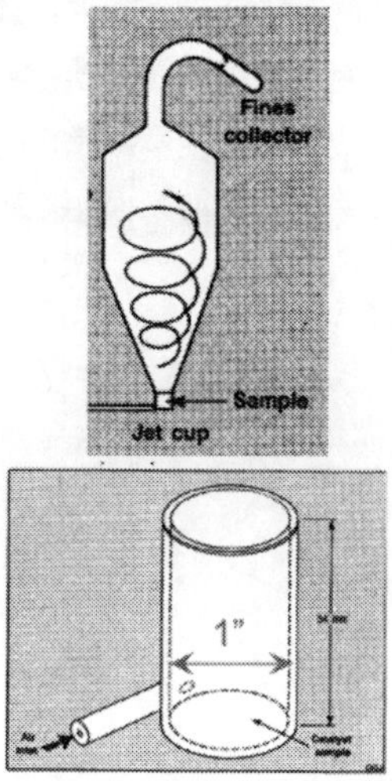

PSRI Conical Jet Cup

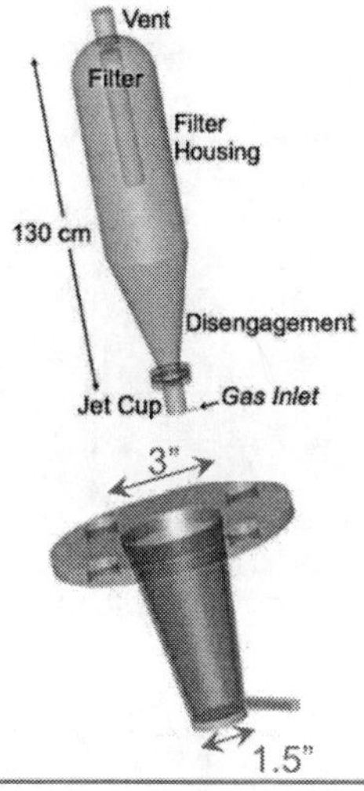

Weeks S., Dumbill P. (1990), Method speeds FCC catalyst attrition resistance determinations, Oil & Gas Journal, 38, 38-40

5

Air Jet and Davison Index
Abrasion vs Fracture

BASF We create chemistry

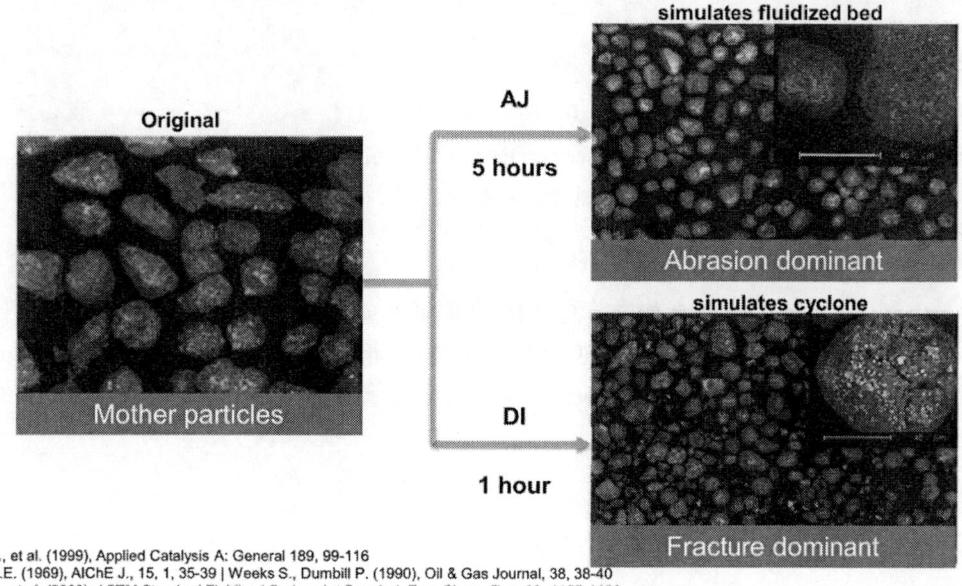

Zhao R., et al. (1999), Applied Catalysis A: General 189, 99-116
Gwyn J.E. (1969), AIChE J., 15, 1, 35-39 | Weeks S., Dumbill P. (1990), Oil & Gas Journal, 38, 38-40
Zhao R., et al. (2000), ASTM Standard Fluidized Bed vs Jet Cup, Ind. Eng. Chem. Res, 39, 1155-1158

6

High Velocity Needed for Cylindrical Jet Cup due to Stagnation Problem

BASF We create chemistry

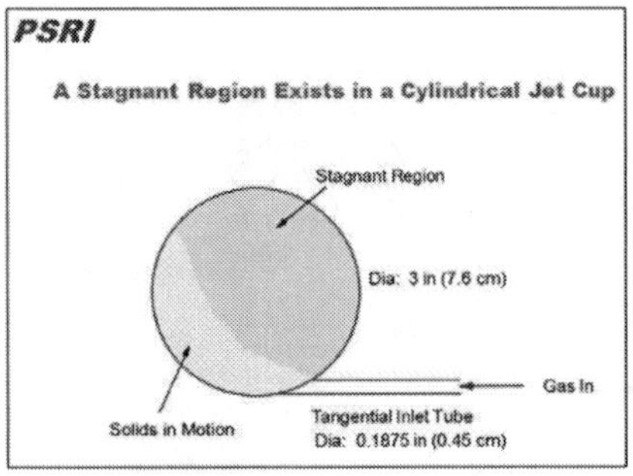

PSRI illustration of stagnant region.
Attempted to fix by introducing conical cup.

Evidence of stagnant regions in cylindrical jet cup

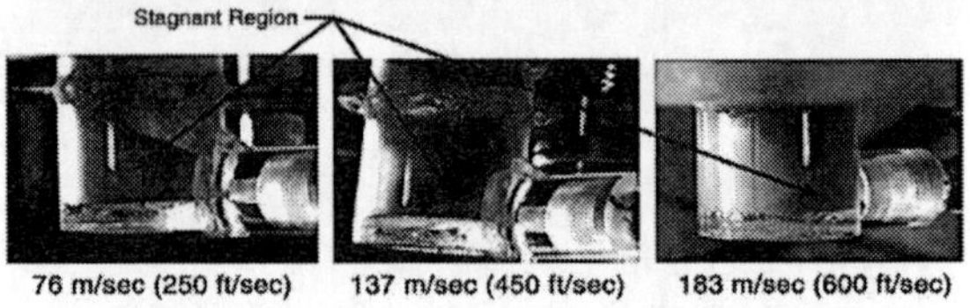

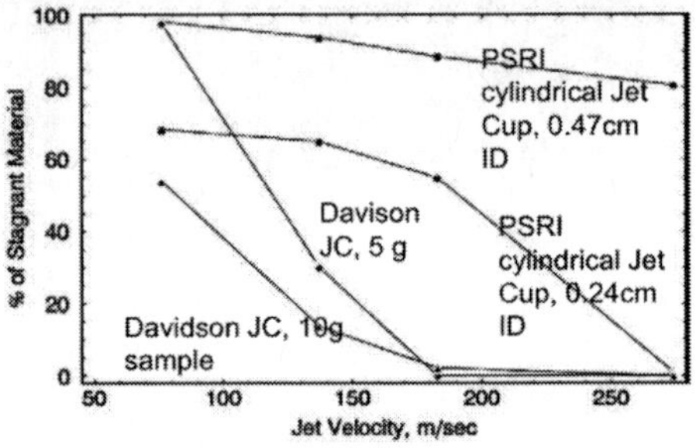

Jet-cup: Stagnation is severe below 550 ft/s (175 m/s). Worse at low loading (5g)

Cocco, R. et al. Powder Technology. 2010

7

Modeling Catalyst Attrition Mechanisms

BASF We create chemistry

Particles can Fracture or Abrade

- Fragments vs. micro-fines generation
- Dependent on catalyst properties and unit forces (cyclone loadings, superficial velocities, wall collisions, gas jet impingement, etc.)

Population Balance Model (PBM)

- Measures particle ***breakage rate*** and ***probability*** to break into fragments versus fines as a function of particle size and time
- Elucidates the dominating attrition mechanism

8

Importance of testing length

BASF
We create chemistry

Al-sol technology has characteristic densification at outer surface due to manufacturing
Longer testing needed to reach steady state attrition in abrasion dominated mechanism

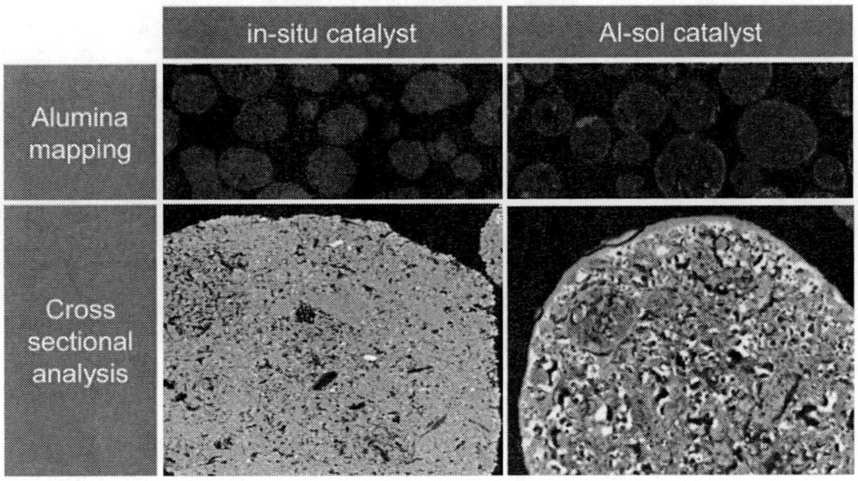

9

What is Population Balance Modeling?

BASF
We create chemistry

- Predicts **degradation rate** (S_i)of particles of a given size to smaller sizes
- Predicts **probability of particle formation** ($b_{i,j}$) from larger size bins

Characterization of Bin Size Movements

Starting size, µ		Product size, µ: >150	80-150	40-80	20-40	10-20	0-10
1	>150		Fracture				
2	80-150			Fracture		Abrasion	
3	40-80				Fracture		
4	20-40		Forbidden			SPT	
5	10-20						SPT
6	0-10						

SPT = small particle transitions

10

Catalyst Attrition Mechanisms

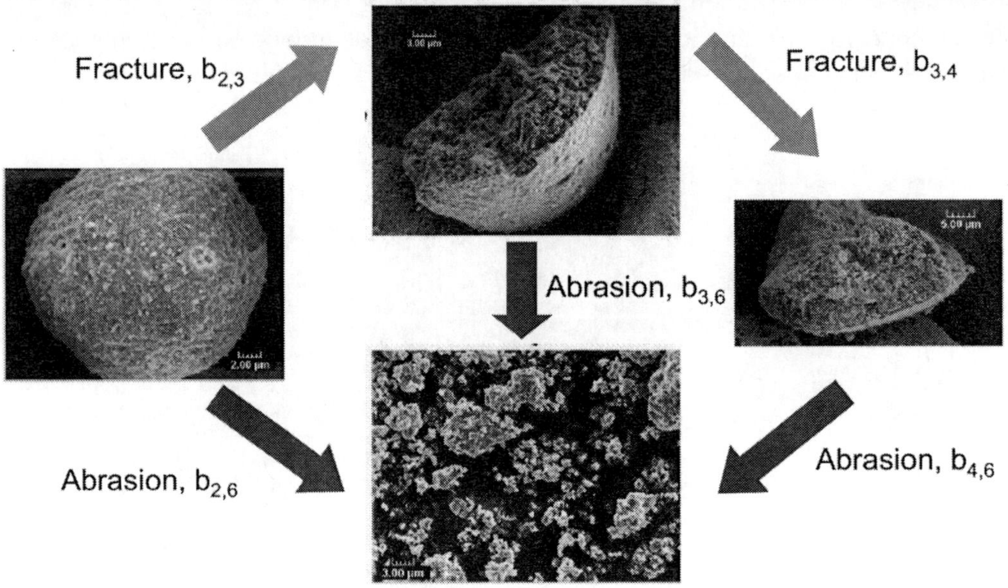

11

PBM Size Dependent Material Balance

$$+\sum_{j=1}^{i-1} b_{i,j} S_j m_j$$

Generation of size *i* from larger size *j*

$$+\frac{w_{fresh} * m_{i,fresh}}{\Delta t}$$

Fresh catalyst additions of size *i*

$$\frac{w_i(t+\Delta t) - w_i(t)}{\Delta t}$$

Rate of change of size *i* in a fluid bed

$$-\frac{w_{losses} * m_{i,losses}}{\Delta t}$$

Rate of size *i* lost: slurry, PM controls, stack

$$-\frac{w_{e-cat} * m_{i,e-cat}}{\Delta t}$$

Rate of size *i* removed from Ecat

$$-S_i m_i$$

Rate of size *i* destroyed by attrition

12

FCCU Data Collection
Complete Catalyst Material Balance

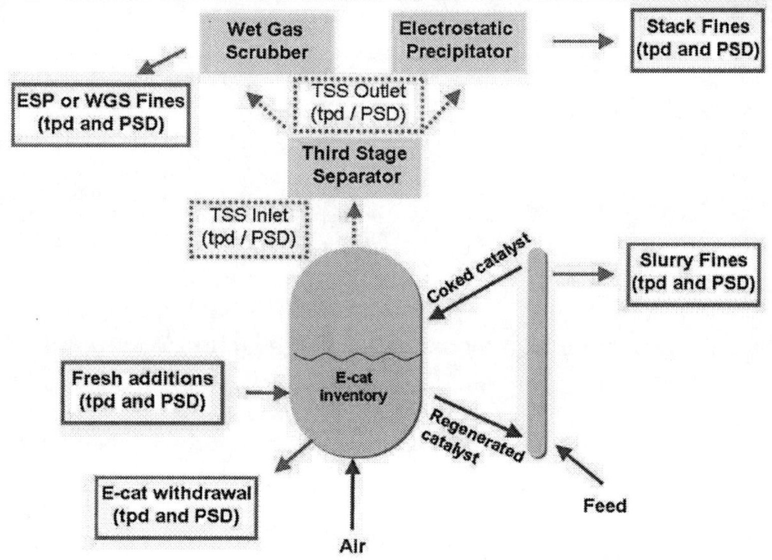

Assumption: unit is under steady state.

13

Catalyst Losses in Six FCCUs

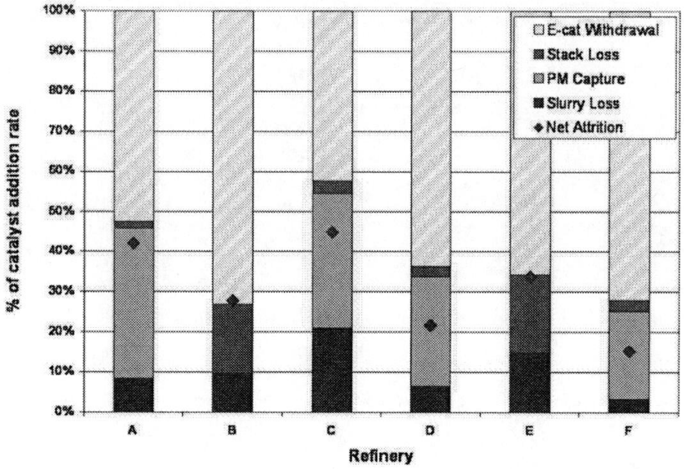

- 27-58% of total catalyst addition is lost to the slurry or out the flue gas train.
- Reducing losses allows a refiner to lower catalyst addition rates
- Net Attrition is a rough measure to assess how much of the cat losses are due to attrition.

$$\frac{\sum_{i=1}^{6}\left|m_{i\_ecat} + m_{i\_losses} - m_{i\_fresh}\right|}{2}$$

14

Attrition Properties of 6 FCCUs

BASF
We create chemistry

Refinery	A	B	C	D	E	F
Total attrition / cat adds, %	84	36	60	25	43	16
Fracture, wt%	43 ± 13	30 ± 9	32 ± 9	24 ± 7	21 ± 5	13 ± 5
Abrasion, wt%	**53** ± 14	**64** ± 13	**65** ± 11	**71** ± 8	**75** ± 7	**86** ± 5
SPT, wt%	4 ± 2	6 ± 4	3 ± 2	2 ± 1	4 ± 2	1 ± 1

- Total attrition accounts for all attrition transitions, even those that occur through several size bins
- Refinery A shows fracture and abrasion to both be important. This unit also shows the highest total attrition.

Abrasion is the most important mechanism in 5 of 6 FCCUs studied

15

Air Jet loss rates: Ecat from Refinery D

BASF
We create chemistry

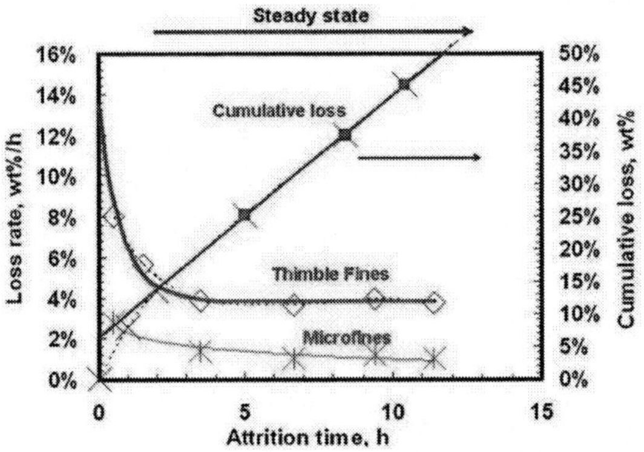

- Tests are run until 25 -50% of fresh catalyst is degraded to fines, matching commercial loss rates
- Initial, transient loss followed by a steady state loss rate
- Single parameter index will not capture all of the attrition properties

16

Air Jet is Abrasion Dominant

BASF We create chemistry

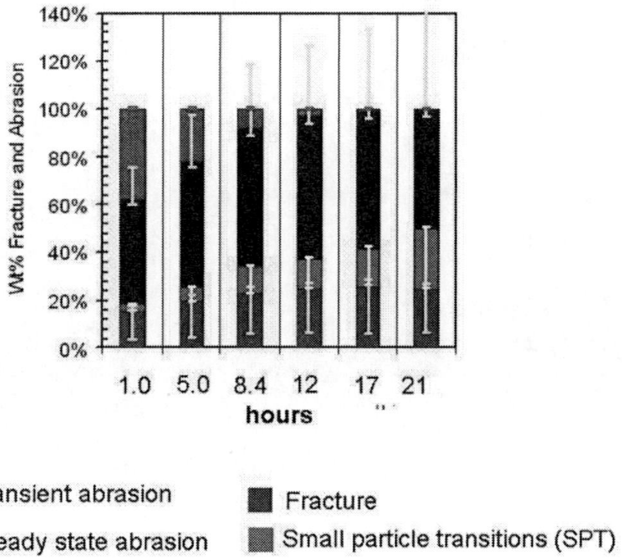

17

Orifice-to-Particle Size Ratio in AJ Testing

BASF We create chemistry

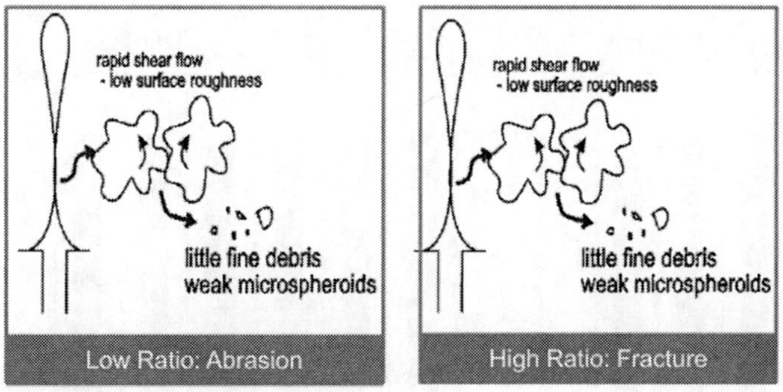

d_{or} / d_p

Air Jet = 4
Jet Cup = 42

Ratios by Test Method

Experimentation on Ecat shows abrasion dominant when ratio < 7.5

- Air velocity at the orifice of AJ testing can be as high as 1400 ft/s
- AJ is abrasion dominant: Low d_{or} / d_p facilitates abrasion despite high air velocity

Boerefijin R., et al. (2000), Powder Technology, 11(2), 145-174

18

PSRI Conical Jet Cup loss rates

BASF
We create chemistry

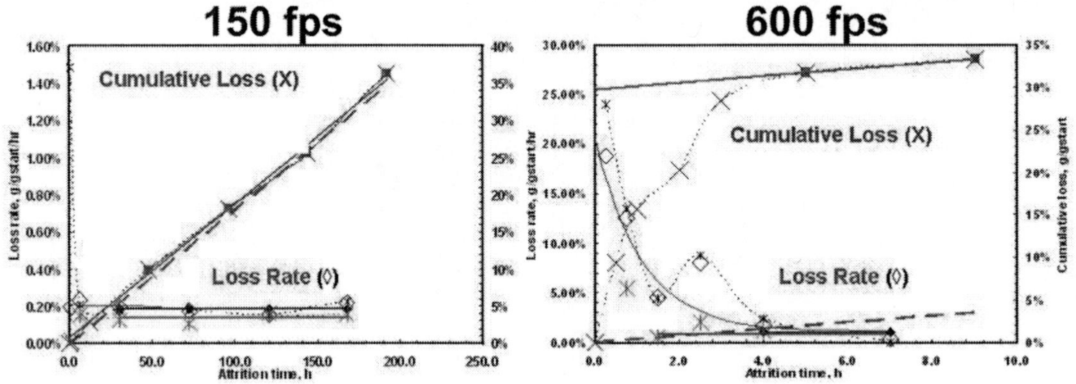

- **Behavior at two velocities is very different.**
  - Steady state loss rate achieved at 150 fps
  - Long transient with a drop to very low attrition rates at 600 fps

19

PSRI Conical Jet Cup

BASF
We create chemistry

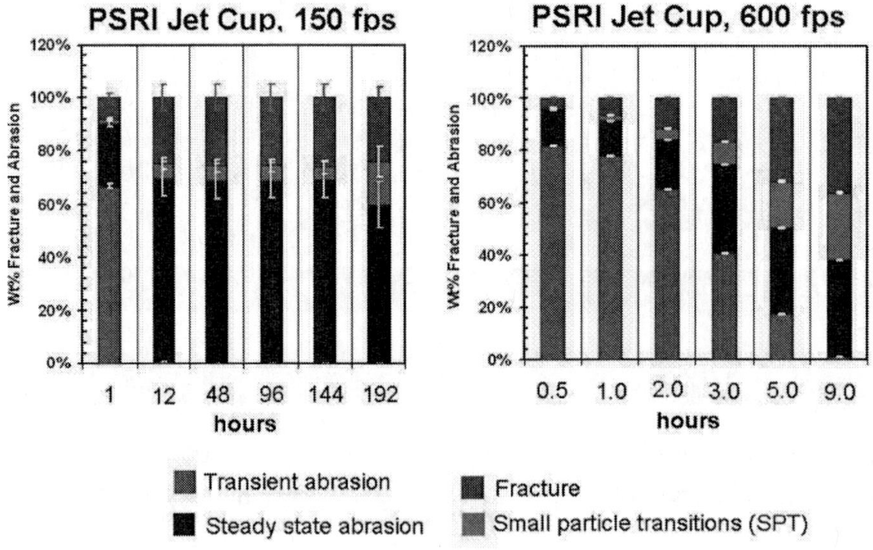

20

Davison Jet Cup (Davison Index) Measurements of Ecat from Refinery D

BASF
We create chemistry

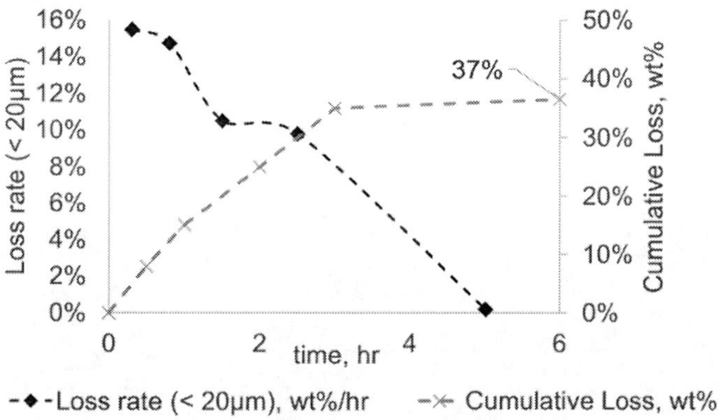

- Independent lab measurements show total loss <20%, lower than refinery level
- BASF measurements: attrition rate drops to zero after 5h
  - Insufficient sample remaining remains in the cup
  - Sample floating up into the separation chamber

21

Summary: Laboratory Evaluation of Attrition

- Attrition testing should be a multi-faceted approach to best mimic forces and effects in a refinery FCC unit
- Population Balance Model enables the understanding of catalyst attrition mechanisms
- Early results show abrasion-based attrition as the most important mechanism in commercial FCC units
- Air Jet and PSRI conical jet cup are abrasion dominant and can match the loss levels in commercial units

22

Questions?

BASF

We create chemistry

Combined High-Energy Synchrotron X-Ray Diffraction and Computed Tomography to Characterize Fracture Behavior of Sand

Khalid Alshibli[*], Mehmet Cil[+], Peter Kenesei[++], Ulrich Lienert[++]

[*] University of Tennessee, Knoxville, Tennessee, USA

[+] Northwestern University, Evanston, IL

[++] Advanced Photon Source (APS), Argonne National Laboratory, Illinois, USA

[++] Deutsches Elektronen-Synchrotron, DESY Photon Science, Hamburg, Germany

THE UNIVERSITY OF TENNESSEE KNOXVILLE

COLLEGE OF ENGINEERING

Argonne NATIONAL LABORATORY

NSF

Overview of Research

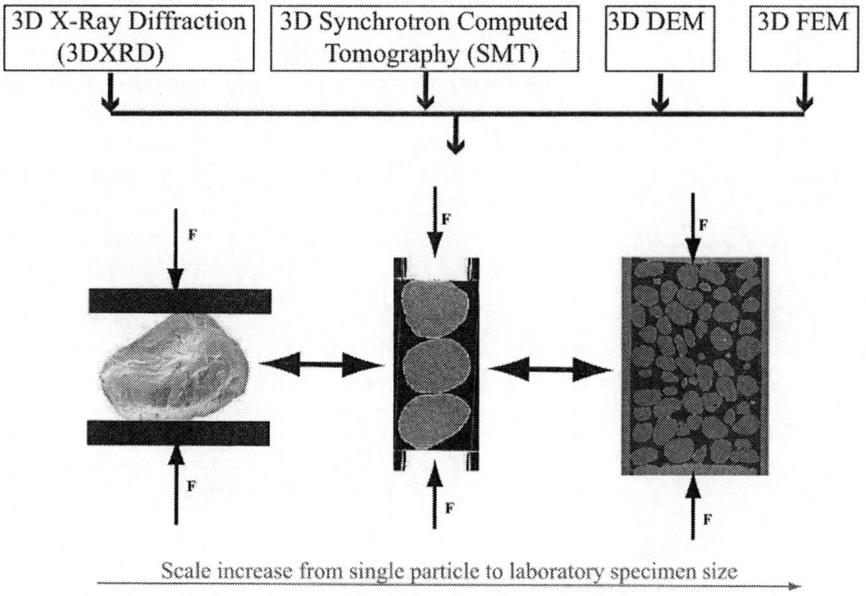

Presentation Outline

- 3D Synchrotron Micro-Computed Tomography (SMT) experiments
- 3D X-ray Diffraction (3DXRD) experiments
- Preliminary results of 3D Finite Element Method (FEM) to model single particle fracture
- Conclusions
- Acknowledgements

3

Schematic of Synchrotron Micro-Computed Tomography (SMT)

- It is a 3D non-destructive technique to map density of materials.
- When an electron travelling at a speed close to the speed of light ***is accelerated***, it emits a ***continuum*** of ***electromagnetic radiation*** (***Synchrotron light***) tangential to the orbit.
- SMT Produces an intense monochromatic highly collimated beam of x-ray which gives a higher resolution than industrial CT systems.

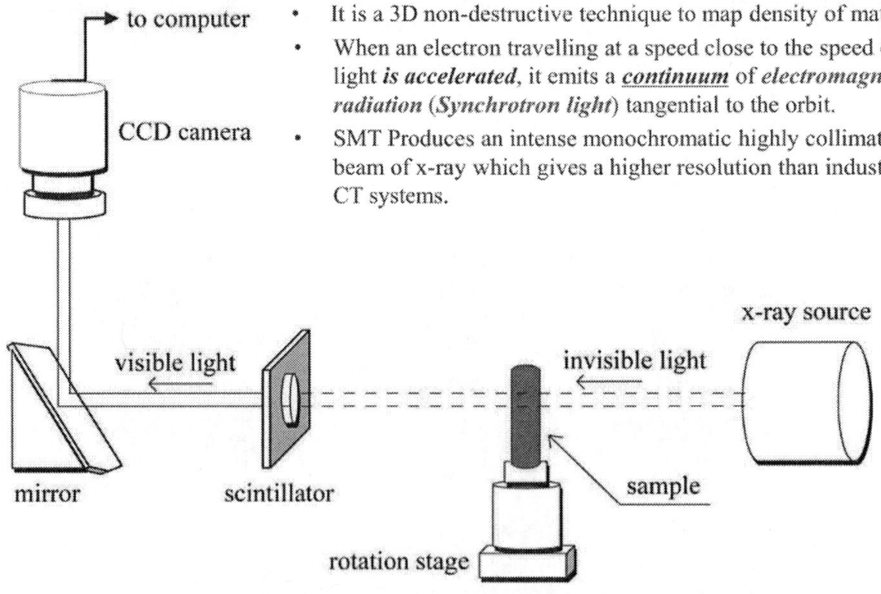

4

Advanced Photon Source (APS)
https://www1.aps.anl.gov/

5

ASTM 20-30 Ottawa Sand
(Grain size: 0.59 mm to 0.84 mm)

6

1D Compression Experiments

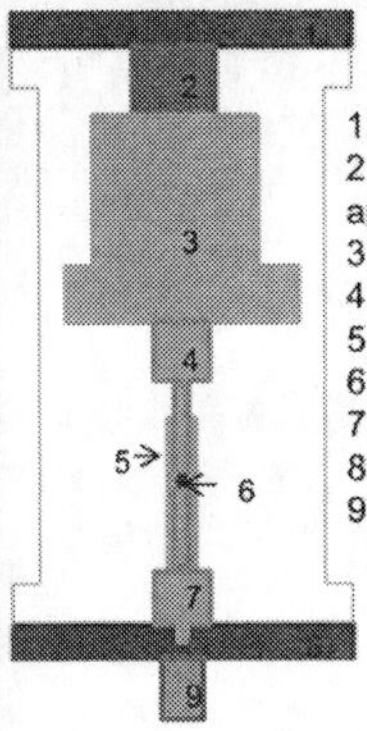

1: Top Plate
2: Stepper motor to apply axial load
3: Load cell
4: Top loading plate
5: Aluminum guide
6: Particles
7: Bottom loading plate
8: Bottom plate
9: Cell-stage connector

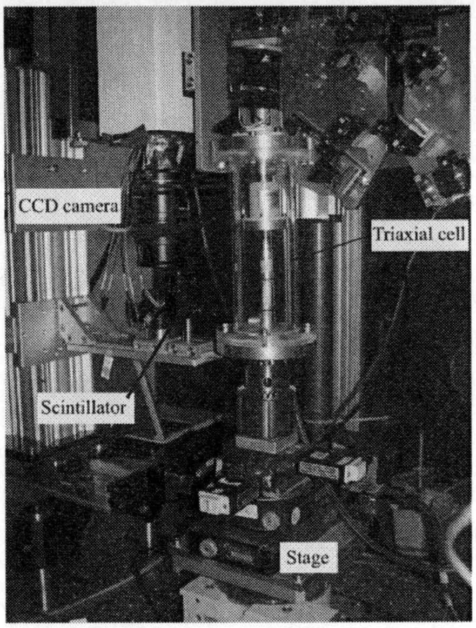

Triaxial test cell on the SMT stage

SMT1 Experiment:

Initial D = 4.32 mm

Initial H = 6.63 mm

H/D = 1.54

Initial void ratio = 0.60

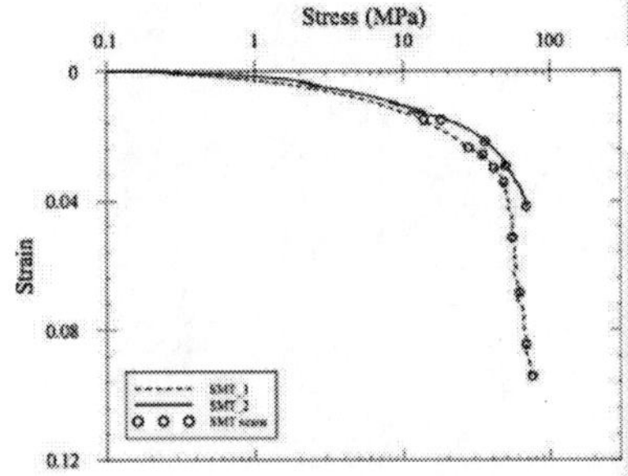

Particle fracture commenced at ~ ε = 2.36% when yielding began

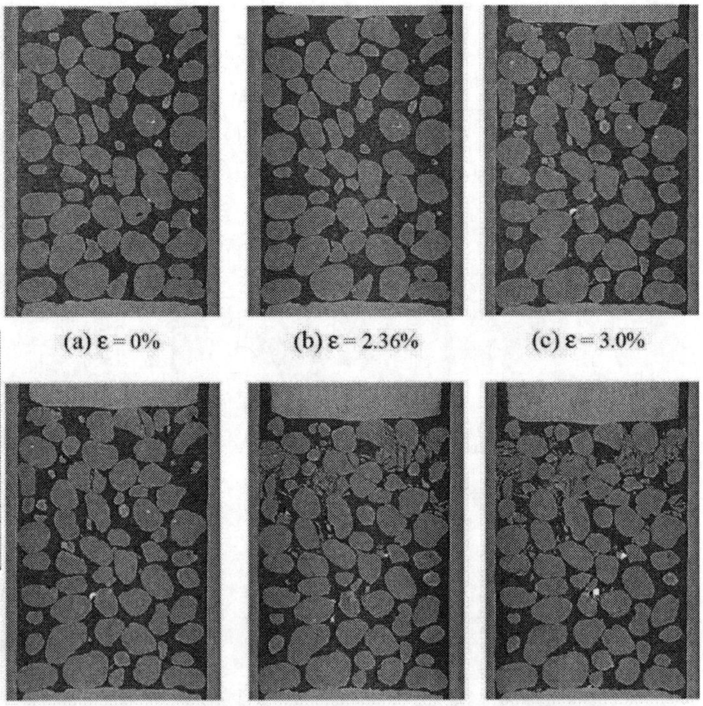

(a) ε = 0% (b) ε = 2.36% (c) ε = 3.0%

(d) ε = 5.14% (e) ε = 6.86% (f) ε = 9.43%

SMT-1

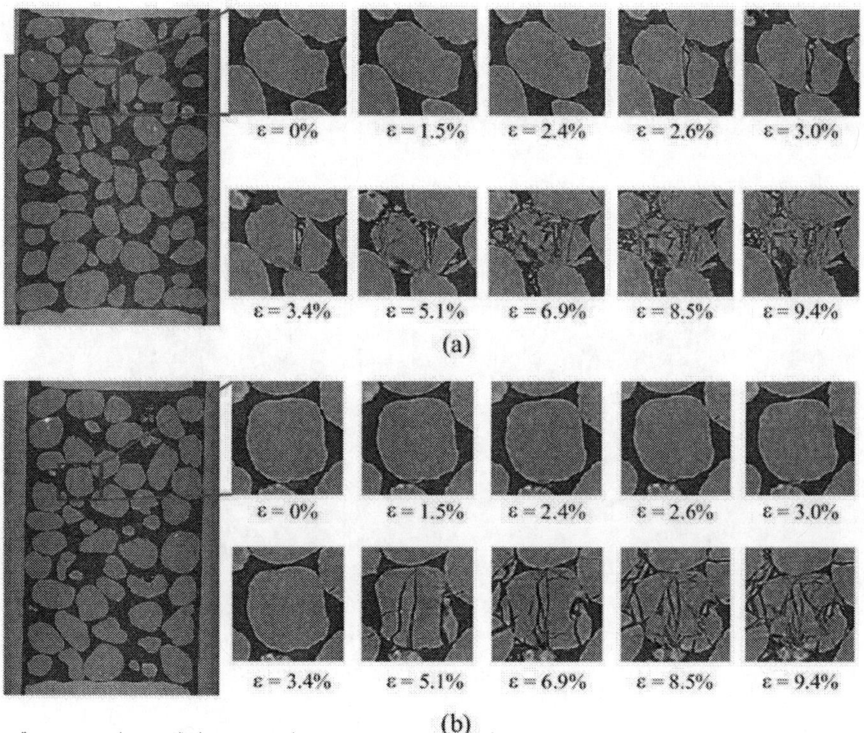

Some fractured particles continue to carry load

SMT 2 Experiment:

Initial D = 4.32 mm

Initial H = 8.51 mm

H/D = 1.97

Initial void ratio = 0.51

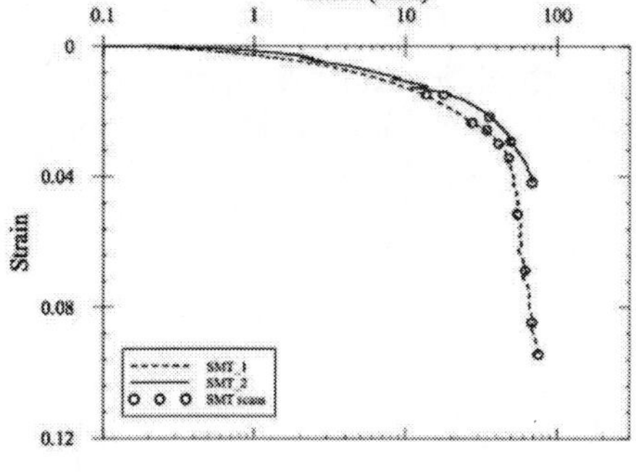

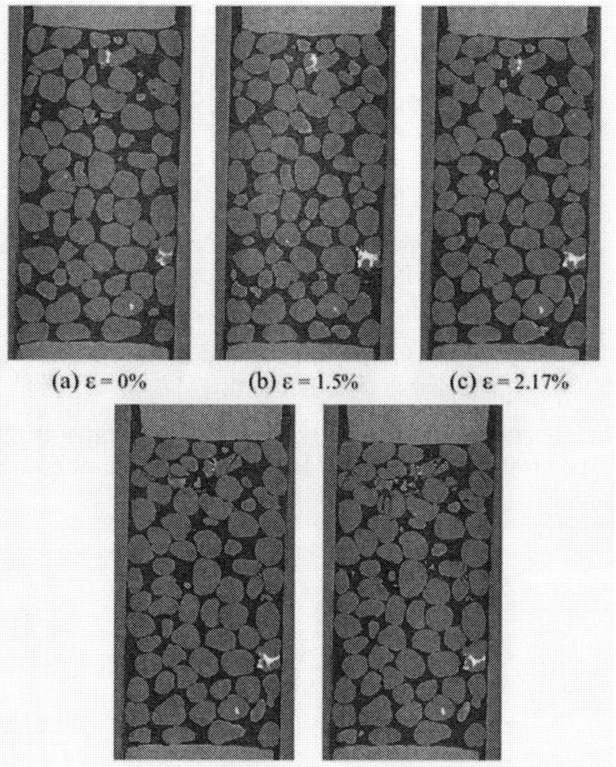

(a) ε = 0% (b) ε = 1.5% (c) ε = 2.17%

(d) ε = 2.92% (e) ε = 4.18%

Evolution of particle fracture

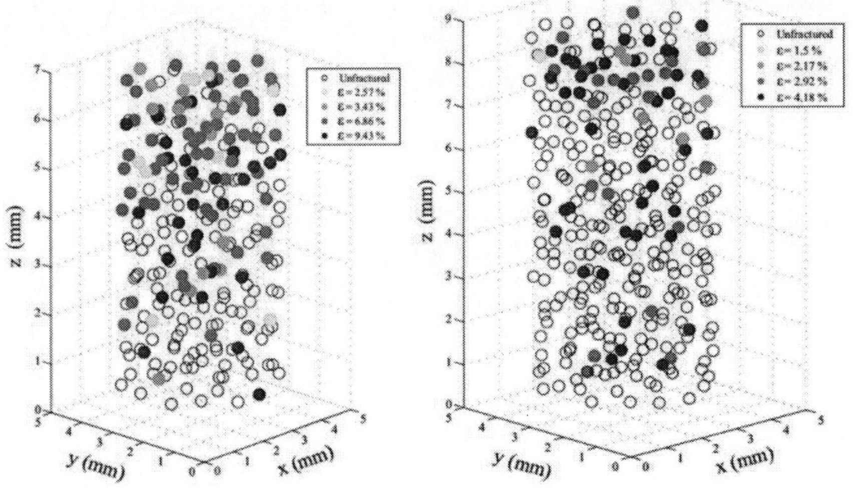

(a) SMT1 (b) SMT2

11

3D DEM Modeling of Particle Fracture

- PFC3D Software was used in the analysis
- Each particle was modeled as an agglomerate of a large number <u>of non-uniform</u> small sub-spheres bonded together according to the Bonded Particle Model (BPM) of Potyondy & Cundall (2004)
- The sub-spheres are bonded at their contact points by parallel bonds that can transfer force and moment.
- The bond strength was chosen randomly from a normal distribution function defined by a mean (475 MPa) and standard deviation (175 MPa) values.

12

DEM simulations

$R_{max}/R_{min} = 1.2$

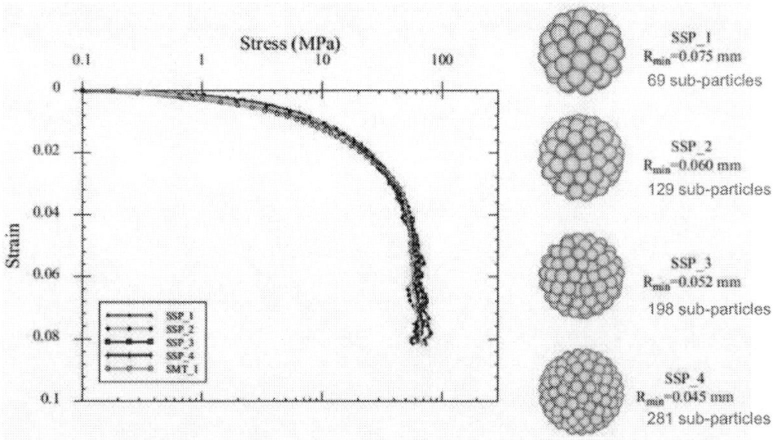

DEM simulations of 4 agglomerate configurations.

~An agglomerate composed of ~69 sub-spheres was used in the following DEM simulations

13

Sequence of Particle generation process

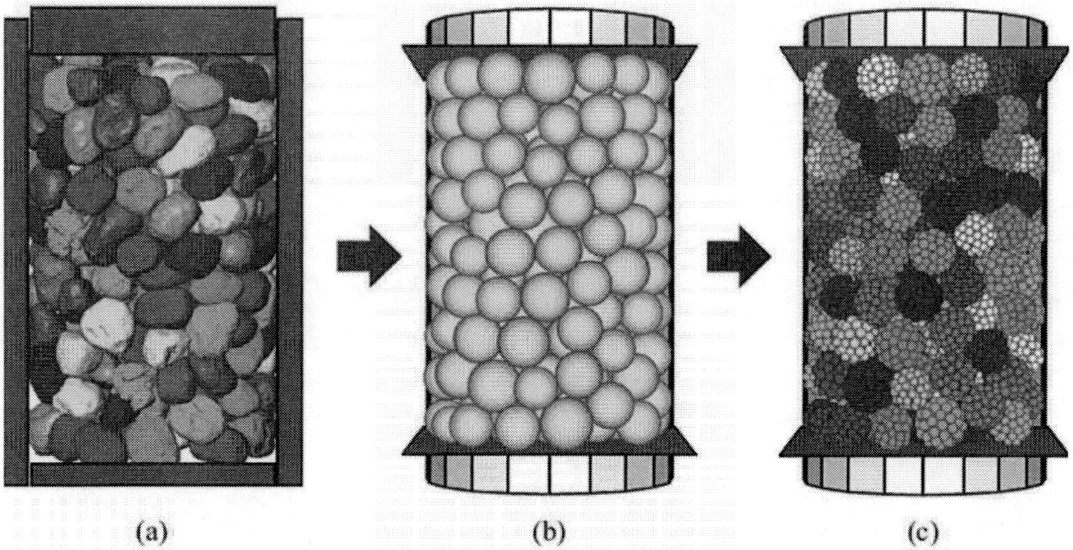

(a) (b) (c)

14

Percentage of fractured particles

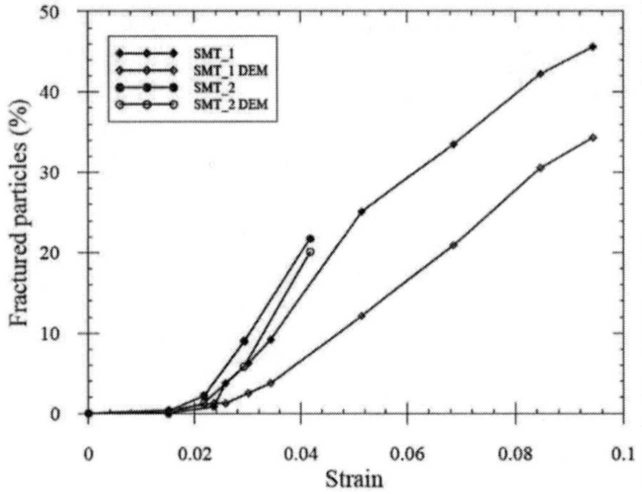

A DEM agglomerate was classified as fractured when the number of bond breakages > 5% of the initial total bonds

15

Bond breakages in DEM agglomerates

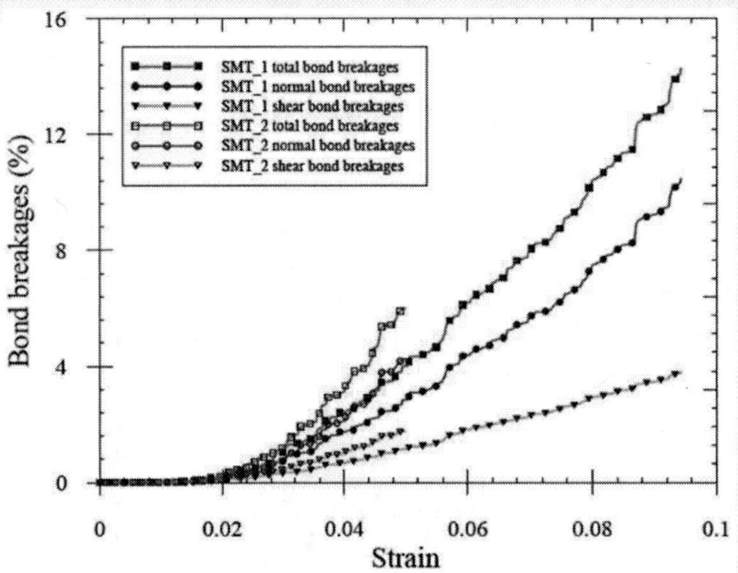

"Asperity" damage commenced at ~1.9% strain
Normal bond breakage is the dominant

16

Strain Measurements within Granular Particles

- Photo-elastic materials
- Neutron Diffraction: Penumadu et al. (2009), Frischbutter et al. (2000), Darling et al. (2004).
- 3D X-ray diffraction (3DXRD):
  - Henning Poulsen, Risø DTU, Denmark
  - Hall *el al.* (2011) presented preliminary experimental measurements for Ottawa sand
- Other techniques?

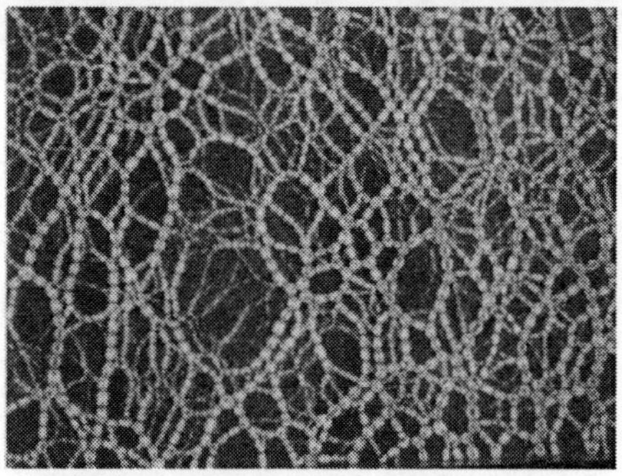

Force chains visualized using photo-elastic discs (Source: Behringer, 2010)

17

Definitions

- α-Quartz ($\alpha\text{–}SiO_2$) is a crystalline material which is composed of uniquely oriented atomic structure that exhibits a repeating pattern.
- A crystal lattice is defined as a network of points that are arranged as a set of identical points called lattice points throughout space.

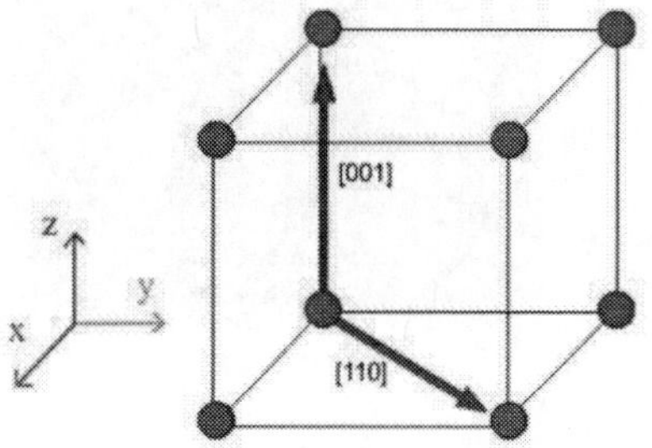

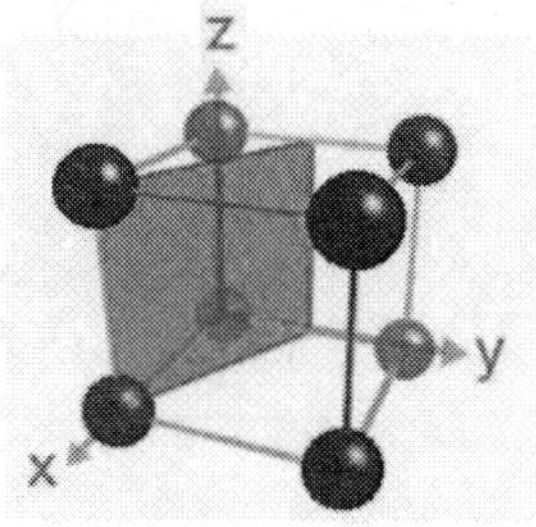

Lattice plane vector (left) and lattice plane [120] (right) are represented by Miller indices (DoITPoMS, 2011).

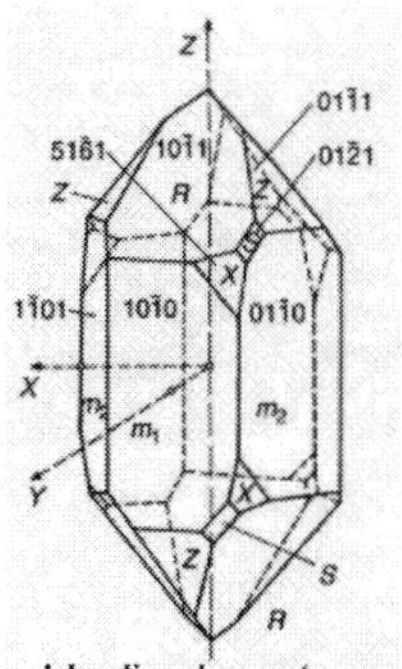

Idealized quartz crystal showing natural facets (Source: Ballato, 2001)

18

Principle of X-Ray Diffraction

The incident monochromatic synchrotron x-ray beam is scattered on the orderly arranged atoms of a crystalline material. The diffracted x-ray beams obey Bragg's law:

$$2d_{hkl}\sin\ _{hkl} = n \quad \Rightarrow \quad = \frac{d_{hkl(0)} - d_{hkl}}{d_{hkl(0)}}$$

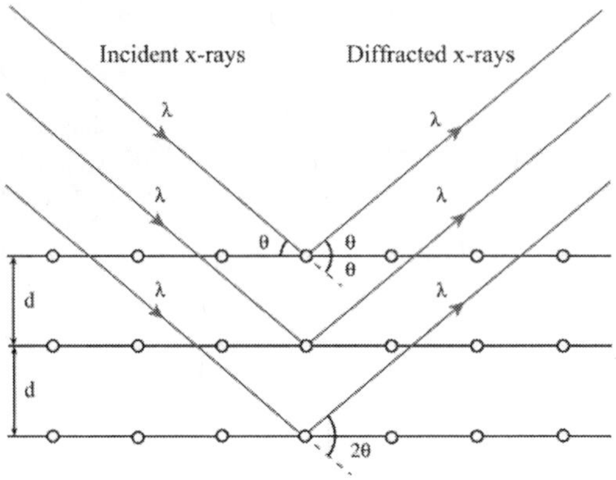

Principle of 3DXRD

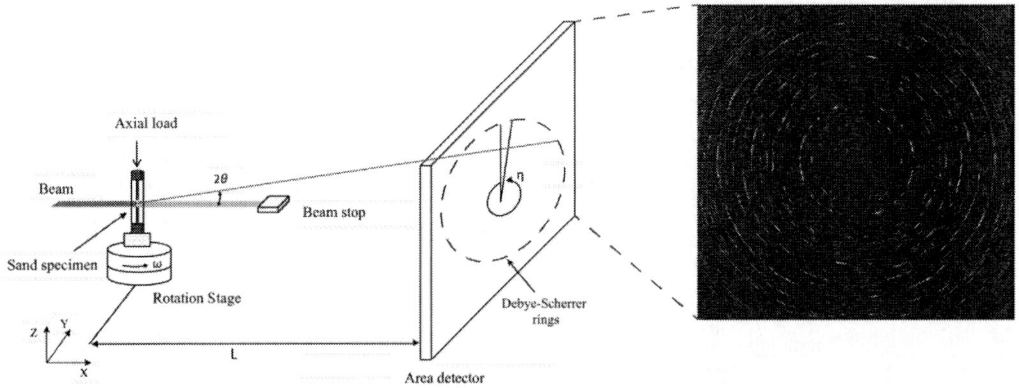

HEXD Beamline (1-ID) at APS

21

Example Diffraction Spots

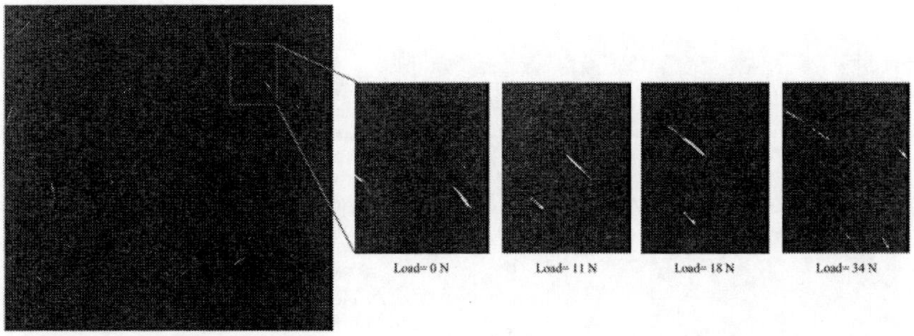

The translation of the diffraction spots as axial load increases. The spots began to move along the azimuthal direction η with increasing load in different directions. The spots broaden in two directions and new reflections also appear with increasing load due to new fragments generated after particle crushing. Diffraction spots move along the azimuthal direction η on the detector plane and broadened in two directions due to grain orientation change and applied deformation.

22

Analysis of Diffraction Spots

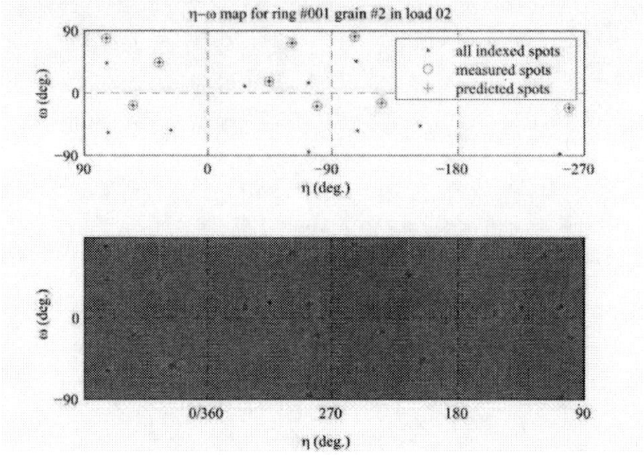

Visual comparison of the diffraction spots that were measured and predicted by Grainspotter for a reference grain for load steps 0 to 2.

Test Cell

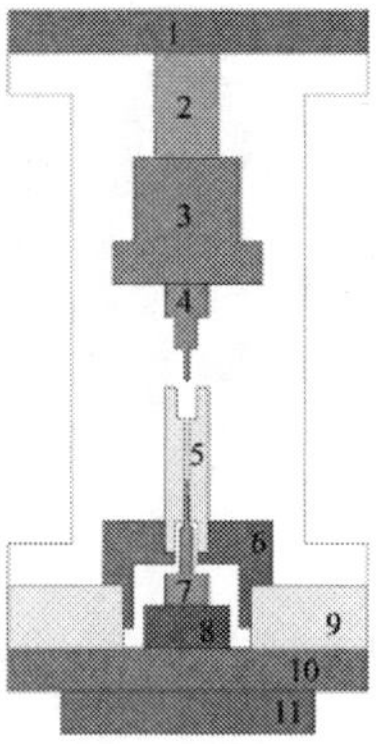

1: Top plate
2: Stepper motor to apply axial load
3: Top load cell
4: Top loading plate
5: Acrylic mold with sand particles
6: Aluminum spacer
7: Bottom loading plate
8: Bottom load cell
9: Acrylic spacer
10: Bottom plate
11: Cell-stage connection plate

Loading and corresponding Particle Behavior

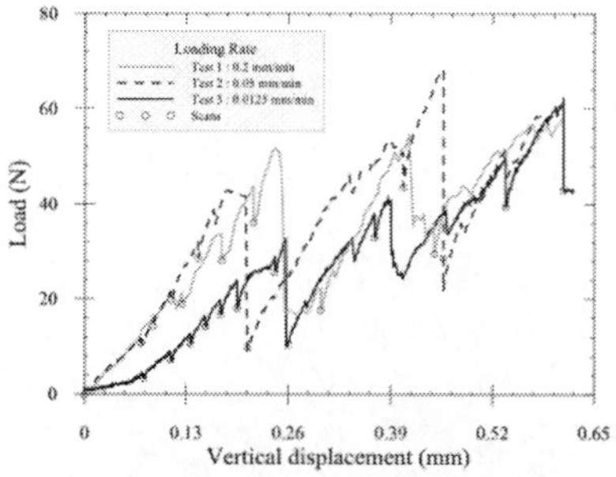

Global load-displacement
For 3DXRD experiments

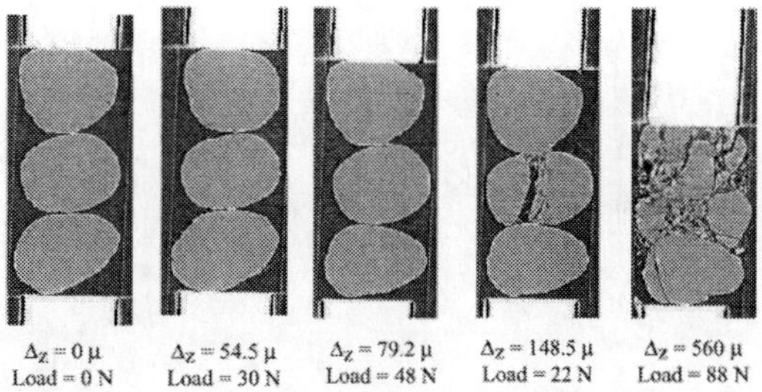

SMT Images

Track grain orientation using 3DXRD

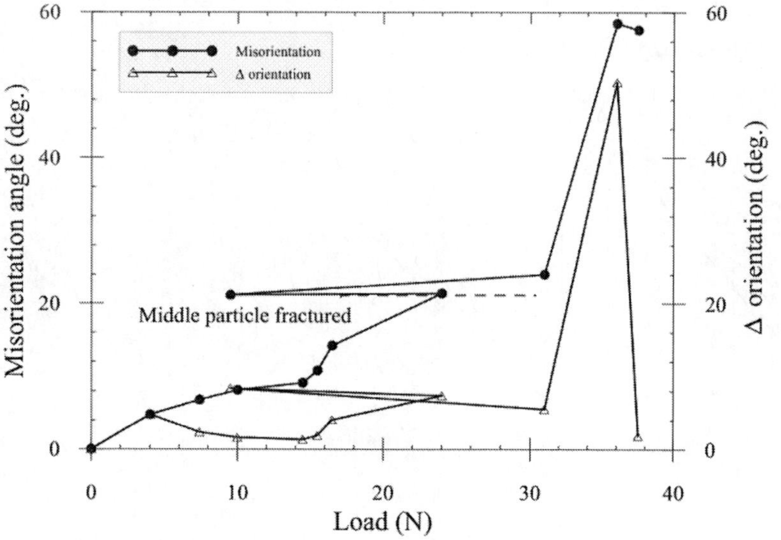

Misorientation angle references rotation to the initial stage

Lattice Strain Measurements

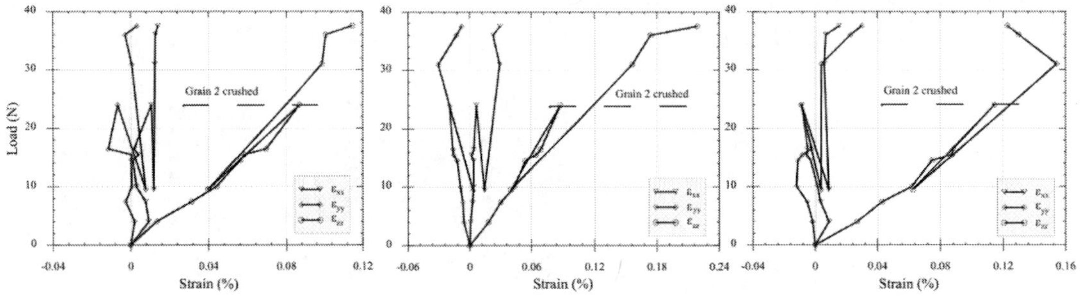

(a) Bottom grain (b) Middle grain (c) Top grain

Load versus lattice strain curves for for normal strain components of Test 3.

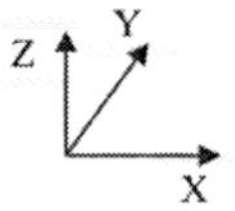

27

Lattice Volumetric Strain for Test 3

Volumetric strain $\varepsilon_v = \varepsilon_{xx} + \varepsilon_{yy} + \varepsilon_{zz}$

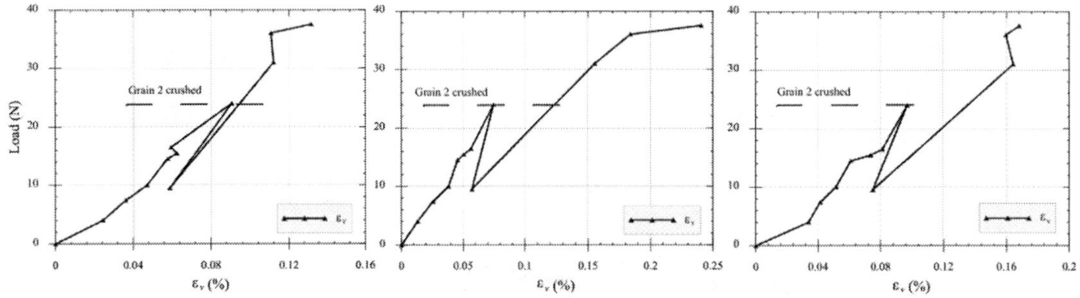

(a) Bottom grain (b) Middle grain (c) Top grain

29

Stress-strain relationship

- Silicone dioxide (SiO_2) is the form of α-quartz in Earth's most common surface mineral
- Strain was calculated from: $= \frac{d_{hkl(0)} - d_{hkl}}{d_{hkl(0)}}$
- α– Quartz possesses anisotropic elastic behavior and trigonal symmetry with 6 independent elastic constants (Ballato 2001; Heylinger *et al.* 2003):

$$\begin{Bmatrix} x \\ y \\ z \\ xy \\ yz \\ zx \end{Bmatrix} = \begin{bmatrix} C_{11} & C_{12} & C_{13} & C_{14} & 0 & 0 \\ & C_{11} & C_{13} & -C_{14} & 0 & 0 \\ & & C_{33} & 0 & 0 & 0 \\ & & & C_{44} & 0 & 0 \\ & & & & C_{44} & C_{14} \\ & & & & & \frac{1}{2}(C_{11} - C_{12}) \end{bmatrix} \begin{Bmatrix} x \\ y \\ z \\ xy \\ yz \\ zx \end{Bmatrix}$$

C_{11} = 87.26 GPa
C_{12} = 6.57 GPa
C_{13} = 11.95 GPa
C_{14} = -17.18 GPa
C_{33} = 105.80 GPa
C_{44} = 57.15 GPa

30

Lattice Strain Measurements (Test 1)

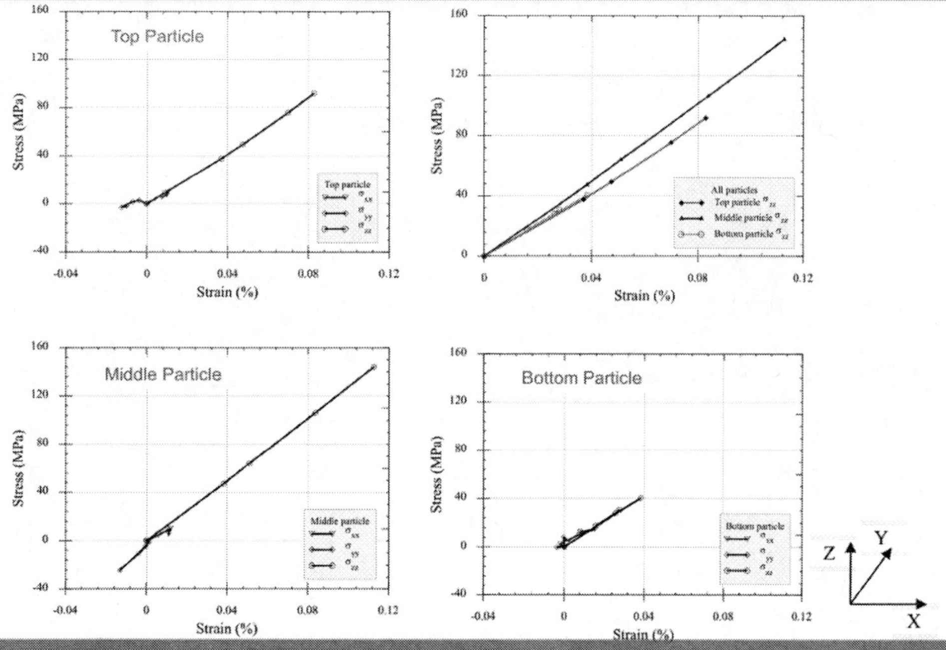

31

Lattice Strain Measurements (Test 2)

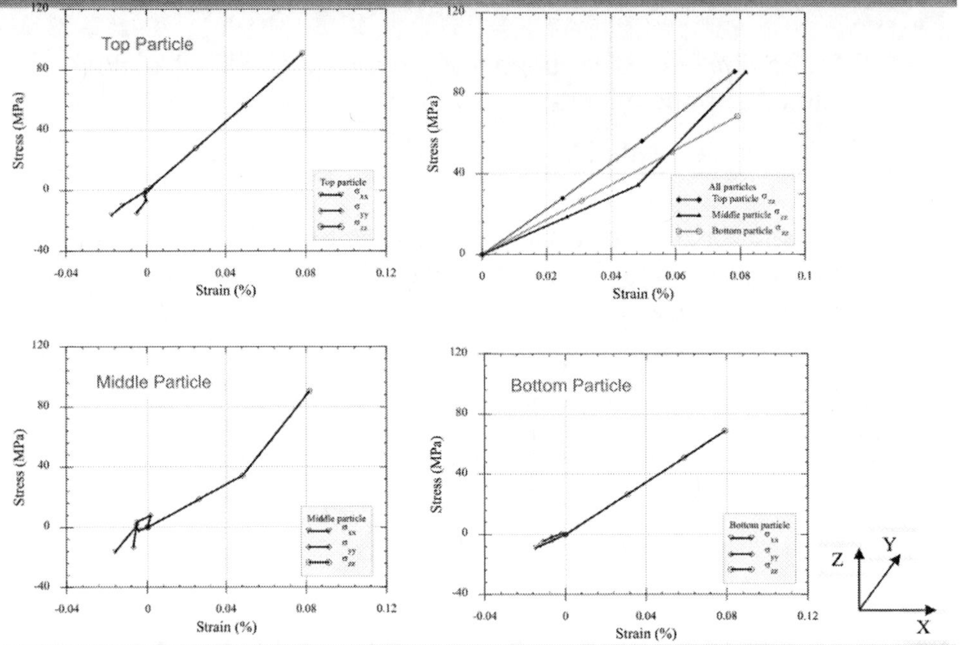

32

Lattice Strain Measurements (Test 3)

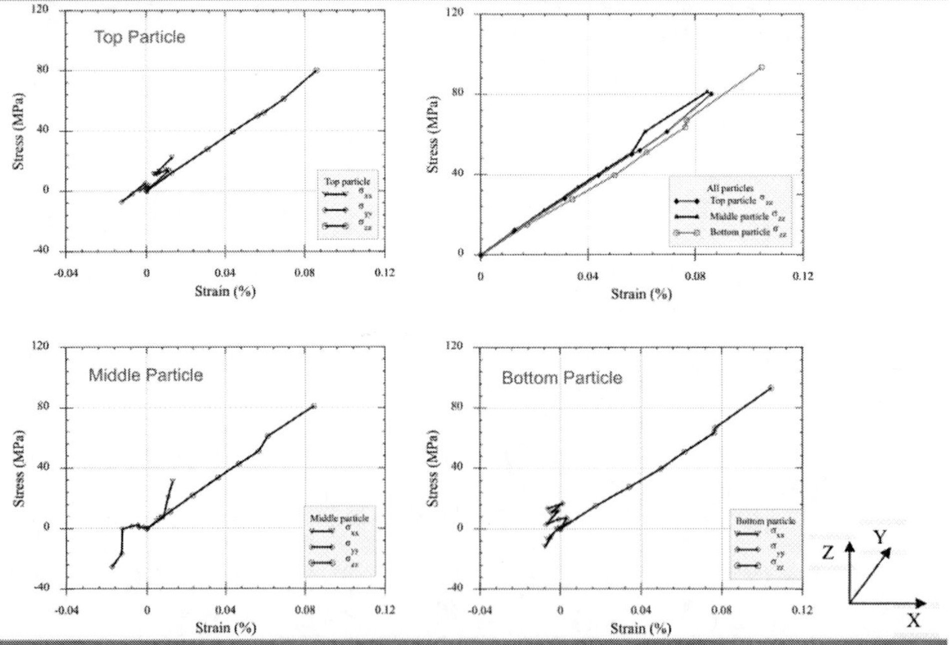

33

σ_{zz} versus ε_{zz} for all particles

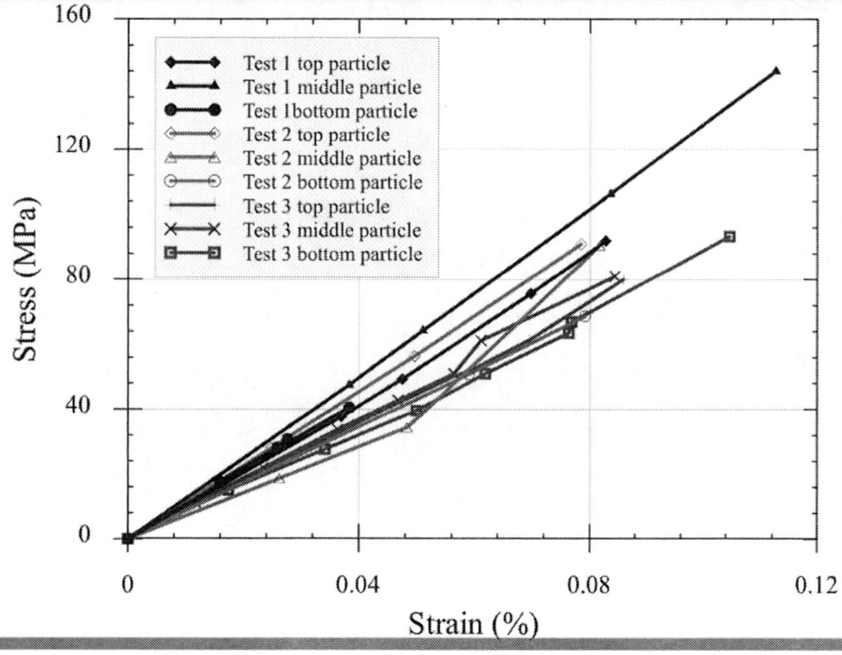

34

Lattice Shear Strains for Middle Particle (Test 3)

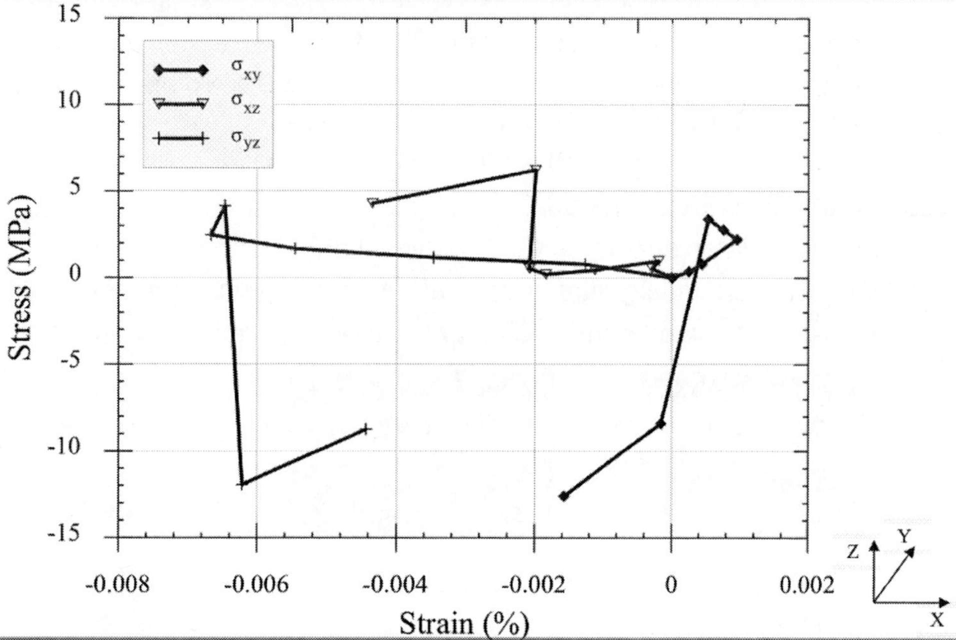

35

Presentation Outline

- 3D Synchrotron Micro-Computed Tomography (SMT) experiments
- 3D X-ray Diffraction (3DXRD)
- → Preliminary results of 3D Finite Element Method (FEM) to model single particle fracture
- Conclusions
- Acknowledgements

36

Image Processing & FE Analysis

- Segment and Binarize grayscale CT data
- Generate surface: optimize surface for tetrahedral volume element development
- Particle volume filled with tetrahedral elements
- Assumed isotropic material
- Loading plates simulated with analytical rigid surfaces
- General contact algorithms applied between particle and plates
- Constant displacement applied to top plate to mimic experiment

Original CT Image

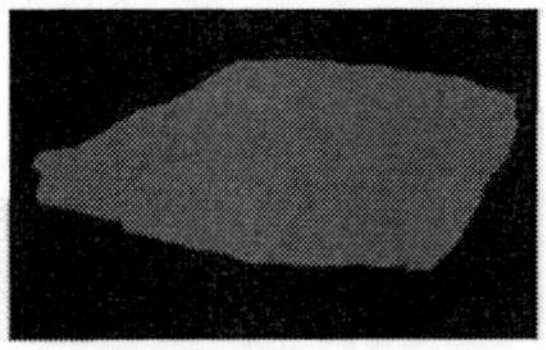

Segmented Image

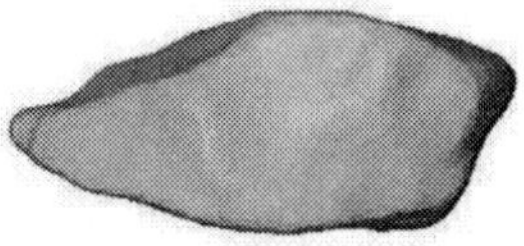

Particle Surface

37

Results

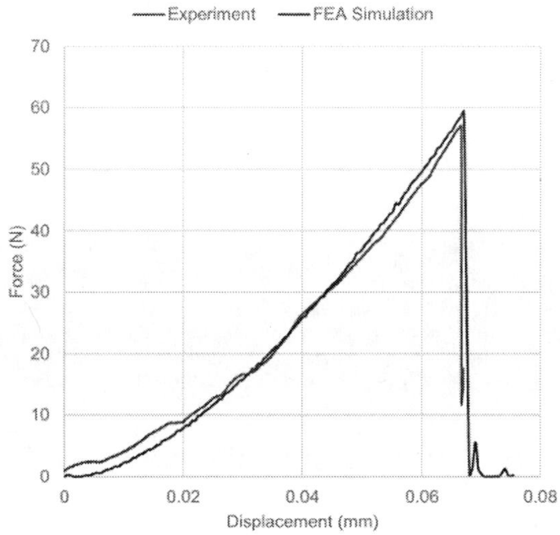

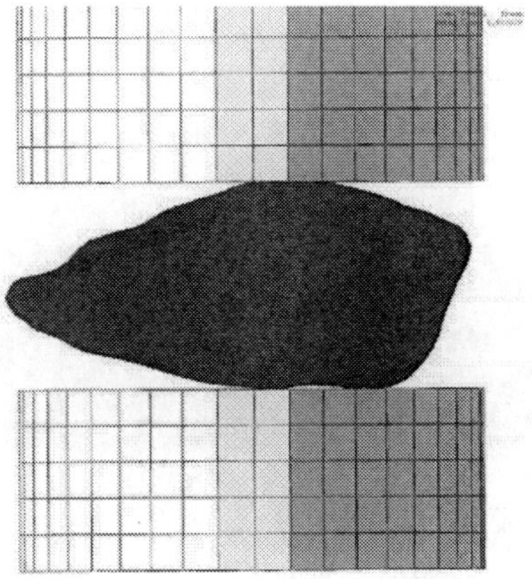

Conclusions

- SMT and 3DXRD complement each other to fully characterize the constitutive behavior of granular materials.
- The fracture mode of sand particles under compression was monitored experimentally using 3D SMT. Particles are susceptible to fracture if their neighbors have already experienced fracture
- The onset of major particle failure in the form of splitting coincides with the specimen yielding level.
- Particles continue to carry load after first major facture.
- 3DXRD technique was successfully used to measure the evolution of particle orientation and to calculate the elastic lattice strain and stress tensors of silica sand at the particle level. Such experimental measurements can be used to develop, validate, or calibrate micromechanics-based finite element and discrete element methods to predict the constitutive behavior of granular materials.

41

Acknowledgments

- This material is based upon work supported by the National Science Foundation under Grant No. (CMMI-1156436). Any opinions, findings, and conclusions or recommendations expressed in this material are those of the authors and do not necessarily reflect the views of the National Science Foundation.

- The 3DXRD and SMT data was collected using the x-ray Operations and Research Beamlines 1-ID and 13D at the Advanced Photon Source (APS), Argonne National Laboratory. Use of the Advanced Photon Source was supported by the U. S. Department of Energy, Office of Science, Office of Basic Energy Sciences under Contract No. DE-AC02-06CH11357.

THE UNIVERSITY OF TENNESSEE KNOXVILLE
COLLEGE OF ENGINEERING

NSF

Argonne NATIONAL LABORATORY

42

Plasmonic Pervaporation for Sustainable Mixture Separation

Austin, T., University of Arkansas Fayetteville

Originally presented on: 4/12/2016 17:30:00 - 17:45:00

Membrane separation provides a potential alternative to distillation, which consumes roughly 40% of the energy used in the chemical and refining industries according to the United States Department of Energy, for separating liquid-liquid mixtures. While providing better separation, current membrane separation techniques are not as energy efficient as distillation. This work seeks to improve the energy efficiency of purifying solvents by coupling the effects of localized plasmonic heating on a membrane with pervaporation, a common membrane separation technique, using polydimethylsiloxane (PDMS) thin-films. Localized plasmonic heating results from the decay of a plasmon that was created when the resonant wavelength of light irradiates gold nanoparticles (AuNP). Preliminary results in a lab-scale, pervaporation system suggest that the combination of these two techniques may be able to facilitate a higher level of mass transfer or flux across membranes.

Drop Impact, Granule Formation, and Granule Morphology: A Comprehensive Look at Single Drop Granulation

Gao, T., Arizona State University

Originally presented on: 4/12/2016 17:45:00 - 18:00:00

Single drop granulation by a drop impacting onto a powder bed is of significant interest since it is the first step in any powder wetting process, including wet granulation. Particle parameters like particle size distribution and density; liquid parameters like viscosity and surface tension; powder-liquid properties like hydrophobicity and solubility; and operating parameters like bed porosity and drop height can all influence the formation mechanism and resulting granule morphology. Breakage and attrition behavior of dry granules is also important for further handling and processing, and we hypothesize that the morphology of granules, which is dependent upon the different formation mechanisms, will also influence the breakage behavior of granules.

Comparison of DEM Force Models for Static and Dynamic Assemblies of Rod-Shaped Particles

Kumar, R., Purdue University

Originally presented on: 4/12/2016 18:00:00 - 18:15:00

Several DEM normal force models for cohesion-less rod-shaped particles are compared under static and dynamic conditions. Comparisons of the force, overlap, and contact duration frequency distributions are presented. Differences are observed in the overlap and contact duration distributions, but the force distributions are nearly independent of the choice of force model.

Solvent-Free Particle Surface Modification using Amino Acid L-Leucine during Milling for Improved Dry Powder Dispersion

Kunnath, K., New Jersey Institute of Technology

Originally presented on: 4/12/2016 18:15:00 - 18:30:00

Inhalation therapy requires the efficient powder de-agglomeration and dispersion in addition to a very fine particle size (<5 μm). Unfortunately, milling of pharmaceutical powders, which is required to achieve the right size range, leads to formation of large agglomerates with poor dispersion after inhalation. In this work, using KCl as a surrogate material, milling and surface coating was done in a Fluid Energy Mill (FEM) to produce particles suitable for inhalation. Several different coating materials were investigated and it was found that L-Leucine, a naturally occurring amino acid, offers similar powder dispersibility performance even at the fine particle sizes used in inhalation.

A Model-Guided Process Intensification Strategy for the Preparation of Sub-100 nm Drug Nanoparticles via Wet Stirred Media Milling

Li, M., New Jersey Institute of Technology

Originally presented on: 4/12/2016 18:30:00 - 18:45:00

There is sustained interest in sub-100 nm particles of poorly water-soluble drugs as such small particles offer improved permeation through various biological barriers and result in rapid onset of therapeutic action. For fast production of sub-100 nm drug particles, yttrium-stabilized zirconia beads with a nominal size ranging from 50 µm to 800 µm were used in a baseline wet stirred media milling (WSMM) process, which was subsequently intensified with the optimal bead size by increasing rotor tip speed, bead loading, and suspension flow rate step-wise, as guided by a microhydrodynamic model. This study has demonstrated, for the first time, the use of small beads in an intensified WSMM process for faster production of sub-100 nm particles of two BCS Class II drugs while achieving reduced energy consumption and keeping metal (Zr) contamination low.

Impact Dynamics of Low-Velocity Projectiles on Wet Granular Media

Vajrala, S., Arizona State University

Originally presented on: 4/12/2016 18:45:00 - 19:00:00

In this work, we use the Discrete Element Modelling (DEM) technique to study the impact of low velocity projectiles on wet, cohesive granular media with different packing fractions. Our motivation is to understand animal interaction on various sand terrains, and we aim to achieve this by observing different stages of interaction of solid projectiles with granular beds.

JENIKE & JOHANSON
SCIENCE ENGINEERING DESIGN

Frontiers in Particle Science & Technology
FPST

Particle attrition: The bane of many industrial plants Problems, solutions and red flags

John Carson, President, Jenike & Johanson
Eric Maynard, Senior Consultant, Jenike & Johanson

Houston, April 13, 2016

Frontiers in Particle Science & Technology Conference:
Mitigating & Application of Particle Attrition

Outline

- Definition
- Examples of attrition
- Results of attrition
- Locations and causes of attrition
- Solutions to minimize attrition
- Some unintended consequences of attrition solutions

Definition

Attrition is defined here as the "unwanted" breakdown, degradation, fragmentation, reduction, and/or accidental damage of large particles into smaller fractions or "daughter particles".

British Materials Handling Board, 1987

JENIKE & JOHANSON SCIENCE | ENGINEERING | DESIGN 3

Attrition

- Intentional
  - Blasting
  - Crushing and grinding
- Unintentional
  - Loading and discharging
  - Handling and storing

JENIKE & JOHANSON SCIENCE | ENGINEERING | DESIGN 4

Outline

- Definition
- Examples of attrition
- Results of attrition
- Locations and causes of attrition
- Solutions to minimize attrition
- Some unintended consequences of attrition solutions

JENIKE & JOHANSON SCIENCE | ENGINEERING | DESIGN 5

Attrition occurs in virtually every industry

- Plastics
- Foods
- Pharmaceuticals

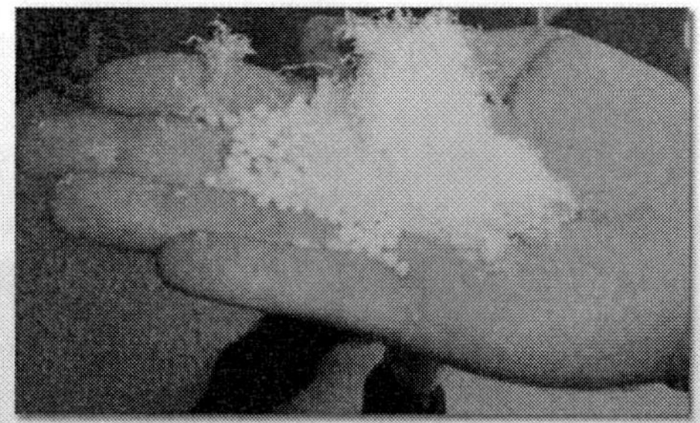

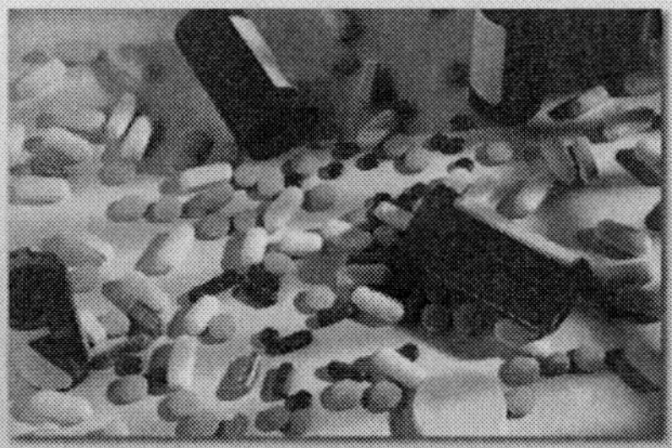

JENIKE & JOHANSON SCIENCE | ENGINEERING | DESIGN 6

Attrition occurs in virtually every industry

- Power generation
- Minerals
- Detergents

JENIKE & JOHANSON SCIENCE | ENGINEERING | DESIGN 7

Salt pellets

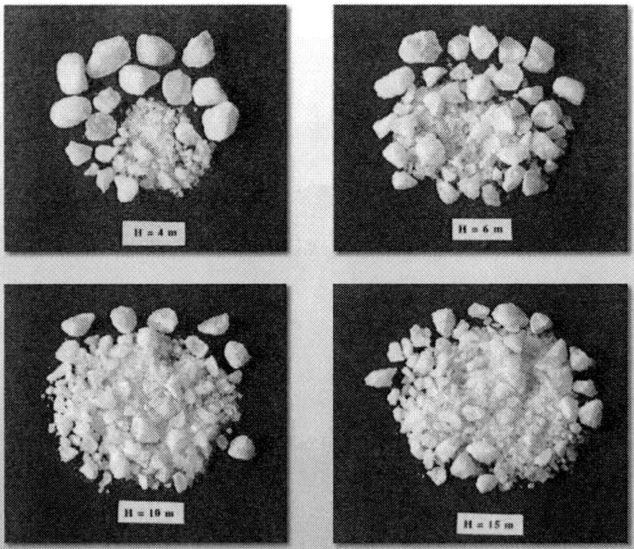

JENIKE & JOHANSON SCIENCE | ENGINEERING | DESIGN 8

Chocolate candies

JENIKE & JOHANSON SCIENCE | ENGINEERING | DESIGN 9

Cat litter

JENIKE & JOHANSON SCIENCE | ENGINEERING | DESIGN 10

Prilled urea and crystalline fines

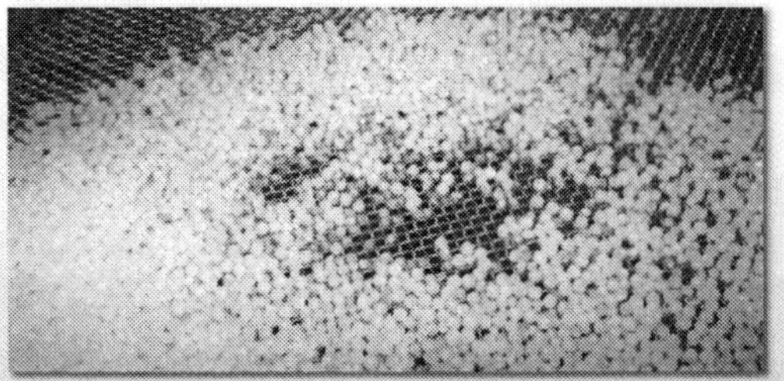

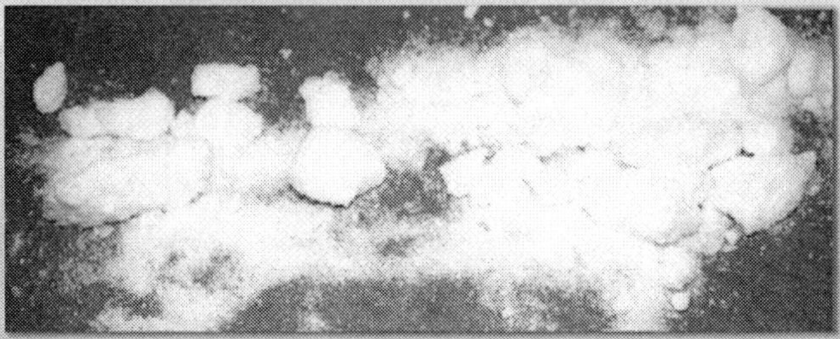

JENIKE & JOHANSON SCIENCE | ENGINEERING | DESIGN 11

Carbon fiber bundles

JENIKE & JOHANSON SCIENCE | ENGINEERING | DESIGN 12

Fuel cell pellets

JENIKE & JOHANSON SCIENCE | ENGINEERING | DESIGN 13

Outline

- Definition
- Examples of attrition
- Results of attrition
- Locations and causes of attrition
- Solutions to minimize attrition
- Some unintended consequences of attrition solutions

JENIKE & JOHANSON SCIENCE | ENGINEERING | DESIGN 14

Results of attrition

- Unacceptable dust levels
- Customer complaints and rejection
- Flow stoppages (arching, ratholing)
- Flow rate limitations
- Off-spec product - too fine or too lumpy
- High pressure drop in packed beds
- Increased segregation
- Equipment performance issues

JENIKE & JOHANSON SCIENCE | ENGINEERING | DESIGN 15

Dust generation at minerals plant

JENIKE & JOHANSON SCIENCE | ENGINEERING | DESIGN 16

Caking - fertilizer from 50 lb (23 kg) bag

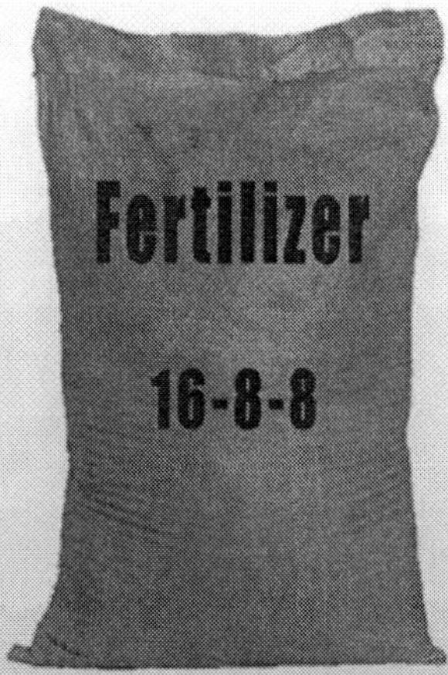

JENIKE & JOHANSON SCIENCE | ENGINEERING | DESIGN 17

Flow stoppages and flow rate limitations

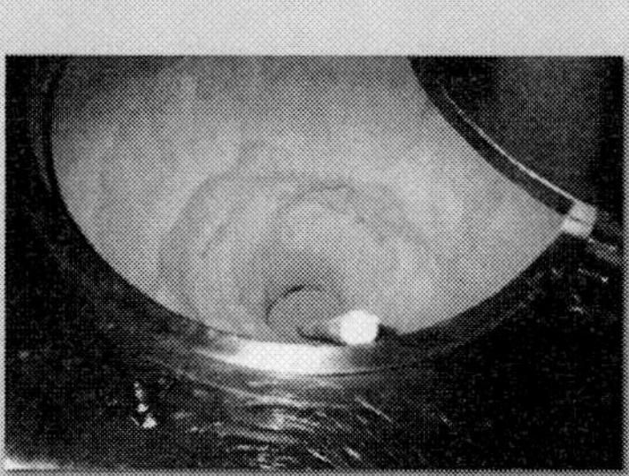

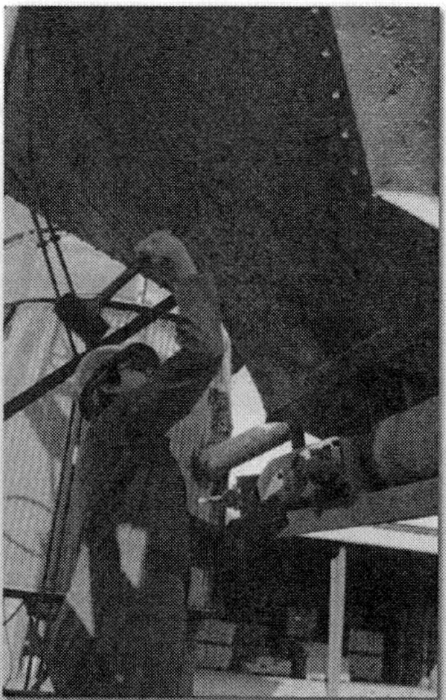

JENIKE & JOHANSON SCIENCE | ENGINEERING | DESIGN 18

Reduced solubility – ice tea powder

- Attrited material (right) more difficult to "wet"

JENIKE & JOHANSON SCIENCE | ENGINEERING | DESIGN 19

Decreased product performance

JENIKE & JOHANSON SCIENCE | ENGINEERING | DESIGN 20

Degraded appearance - cereal

- Breakage and "scuffing" of sugar coating are issues

JENIKE & JOHANSON SCIENCE | ENGINEERING | DESIGN 21

Dust explosivity hazards at board plant

JENIKE & JOHANSON SCIENCE | ENGINEERING | DESIGN 22

Fluidized bed collapse of agglomerates

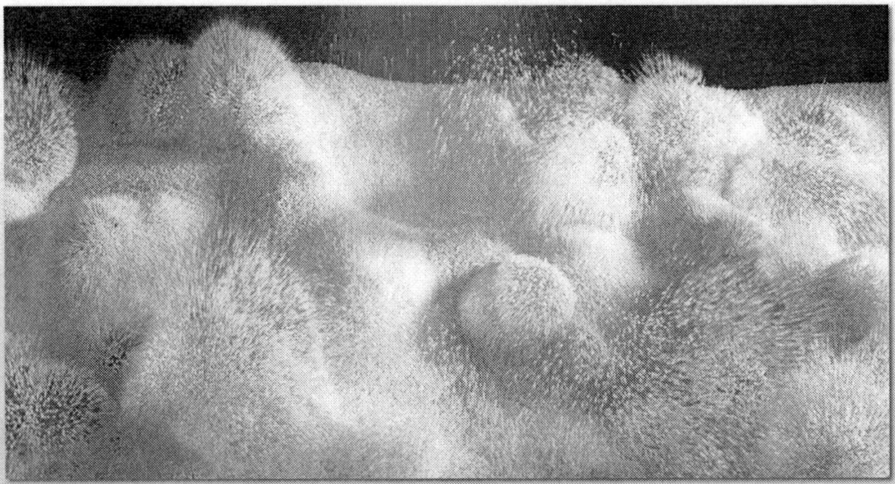

Ref.: http://www.gea.com/global/en/products/vibro-fluidizer.jsp

JENIKE & JOHANSON SCIENCE | ENGINEERING | DESIGN 23

Equipment performance issues

Screw jamming

JENIKE & JOHANSON SCIENCE | ENGINEERING | DESIGN 24

Equipment performance issues

- Finer material increases exothermic reaction

JENIKE & JOHANSON SCIENCE | ENGINEERING | DESIGN 25

Outline

- Definition
- Examples of attrition
- Results of attrition
- Locations and causes of attrition
- Solutions to minimize attrition
- Some unintended consequences of attrition solutions

JENIKE & JOHANSON SCIENCE | ENGINEERING | DESIGN 26

Transfer chutes

JENIKE & JOHANSON SCIENCE | ENGINEERING | DESIGN 27

Pneumatic conveying systems

JENIKE & JOHANSON SCIENCE | ENGINEERING | DESIGN 28

Rotary valve feeder

JENIKE & JOHANSON SCIENCE | ENGINEERING | DESIGN 29

Attrition during pneumatic conveying

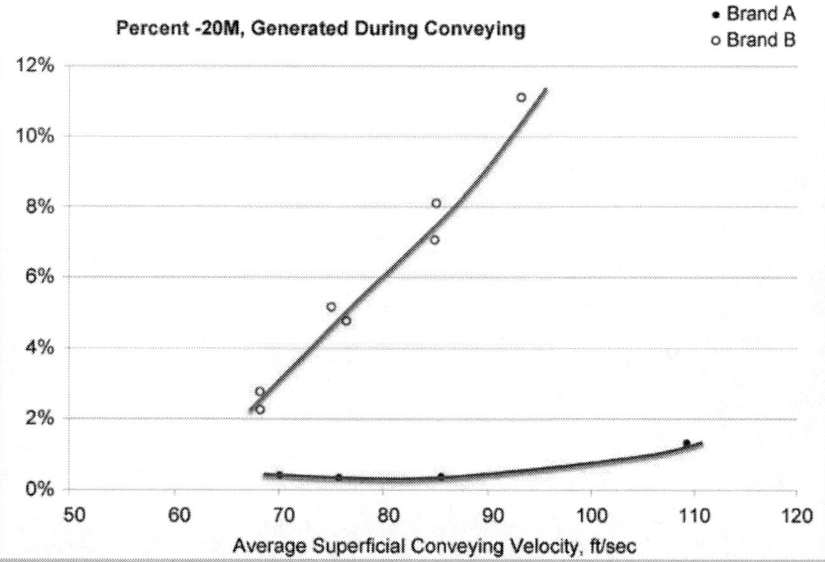

JENIKE & JOHANSON SCIENCE | ENGINEERING | DESIGN 30

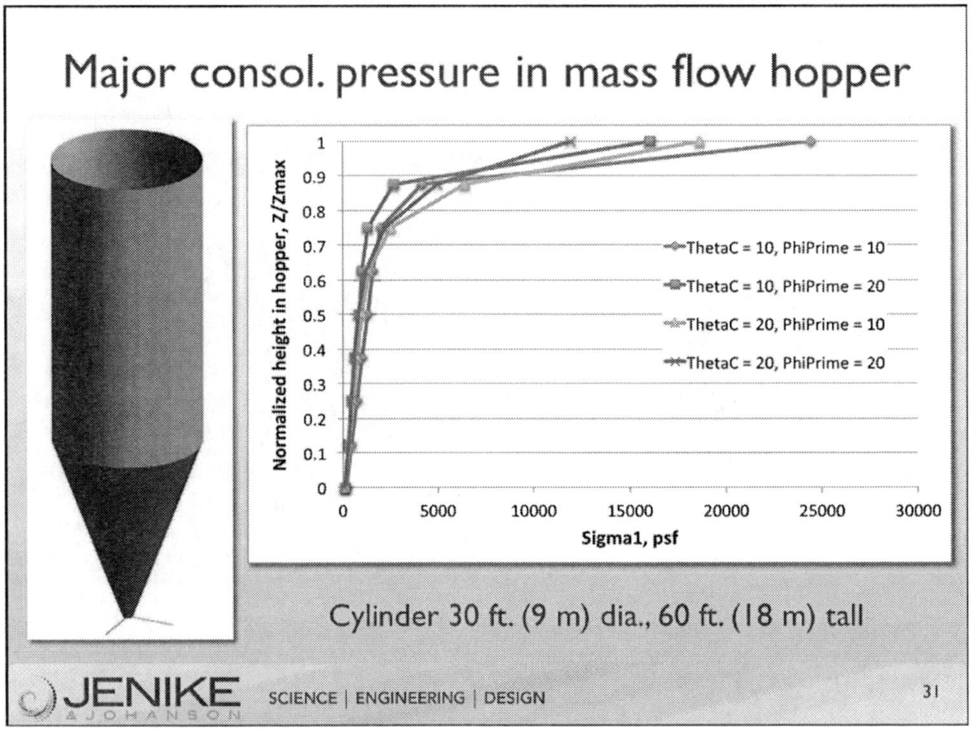
Major consol. pressure in mass flow hopper
Normalized height in hopper, Z/Zmax
1
0.9
0.8
0.7
0.6
0.5
0.4
0.3
0.2
0.1
0
0
5000
10000
15000
20000
25000
30000
Sigma1, psf
ThetaC = 10, PhiPrime = 10
ThetaC = 10, PhiPrime = 20
ThetaC = 20, PhiPrime = 10
ThetaC = 20, PhiPrime = 20
Cylinder 30 ft. (9 m) dia., 60 ft. (18 m) tall
JENIKE & JOHANSON
SCIENCE | ENGINEERING | DESIGN
31

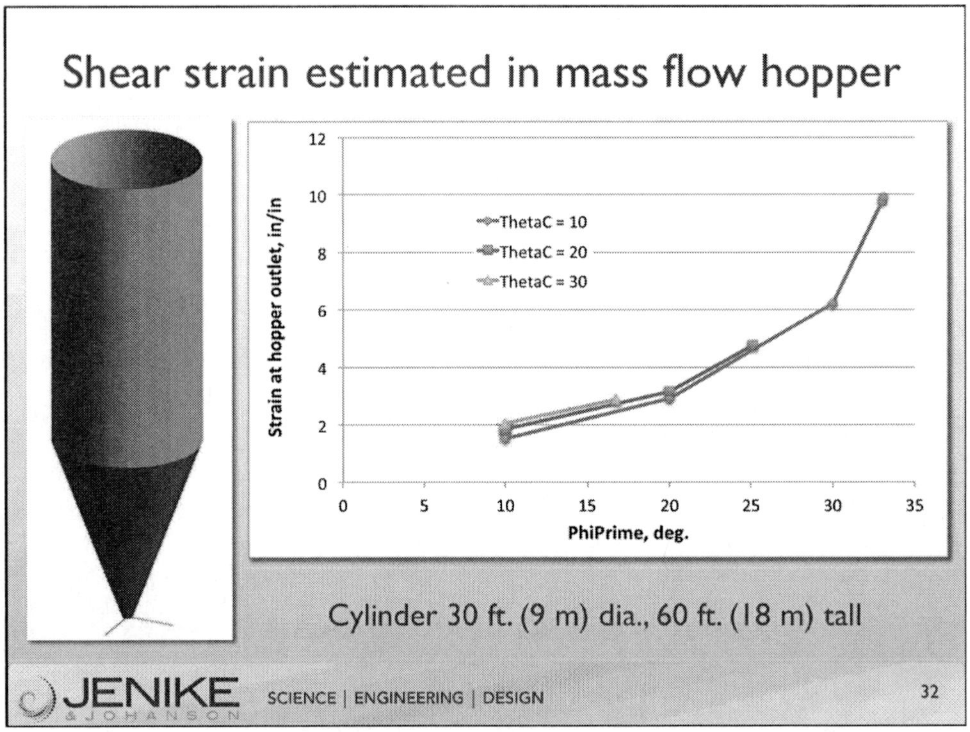
Shear strain estimated in mass flow hopper
Strain at hopper outlet, in/in
12
10
8
6
4
2
0
0
5
10
15
20
25
30
35
PhiPrime, deg.
ThetaC = 10
ThetaC = 20
ThetaC = 30
Cylinder 30 ft. (9 m) dia., 60 ft. (18 m) tall
JENIKE & JOHANSON
SCIENCE | ENGINEERING | DESIGN
32

Attrition effects: mass flow vs. funnel flow

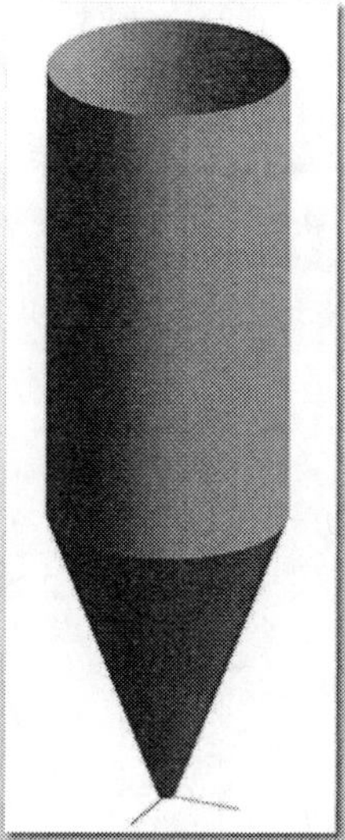

- Mass flow
  - Highest pressure at transition point
  - Lowest pressure at outlet
  - Minimal shear strain, unless large velocity gradient
- Funnel flow
  - High pressure at transition point, but, not as large as mass flow peak value
  - Higher pressure at outlet compared to mass flow
  - Major shear strain due to flow channel

JENIKE & JOHANSON SCIENCE | ENGINEERING | DESIGN 33

Screw feeder

JENIKE & JOHANSON SCIENCE | ENGINEERING | DESIGN 34

Outline

- Definition
- Examples of attrition
- Results of attrition
- Locations and causes of attrition
- Solutions to minimize attrition
- Some unintended consequences of attrition solutions

JENIKE & JOHANSON SCIENCE | ENGINEERING | DESIGN 35

Solutions to minimize attrition

- Increase particle hardness
- Decrease particle size
- Make particles rounded vs. angular or flaky
- Cushion impact
- Use dense phase vs. dilute phase conveying
- Reduce bin size or level
- Use belt feeder vs. screw or rotary valve
- Decrease angles of impact

JENIKE & JOHANSON SCIENCE | ENGINEERING | DESIGN 36

Modifying the particle

Potato flakes

Flake vs pellet

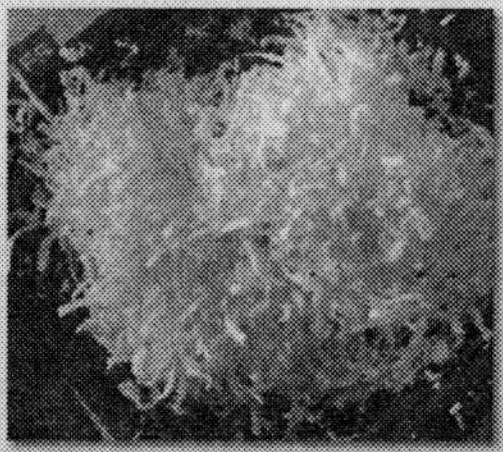

JENIKE & JOHANSON SCIENCE | ENGINEERING | DESIGN 37

Increasing particle hardness

Roller compaction

JENIKE & JOHANSON SCIENCE | ENGINEERING | DESIGN 38

Cushion impact

- Drop on bed vs plate
- Let-down chute
- Minimize impact angle

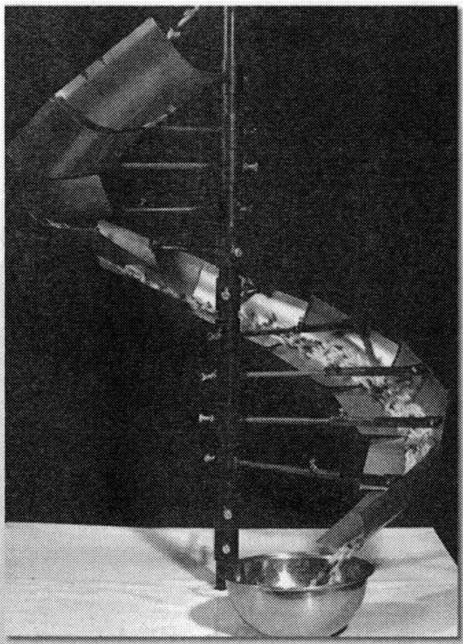

Hood and spoon at transfer point instead of rock box

US patent #4,646,910 A

Pneumatic conveying – "speed kills"

- Lower conveying velocity
- Fewer bends
- Different types of bends

JENIKE & JOHANSON SCIENCE | ENGINEERING | DESIGN 41

Different types of bends

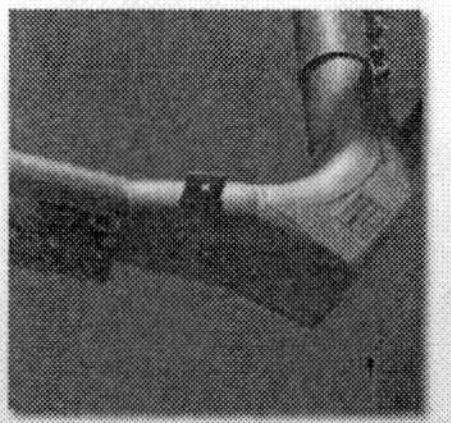

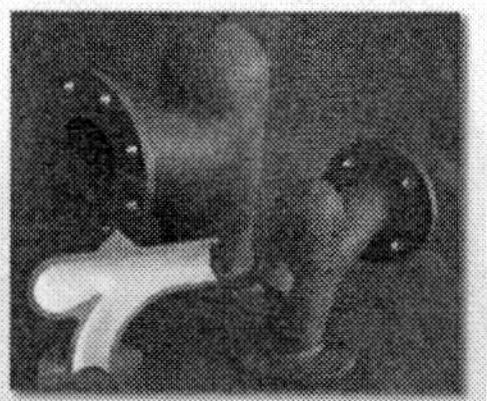

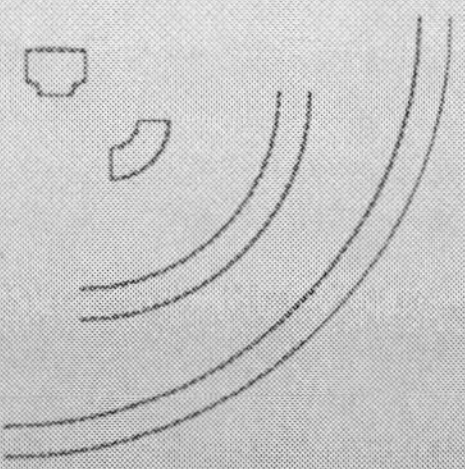

JENIKE & JOHANSON SCIENCE | ENGINEERING | DESIGN 42

Converging/diverging bins

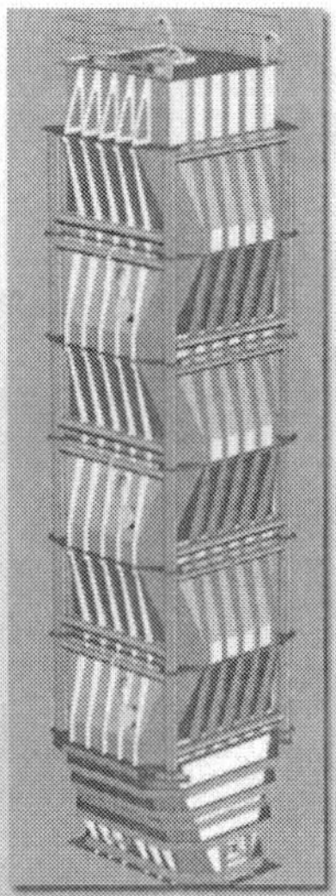

US patent
#6139241 A

JENIKE & JOHANSON SCIENCE | ENGINEERING | DESIGN 43

Use different feeder

Top feed

Side feed

JENIKE & JOHANSON SCIENCE | ENGINEERING | DESIGN 44

Outline

- Definition
- Examples of attrition
- Results of attrition
- Locations and causes of attrition
- Solutions to minimize attrition
- Some unintended consequences of attrition solutions

JENIKE & JOHANSON SCIENCE | ENGINEERING | DESIGN 45

Unintended consequence: let-down slide

- Particle attrition
- Dust explosion
- Structural failure of let-down chute

JENIKE & JOHANSON SCIENCE | ENGINEERING | DESIGN 46

Structural failure of let-down slide

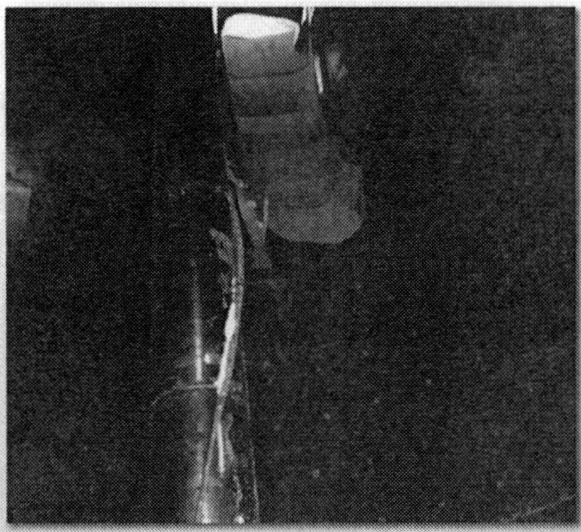

JENIKE & JOHANSON SCIENCE | ENGINEERING | DESIGN 47

Unintended consequence: chute design

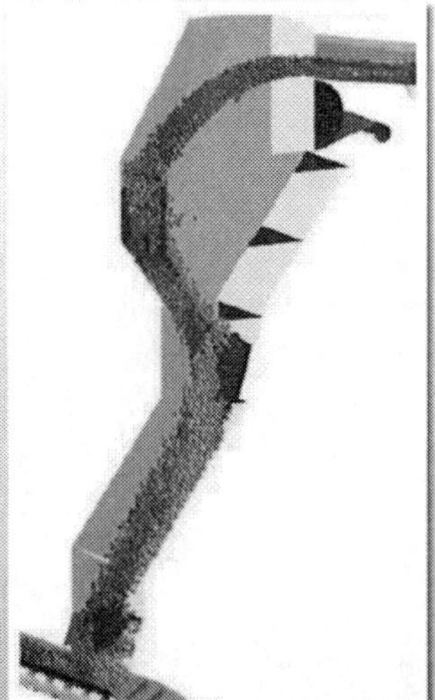

JENIKE & JOHANSON SCIENCE | ENGINEERING | DESIGN 48

Unintended consequence: silo/bin inserts

JENIKE & JOHANSON SCIENCE | ENGINEERING | DESIGN 49

Unintended consequence: silo/bin inserts

- Tremendous material-induced loads on silo inserts

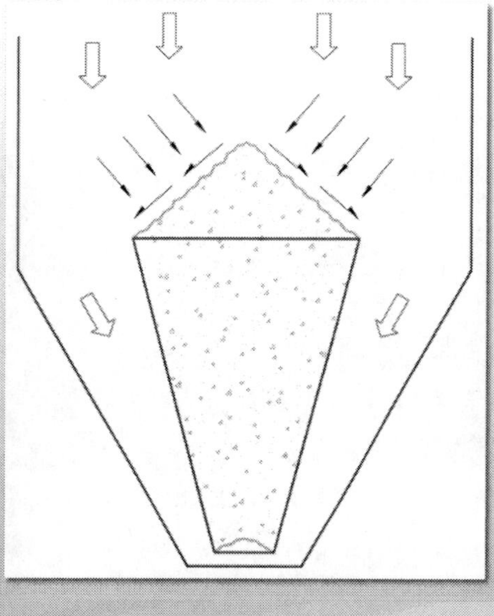

JENIKE & JOHANSON SCIENCE | ENGINEERING | DESIGN 50

Unintended consequence

- Bin ≈ 300 ton cap.
  - 14 ft. diameter
  - 36 ft. tall cylinder
  - Mass flow Binsert®
- Limestone
  - 90 pcf (1440 kg/m^3)

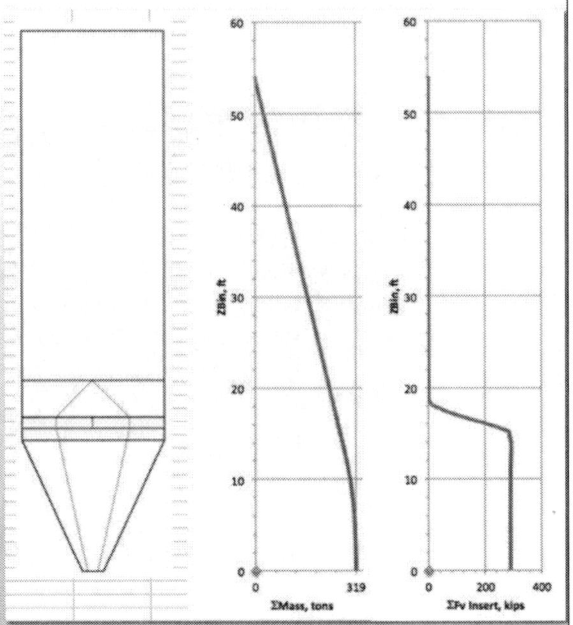

JENIKE & JOHANSON SCIENCE | ENGINEERING | DESIGN 51

Unintended consequence

- Insert failure

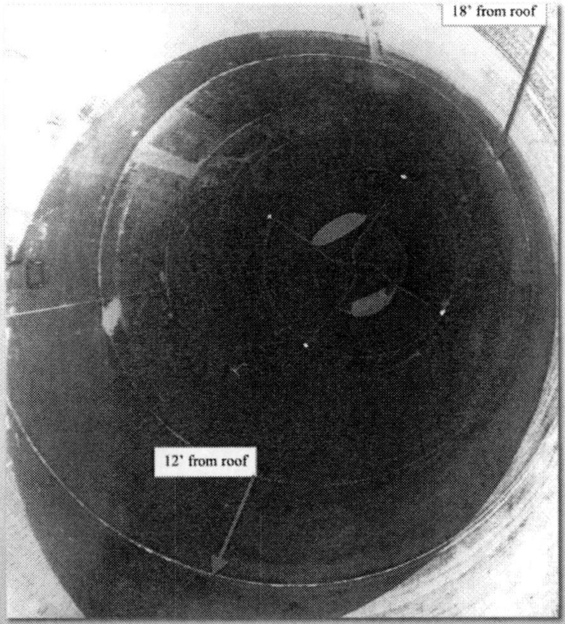

JENIKE & JOHANSON SCIENCE | ENGINEERING | DESIGN www.jenike.com 52

Unintended consequence: silo/bin inserts

- Inserts used in bin induce high loads on roof and break-off periodically breaking equipment

JENIKE & JOHANSON SCIENCE | ENGINEERING | DESIGN 53

Unintended consequence: unique silo/bin design

- Converging/diverging bin for handling compression-sensitive bulk solids
  - Sweet bran
  - Fly ash green pellets

JENIKE & JOHANSON SCIENCE | ENGINEERING | DESIGN 54

Unintended consequence: unique silo/bin design

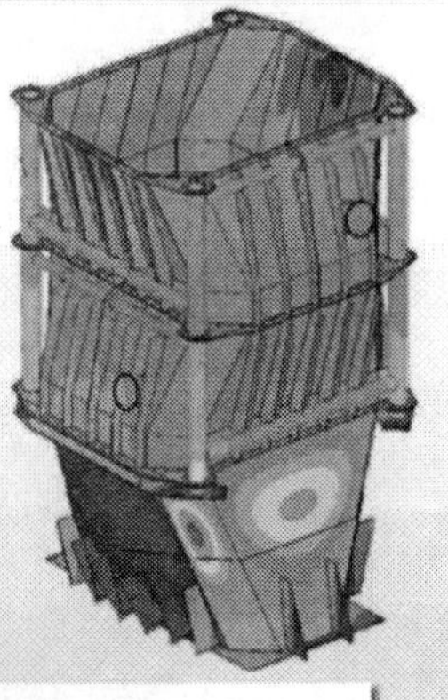

- Structural issues
  - Bending moments in flat walls
- Feed issues
  - Poor feeder interface

JENIKE & JOHANSON SCIENCE | ENGINEERING | DESIGN 55

Unintended consequence: pneu. conveying

- Lowering conveying velocity in pneumatic conveying system
  - Attrition sensitive to velocity power law
- Can induce saltation if operating in dilute phase
- Not all materials can be conveyed in dense phase

JENIKE & JOHANSON SCIENCE | ENGINEERING | DESIGN 56

Unintended consequence: pneu. conveying

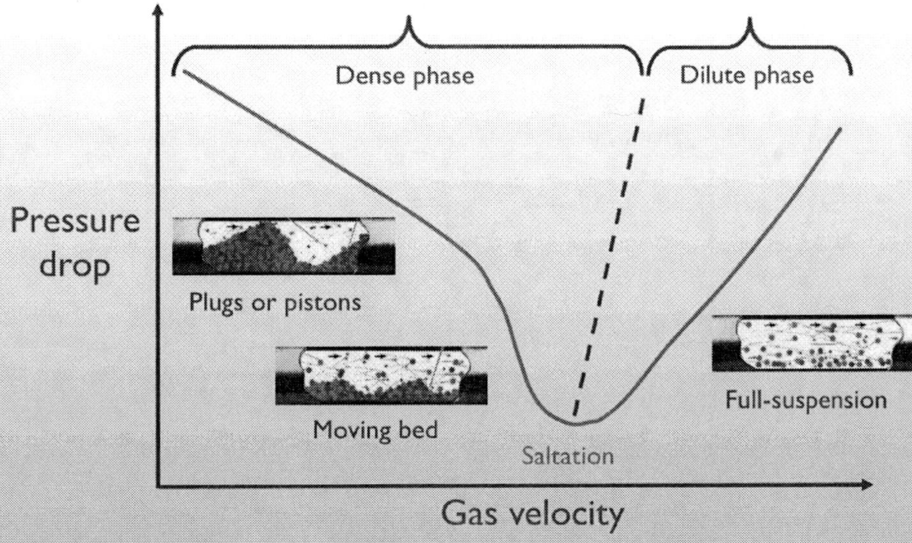

JENIKE & JOHANSON SCIENCE | ENGINEERING | DESIGN 57

Unintended consequence: pneu. conveying

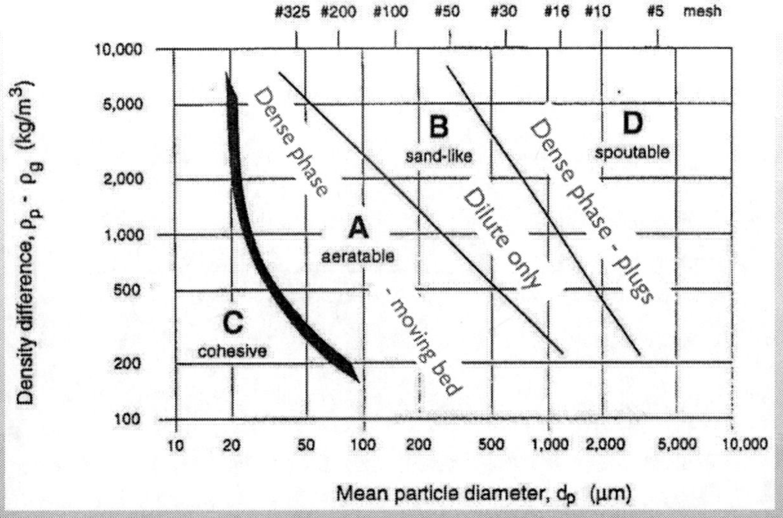

Geldart, D (1973) *Powder Technology*, 7, 285-292

JENIKE & JOHANSON SCIENCE | ENGINEERING | DESIGN 58

Unintended consequence: pneu. conveying

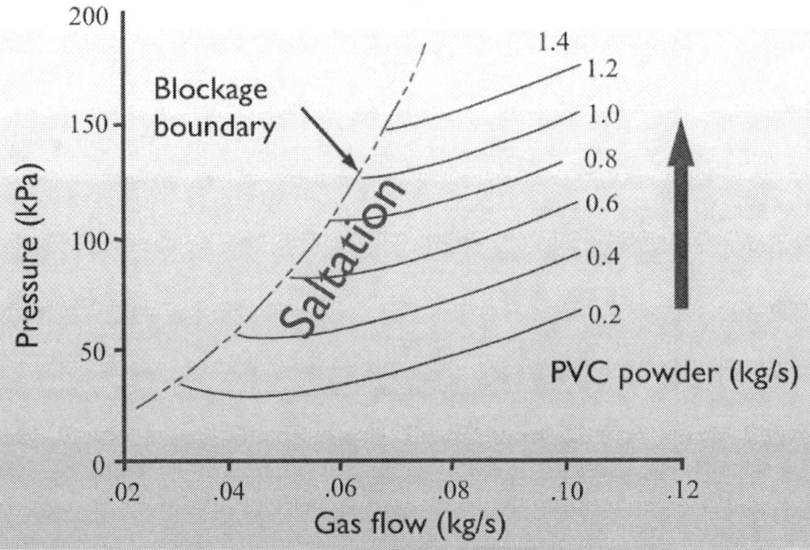

JENIKE & JOHANSON SCIENCE | ENGINEERING | DESIGN 59

Unintended consequence: pneu. conveying

- Lowering conveying velocity using "stepped line"

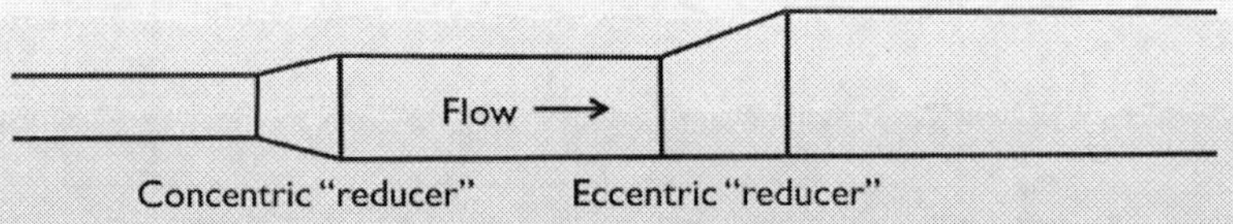

Watch out for too large of area expansion leading to saltation, instabilities, and possible plugging

JENIKE & JOHANSON SCIENCE | ENGINEERING | DESIGN 60

Unintended consequence: flow pattern

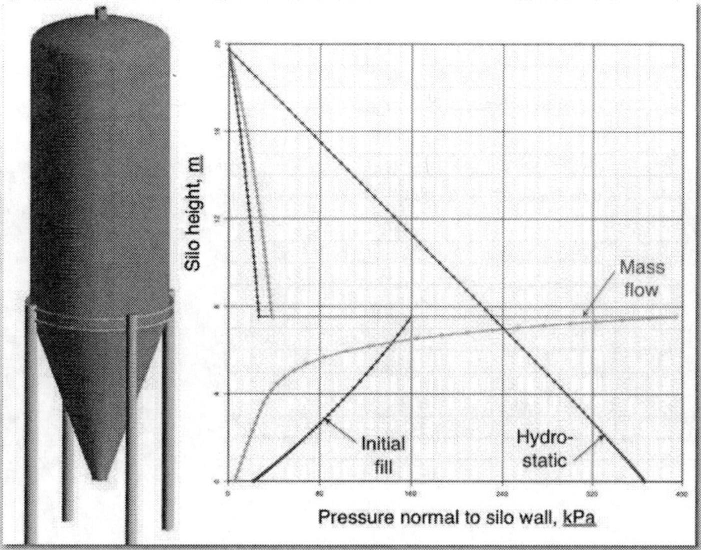

JENIKE & JOHANSON SCIENCE | ENGINEERING | DESIGN 61

Unintended consequence: flow pattern

- Using funnel flow discharge pattern to reduce extreme consolidation due to mass flow transition point peak pressure
- Consequences
  - Increased inter-particle shear
  - Segregation
  - FILO and reduced "live" capacity

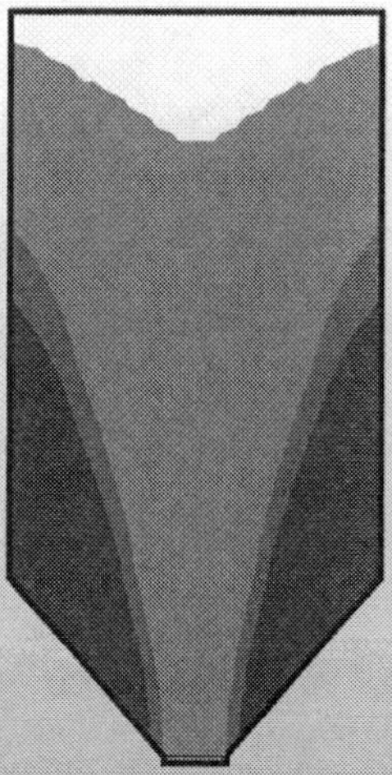

JENIKE & JOHANSON SCIENCE | ENGINEERING | DESIGN 62

Unintended consequence: modified particle

- Reformulate flakes to prills

- Negatively affected reaction rate and processing time

JENIKE & JOHANSON SCIENCE | ENGINEERING | DESIGN 63

JENIKE® & JOHANSON SCIENCE ENGINEERING DESIGN

Thank you

John W. Carson, Ph.D.
jwcarson@jenike.com
1-978-649-3300
Jenike.com

Particle Attrition

S.B. Reddy Karri, Yeook Arrington and Ray Cocco

April 13, 2016

1

Outline

- Attrition sources in a fluidized bed or circulating fluidized bed systems
- Fundamentals of particle attrition
- Characterization of particle attrition using different test apparatus

2

PSRI
Particulate Solid Research, Inc.

Possible Sources of Particle Attrition

- Sources of Attrition
  - Primary cyclones
  - Bubble
  - Riser terminations or bends
  - Jets from spargers and grids
  - Riser feeds
  - Standpipes
  - Attrition at valves or screw feeders

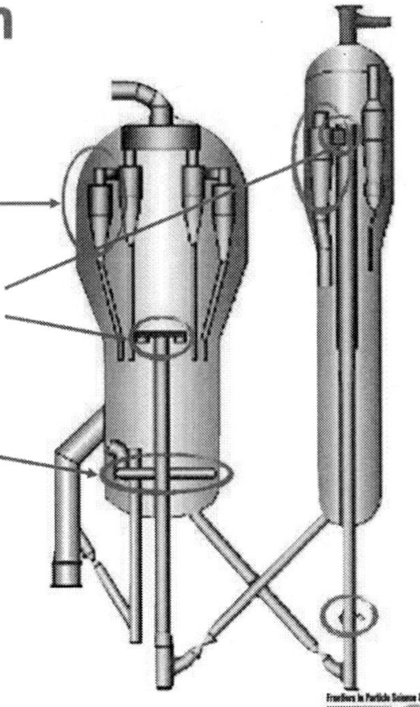

Frontiers in Particle Science & Technology
FPST

3

J1-4

PSRI
Particulate Solid Research, Inc.

Most Attrition in Fluidized Bed Systems is Due to Particle Impact

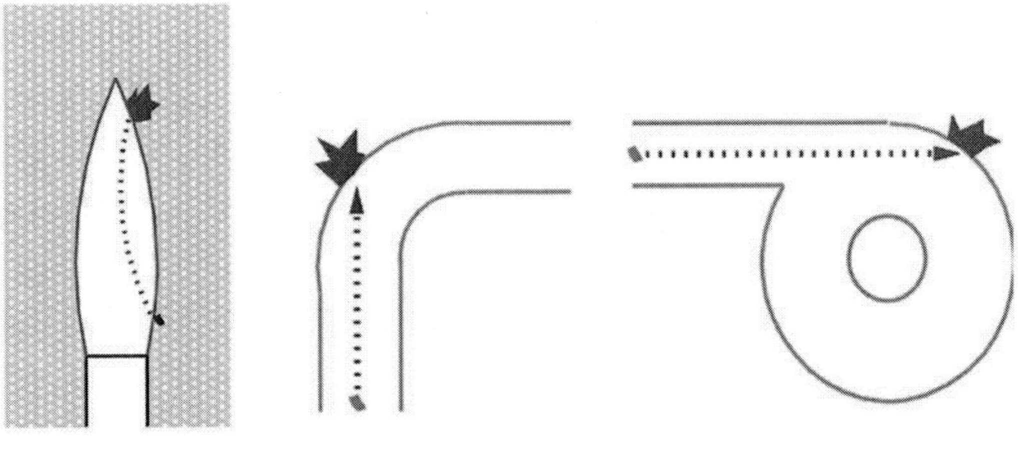

Jet/Nozzle Bend Cyclone

Frontiers in Particle Science & Technology
FPST

Jets in Fluidized Beds

High Angle of Impact

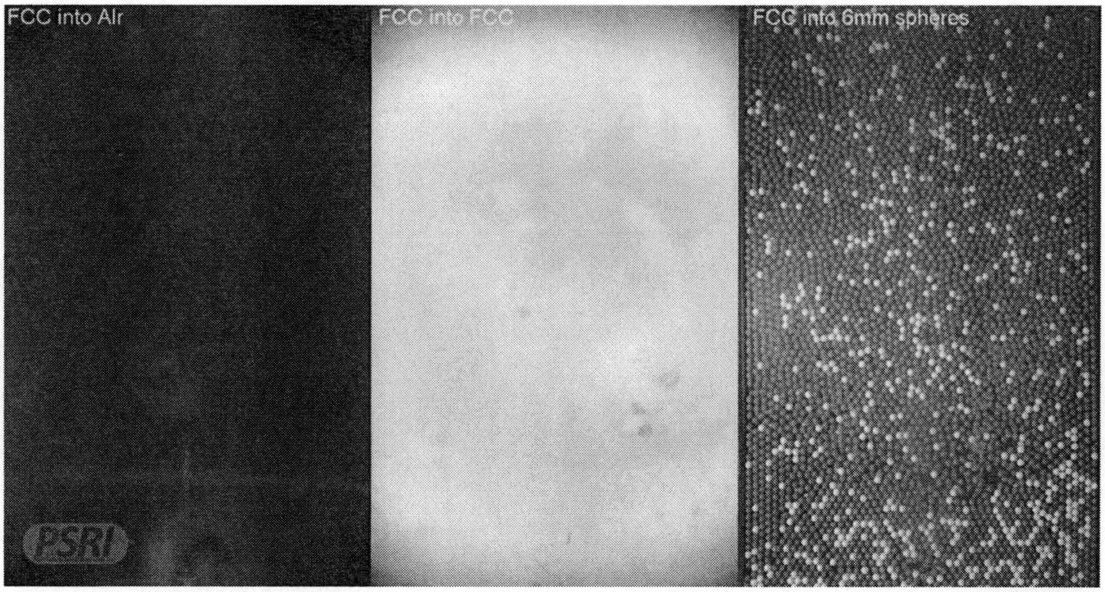

5

Particle Impact in Primary Cyclone

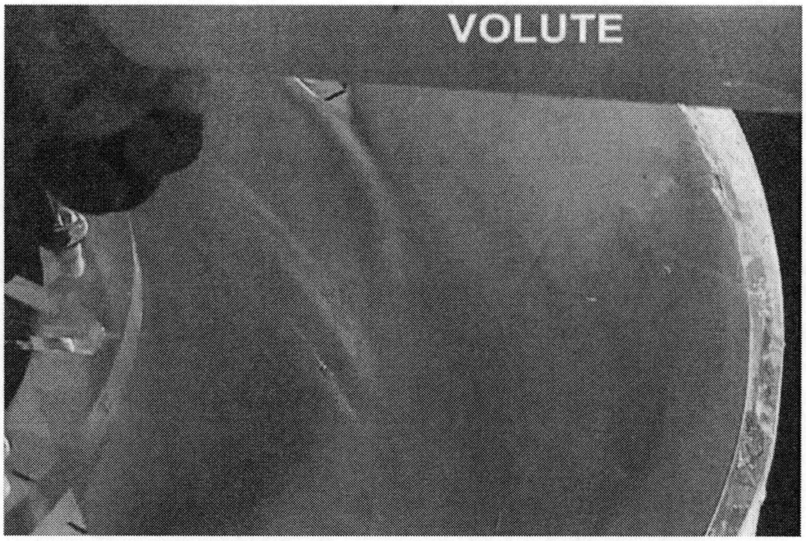

6

Spray Drying and Particle Attrition

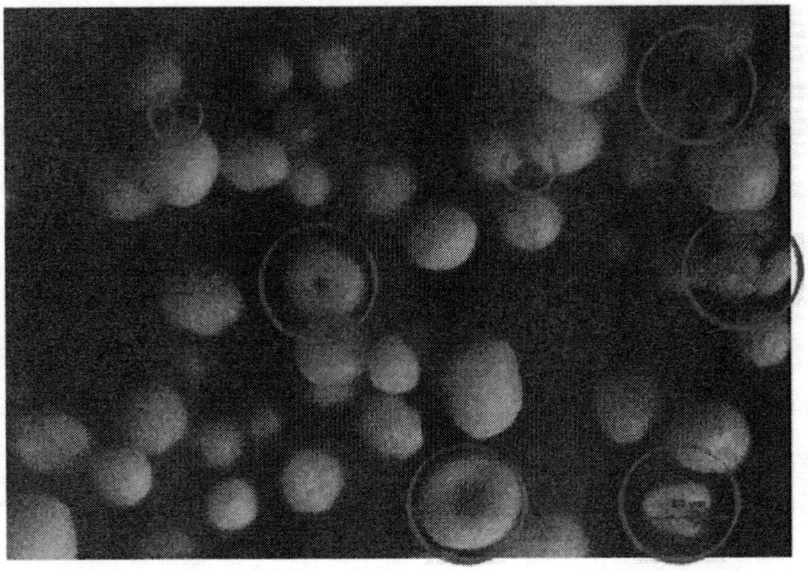

7

Attrition Mechanisms
Particle Fracture vs. Abrasion

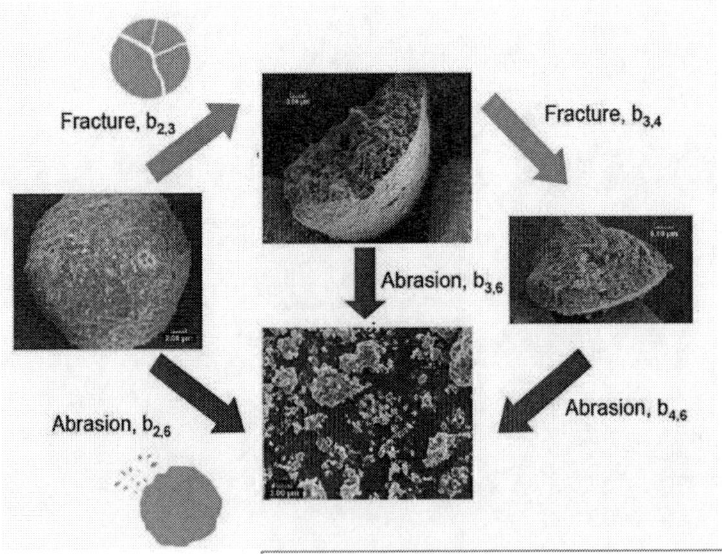

- It is important to understand the attrition mechanisms in commercial units

Source: D. Stockwell, et al., NPRA paper AM-11-39

Types of Particle Attrition

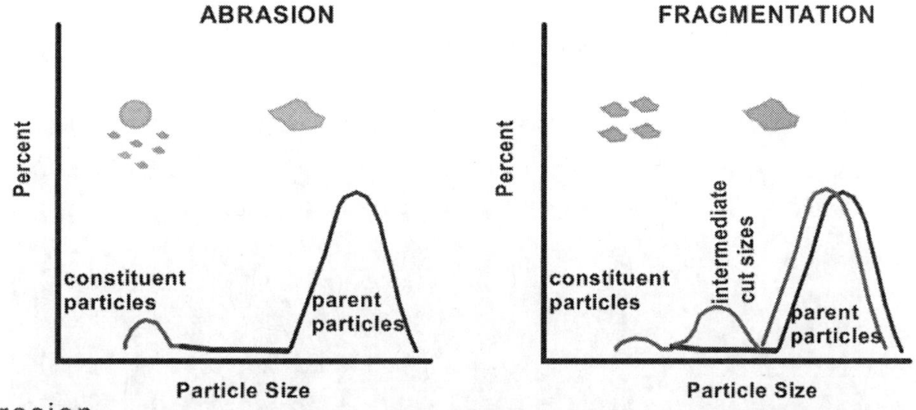

Abrasion

- Abrasion is when constituent particles attrit from the surface of the parent particles
- Typically the removal of edges, corners, protrusions, etc.
- Small, subtle changes in particle size distribution for most cases

Fragmentation

- Fragmentation is when the parent particle breaks down to intermediate sizes and a daughter constituent particl
- Significant change in particle size distribution

Particle Impacts

Abrasion

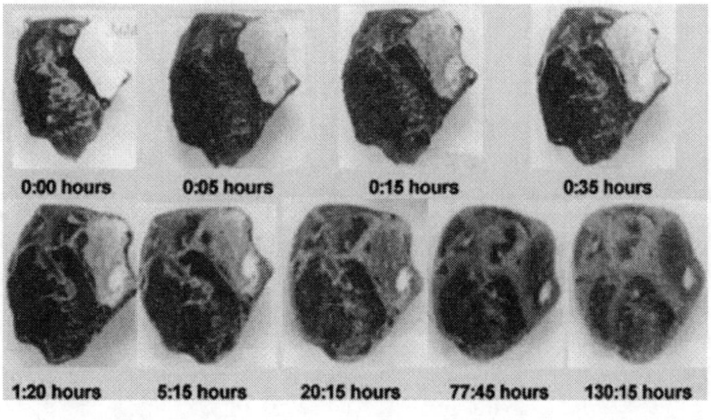

http://www.sussex.ac.uk/geography/researchprojects/BERM/project/tumbling.html

Particle Impacts

Fragmentation

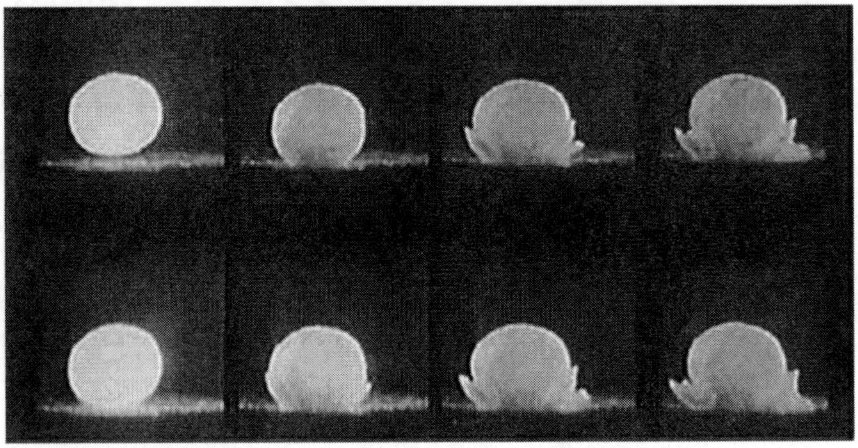

M. Ghadiri, University of Leeds

11

PSRI Conical Jet Cup Attrition

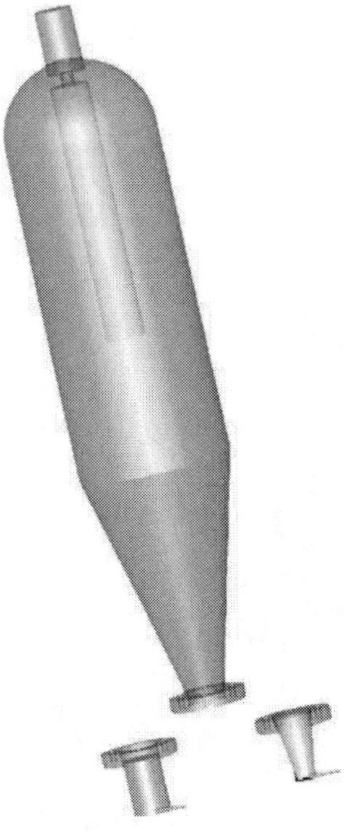

- Jet cup attrition testing is a common method for evaluating particle attrition in fluidized systems
- The jet cup method only provides a relative comparison of attrition rate and does not give a quantitative attrition rate that can be applied towards commercial units

12

PSRI

Threshold Velocity Defines the Transition from Abrasion to Fragmentation for FCC Catalyst

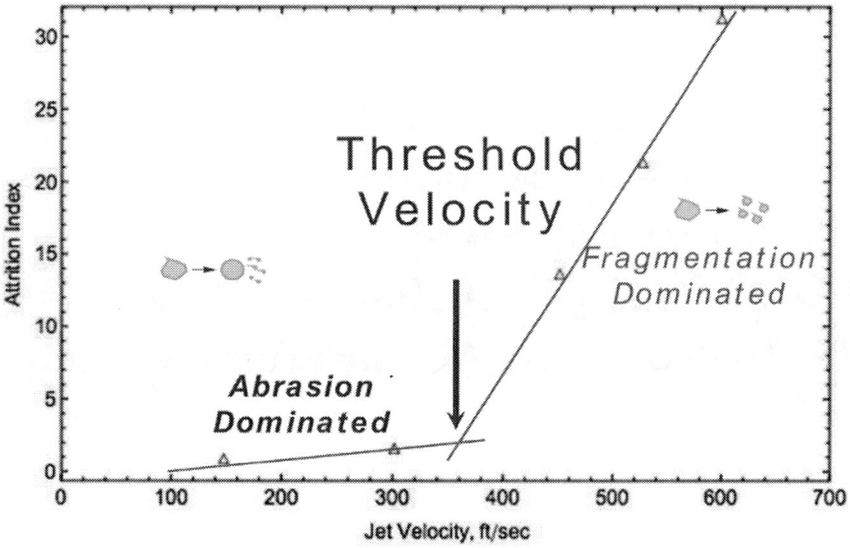

FPST

13

PSRI

Abrasion and Fragmentation for FCC Equilibrium Catalyst

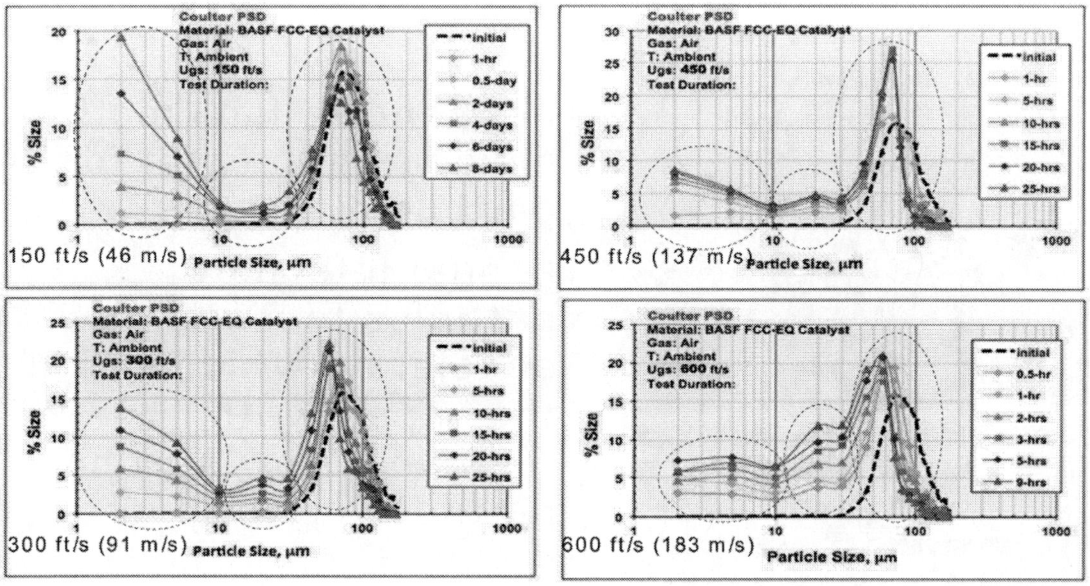

- PSDs of attrited samples at all four jet velocities over time clearly indicated three peaks corresponding to constituent particle sizes, intermediate cut sizes, and parent particle sizes

FPST

14

What Affects Attrition?

- Particle properties
  - Shape factor
  - Strength of material
  - Elasticity of material
  - Structural integrity
  - Surface morphology
- Particle velocity (momentum) and impact angle

15

Jet vs Particles

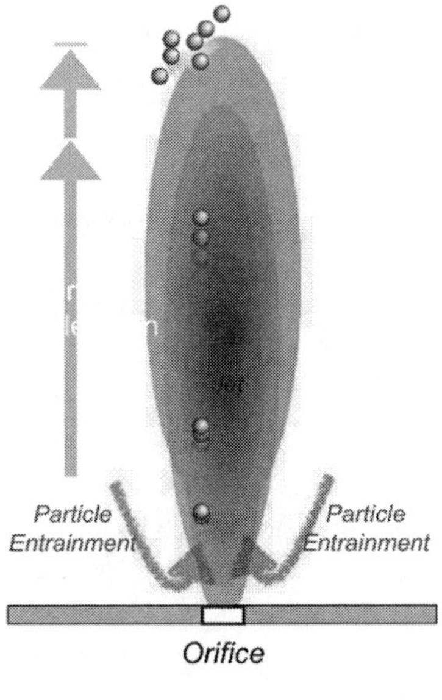

- Jet penetration length is a significant factor for particle attrition
- Longer jets allow particles to more fully accelerate which makes the impact with relatively slow moving particles more severe

16

PSRI

Jets vs Jets

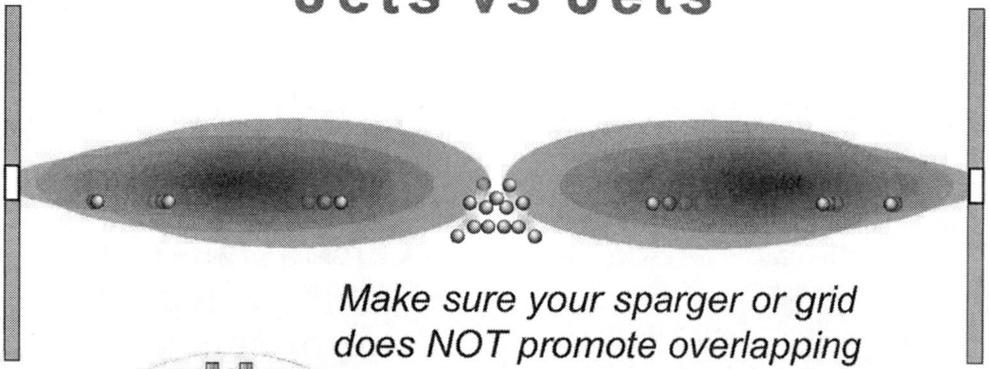

Make sure your sparger or grid does NOT promote overlapping jets

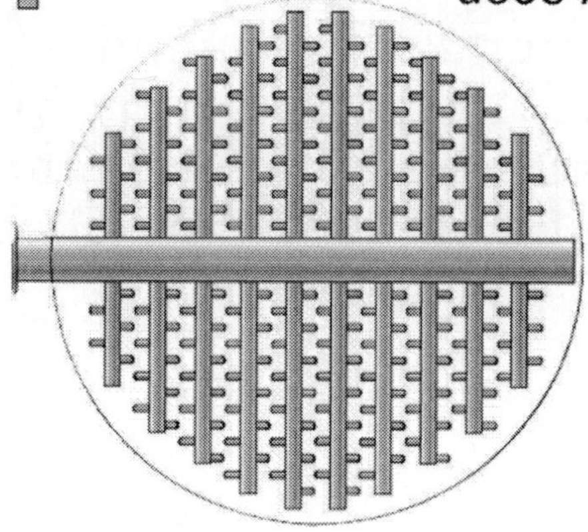

- The collision of opposing jets occurs with double the momentum
  - Best way to ensure attrition

FPST

17

PSRI

Measurement of Relative Particle Attrition Characterisitics

- Drum mill
- Immersed jet
- Impact test unit
- Jet cup

Methods can be used to measure relative attrition characteristics

FPST

18

PSRI

Drum Mill

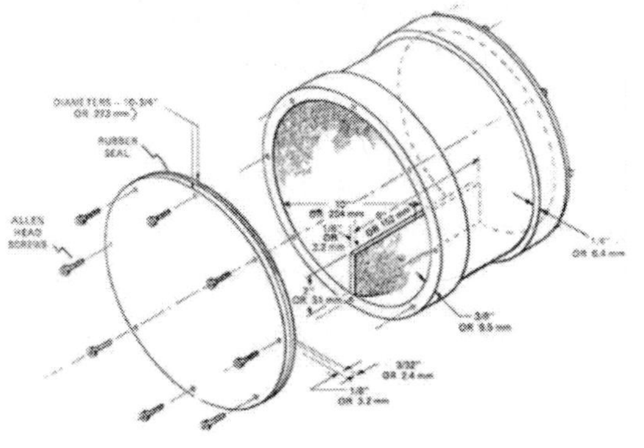

- ASTM 4058
- Typically used for tablets, extrudate, spheres, and irregularly-shaped particles larger than about 1⁄16 in (1.6 mm) and smaller than about 3⁄4 in (19 mm).

FPST

19

PSRI

Immersed Jet

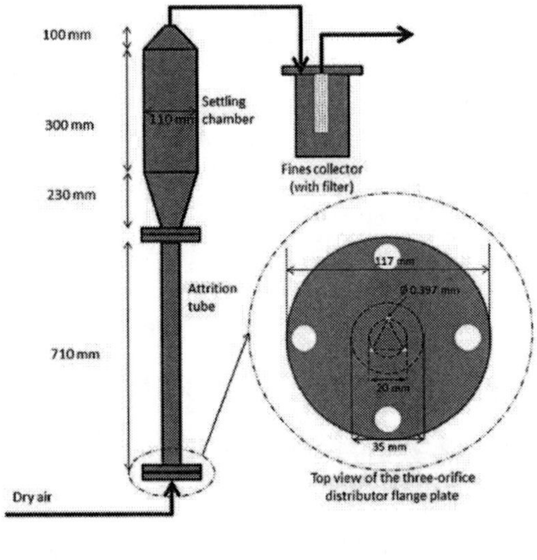

- ASTM D5757
- Applicable to spherical or irregularly-shaped particles which range in size between 10 and 180 µm and have skeletal densities between 2.4 and 3.0 g/cm3 (150 and 187 lb/ft3)
- Particles less than 20 µm are considered fines

FPST

20

PSRI Impact Unit

- PSRI high temperature impact attrition test unit
- Best for Geldart Group A and B particles
- Fines are considered to be 0-20 and 0-44 microns

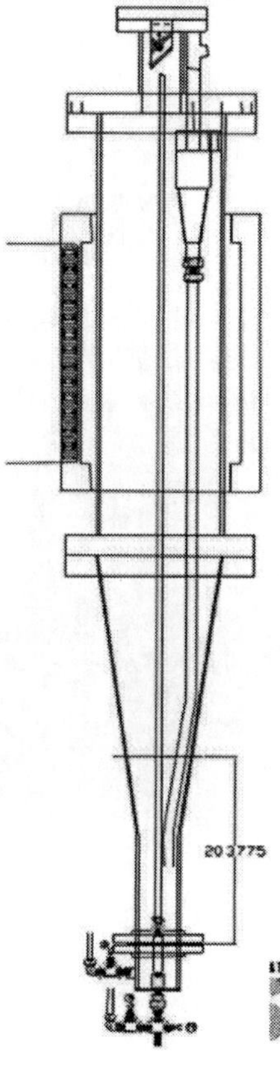

21

Jet Cup Attrition

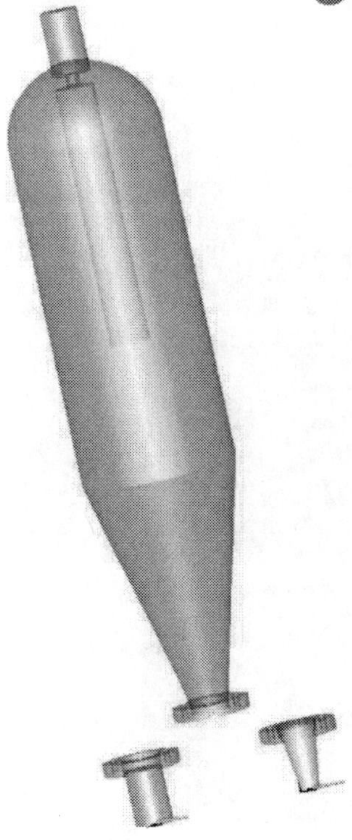

- Jet cup attrition testing is a common method for evaluating particle attrition in fluidized systems
- The jet cup method only provides a relative comparison of attrition rate and does not give a quantitative attrition rate that can be applied towards commercial units,
- Today, there are three types of jet cup test units
  - Davison jet cup
  - PSRI cylindrical jet cup (old method)
  - PSRI conical jet cup (new method)
- The ranking of catalyst attrition is determined by comparing the attrition index for a specific size range

22

Issues with Classical Jet Cup

0.1875 in (0.48 cm) Nozzle with 450 ft/sec (137 m/sec) Jet Velocity

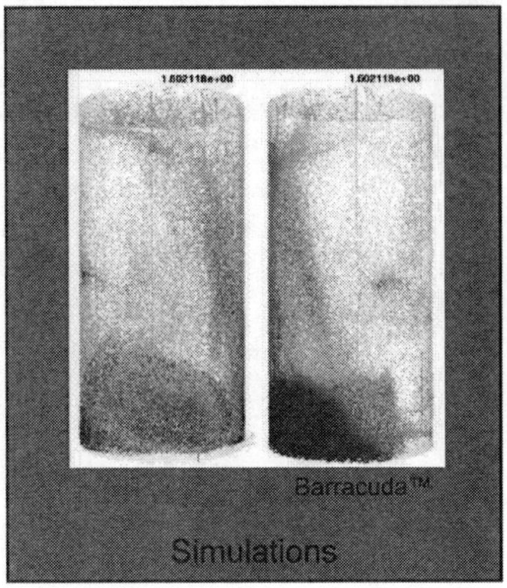

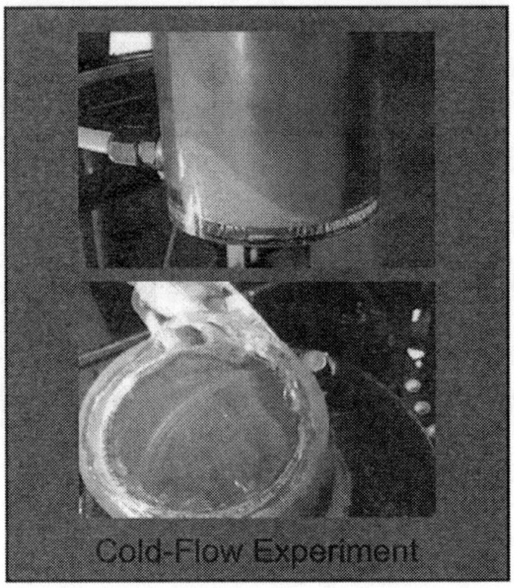

- More than 50% of the bed is stagnant

R. Cocco, Y. Arrington, R. Hays, J. Findlay, S.B. Reddy Karri, T. Knowlton, Jet cup attrition testing, Powder Technology, 200 (2010) 224–23

23

PSRI Conical Jet Cup

- 3 in (7.6 cm) ID on top and 1.5 in (3.8 cm) ID on bottom

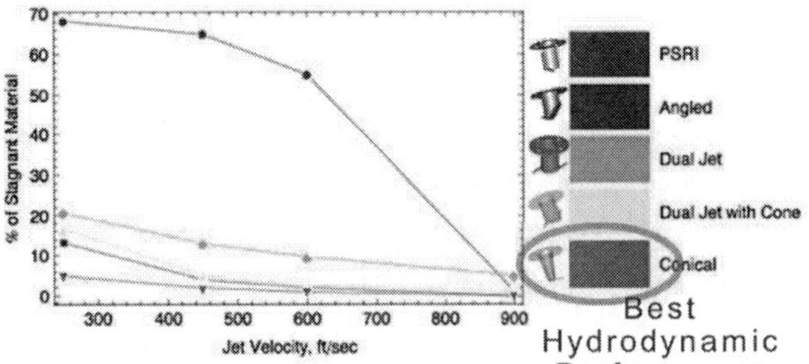

Best Hydrodynamic Performance

250 ft/sec 76.2 m/sec

450 ft/sec 137 m/sec

600 ft/sec 183 m/sec

24

Comparison to Attrition in a Fluidized Bed

PSRI Fluidized Bed Attrition Unit

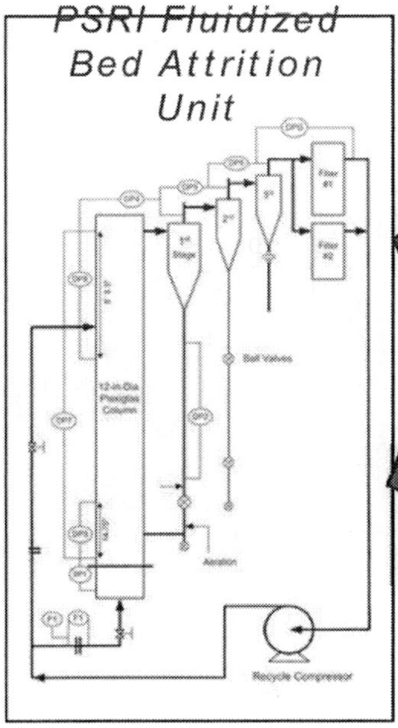

Ranking of Catalyst

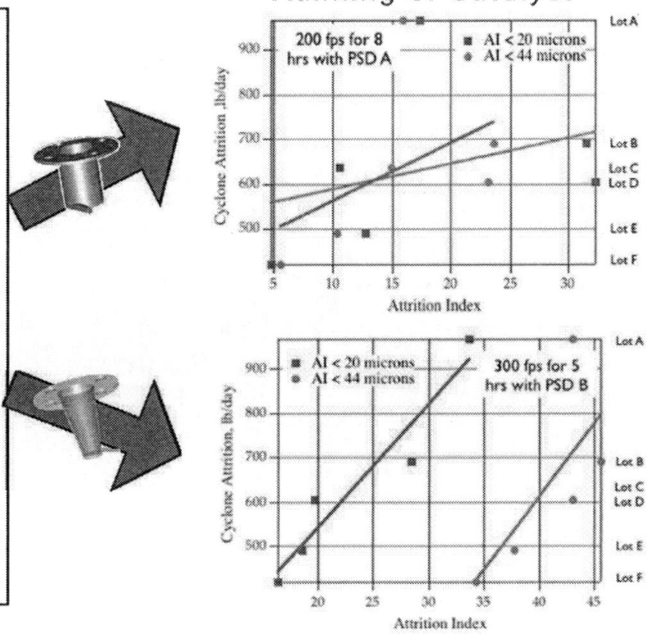

25

Comparison of Different Methods

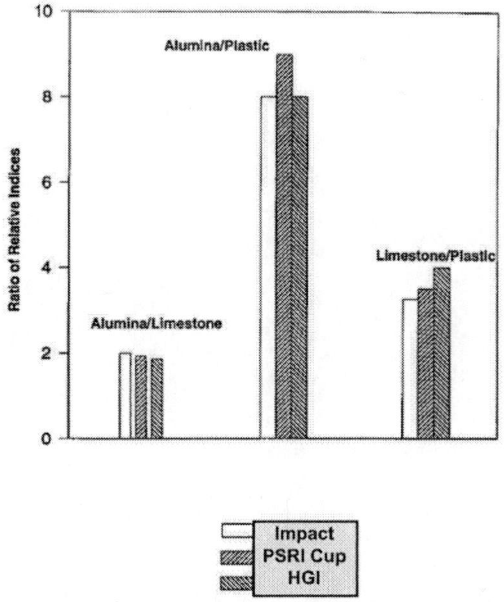

- The ranking of relative attrition indices does not change with the type of method used as long as they are properly designed

26

Summary

- Mitigating particle attrition involves understanding particle exposure to
  - Jets
  - Cyclones
  - Transfer lines
  - Distributor plates
- Attrition testing is helpful but make sure it is a valid test

27

IFP Energies Nouvelles

- **French research center active in the fields of energy, transport and environment: 1 700 employees on two sites (Paris and Lyon)**

1

Frontiers in particle science and technology – April 13th, 2016

IFP Energies Nouvelles

Validate the industrial feasability
Development and scale-up of processes and technologies
www.ifpenergiesnouvelles.com

2

Frontiers in particle science and technology – April 13th, 2016

Renewable energies | Eco-friendly production | Innovative transport | Eco-efficient processes | Sustainable resources

Attrition testing using a Jet Cup rig for the chemical looping process development

Frontiers in Particle Science and Technology, Houston, 13th of April 2016

B.Amblard, C.Bobin, S.Bertholin, Th. Gauthier

TOTAL

IFP Energies nouvelles

Outline

- Scientific context and objectives
- Jet cup attrition testing method development
- Use of the attrition testing method for the European Project « Success »
- Conclusions and perspectives

SUCCESS

Scientific context and objectives

- Chemical Looping Combustion (CLC)
  - Oxy-combustion like technology
    - A metal oxide is used for combustion
    - Intrinsic CO_2 separation
    - Low energy penalty

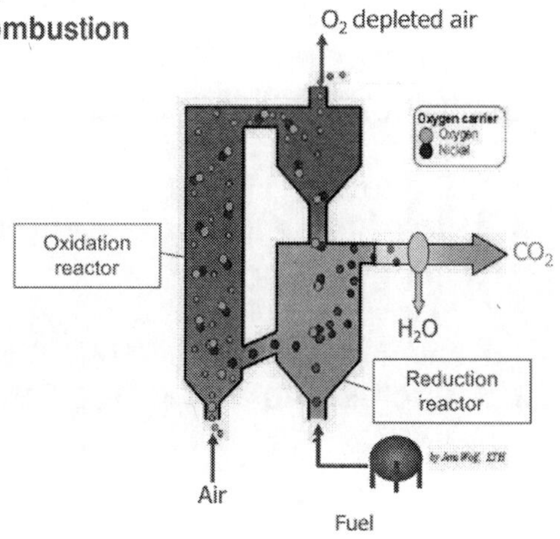

Scientific context and objectives

- Chemical Looping Combustion (CLC)
  - Oxy-combustion like technology
    - A metal oxide is used for combustion
    - Intrinsic CO_2 separation
    - Low energy penalty

- Dense group B particles are used as an oxygen carrier
- Little information is available concerning the attrition of Group B oxygen carriers. However attrition will have an impact on:
  - Economics of the process due to the inventory losses
  - Circulation problems due to the generation of fines
- Objective: evaluate the mechanical resistance to attrition of the oxygen carriers available at the early stage of the process development

Outline

- **Scientific context and objectives**
- **Attrition testing method**
- **Use of the attrition testing method for the European Project « Success »**
- **Conclusions and perspectives**

Jet cup attrition testing method

- **Constraint: Small amount of solid available for the attrition test (50 to 200 grammes)**
- **Objective for the method**
  - **Comparison of material with different properties**

	Mean Diameter from Particle Size Distribution D50 (m)	Particle Bulk Density (kg/m3)
Oxygen Carrier n°1	148	1 590
Oxygen Carrier n°2	169	1 840
Oxygen Carrier n°3	175	980
Oxygen Carrier n°4	136	1 410
FCC particles	*80*	*850*

Jet cup attrition testing method

- **Constraint: Small amount of solid available for the attrition test (50 to 200 grammes)**
- **Objective for the method**
  - **Comparison of material with different properties**
    - **Why comparing the oxgygen carriers with FCC particles ?**

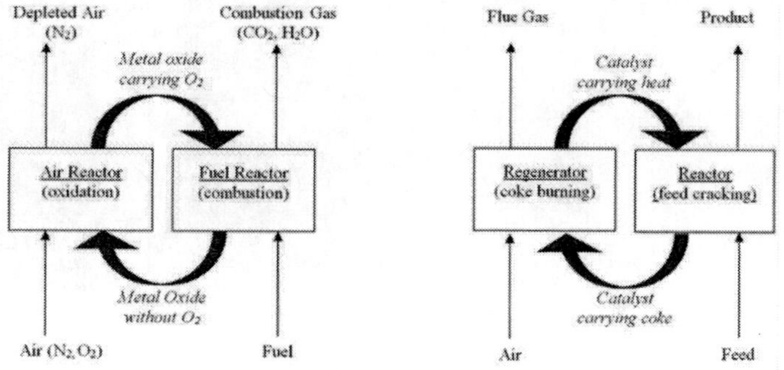

Chemical Looping Combustion (CLC)

Fluid Catalytic Cracking (FCC)

Jet cup attrition testing method

- **Constraint: Small amount of solid available for the attrition test (50 to 200 grams)**
- **Objective for the method**
  - **Comparison of material with different properties**
    - **Why comparing the oxgygen carriers with FCC particles ?**
      - In both processes, attrition is proportional to particles circulation rate
      - For FCC particles, attrition data are available both at attrition testing and industrual scales
      - FCC particles can be a good reference to compare the different oxygen carriers
- **Attrition testing procedure developed using a jet cup rig**

Apparatus

- **Testing Rig: Jet cup from PSRI**

R. Cocco, Y. Arrington, R. Hays, J. Findlay, S.B.R. Karri, T.M. Knowlton, Jet cup attrition testing, Powder Technology 200 (2010)

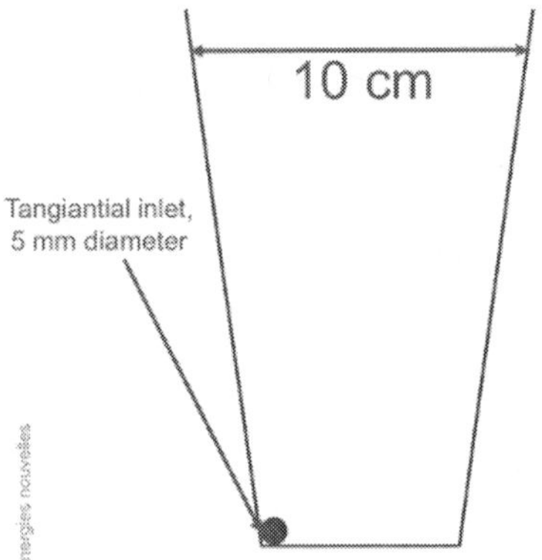

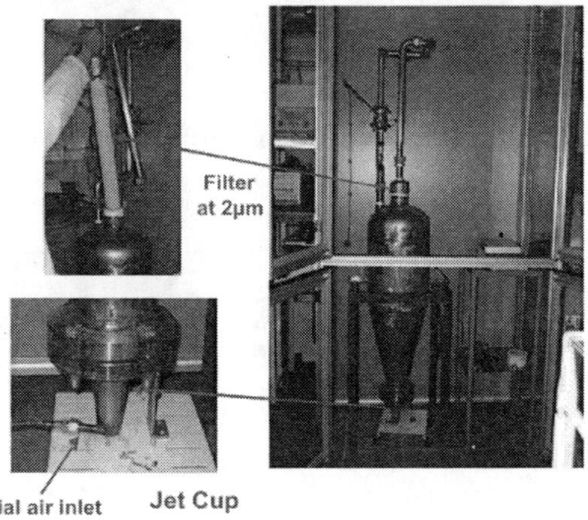

Standard procedure with FCC: Sample of 100g, impact of the test duration

New attrition index

- **Testing Rig: Jet cup from PSRI**
  - **Standard procedure for FCC particles**
    - **100 grams of material**
    - **Attrition indexes (AI 20) and (AI40)**

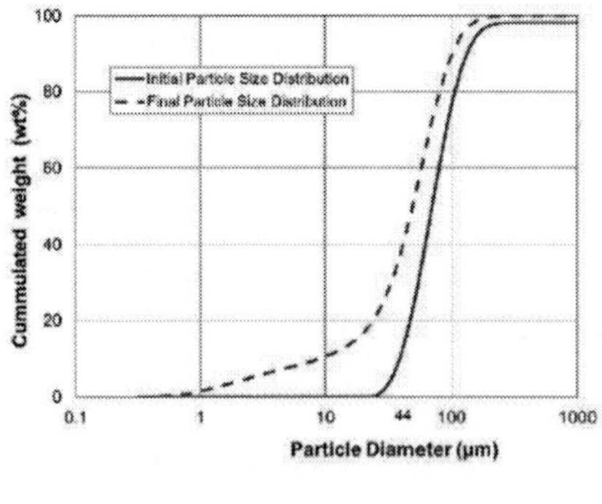

Testing Procedure | Apparatus | New attrition index | Air Jet Velocity | Comparison of solids with different properties | Characterization of the solid resistance

New attrition index

- **Testing Rig: Jet cup from PSRI**
  - **Standard procedure for FCC particles**
    - **100 grams of material**
    - **Attrition indexes (AI 20) and (AI40)**

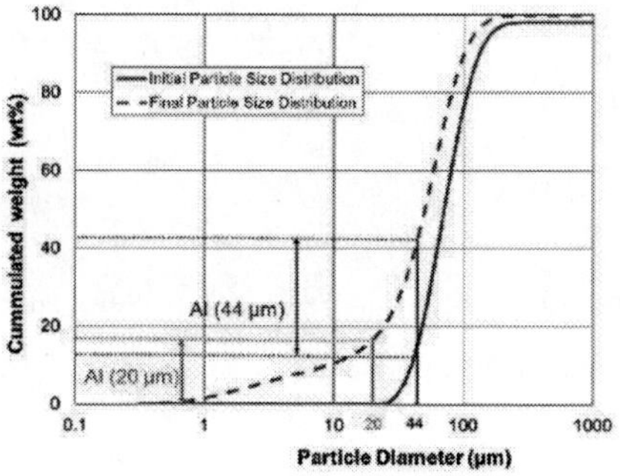

Testing Procedure | Apparatus | New attrition index | Air Jet Velocity | Comparison of solids with different properties | Characterization of the solid resistance

New attrition index

- **When considering large particles such as group B powders, generation of fine particles below 20 or 40 microns is not enough to describe particle attrition.**

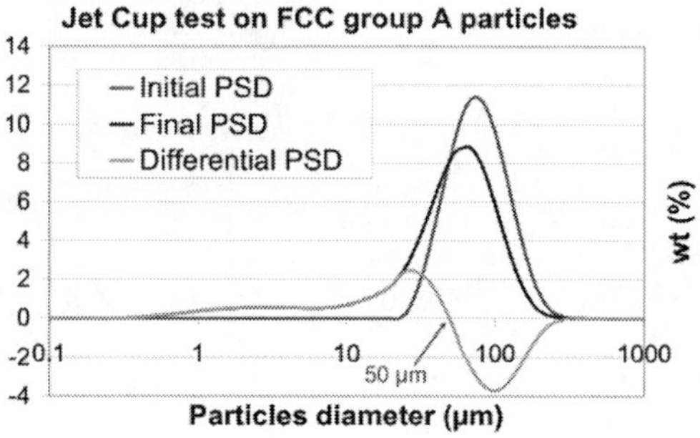

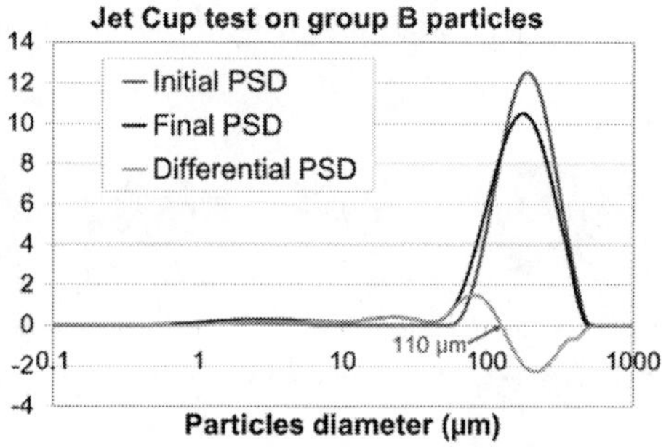

New attrition index

- **When considering large particles such as group B powders, generation of fine particles below 20 or 40 microns is not enough to describe particle attrition.**

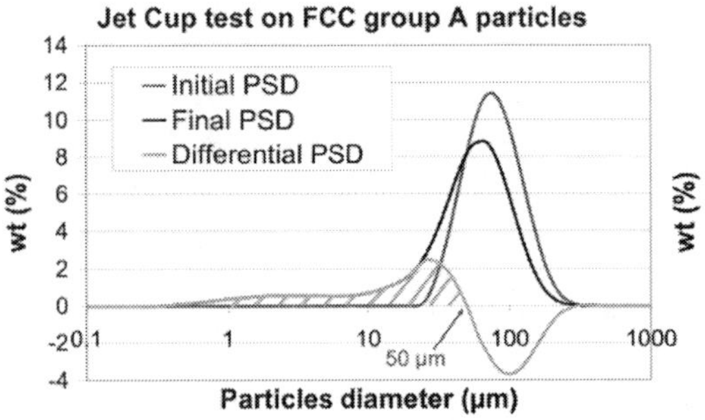

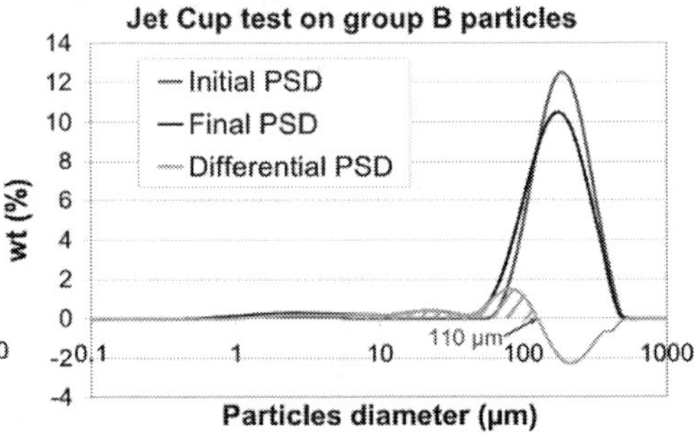

- **New attrition index: Total Particles Generated Index (TPGI)**

	AI(20) (wt%)	AI(44) (wt%)	TPGI (wt%)
Group A particles	15	26	27
Group B particles	3	5	15

Choice of Air Jet velocity (PSRI method)

- **Jet velocity chosen: 90 m/s**
  - **Close to the limit between where fragmentation phenomena are dominant and where abrasion phenomena are dominant.**

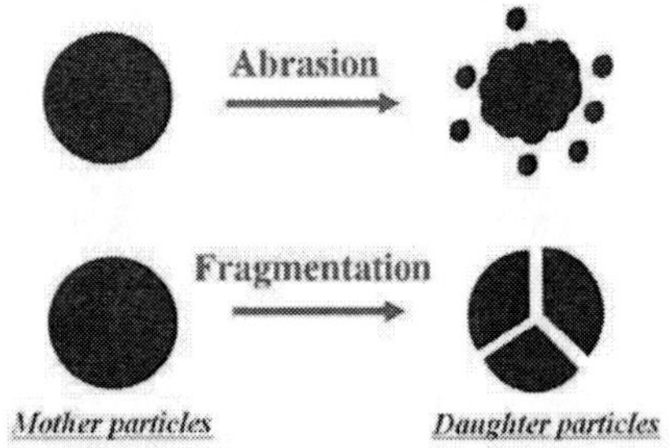

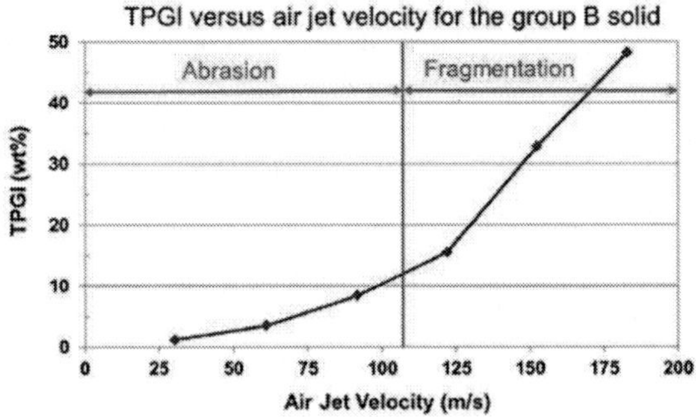

Comparison of solids with different properties

- **How to compare two solids with different size and density ?**
  - **Comparison at iso-volume, iso-mass ?**

	Mean Diameter from Particle Size Distribution D50 (m)	Particle Bulk Density (kg/m3)
Oxygen Carrier n°1	148	1 590
Oxygen Carrier n°2	169	1 840
Oxygen Carrier n°3	175	980
Oxygen Carrier n°4	136	1 410
FCC particles	*80*	*850*

Testing Procedure | Apparatus | New attrition index | Air Jet Velocity | Comparison of solids with different properties | Characterization of the solid resistance

Comparison of solids with different properties

- **How to compare two solids with different size and density ?**
  - **Comparison at iso-volume, iso-mass ?**
  - **How to make sure that particles go through the same mechanical stress?**

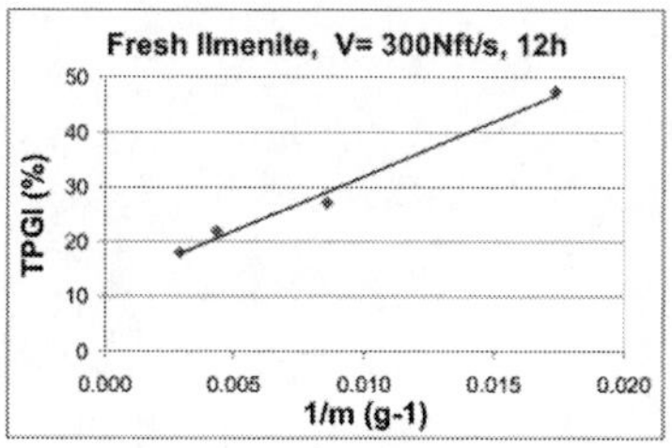

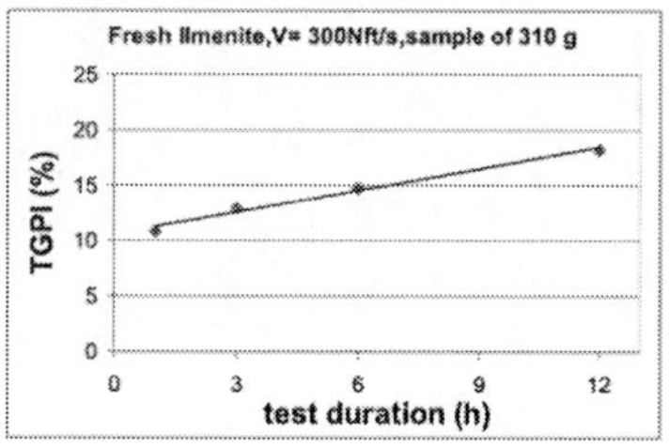

- **Attrition in the Jet Cup is proportional to number of times particles go into the air jet.**
  - **Evaluation of the jet cup internal circulation rate of the different types of particles with CFD**

Comparison of solids with different properties

- **CFD study: Use of the CPFD Barracuda software, Euler-lagrangian approach (MP-PIC method).**
  - **Drag force: Wen-Yu and Ergun models combined**

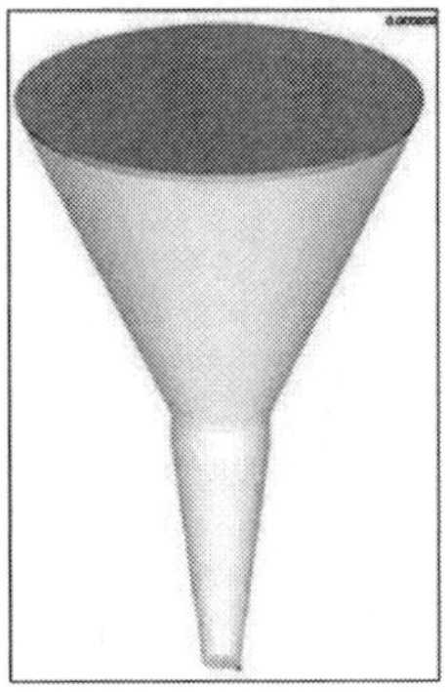

Simulated geometry

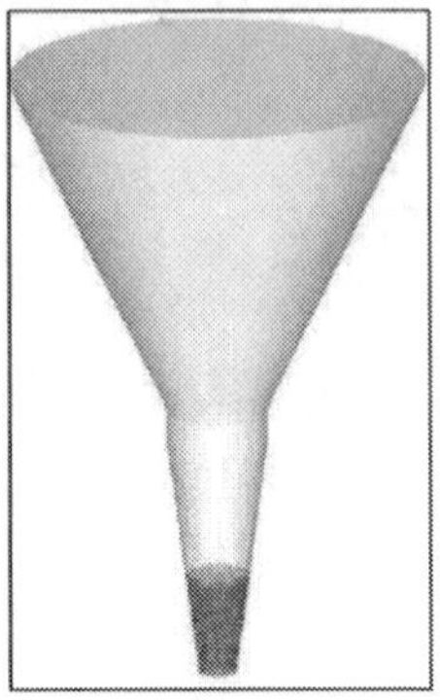

Initial conditions

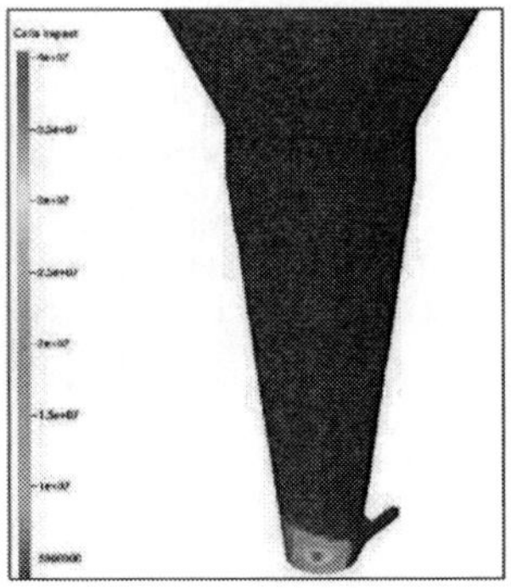

Particles on wall impact number

Comparison of of solids with different properties

- **CFD study: Circulation rate evaluation**

$$CR = \left(\frac{\text{Cumulated solid mass going down through the plane}}{\text{Solid sample mass}} \right) \times \frac{1}{\text{Simulation\_time}}$$

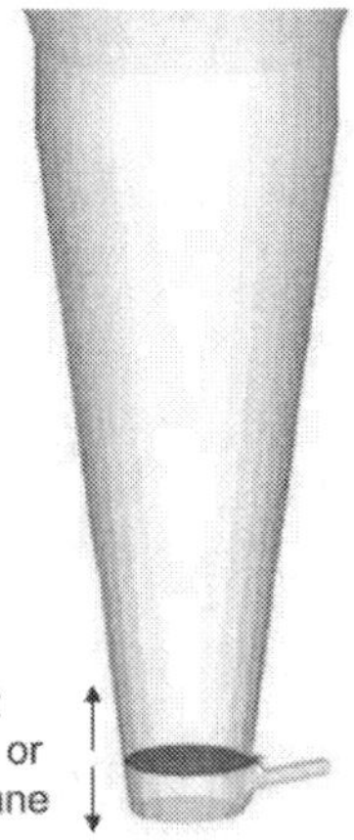

Comparison of of solids with different properties

- **CFD study: Circulation rate evaluation**

Comparison of solids with different properties

- **CFD study: Circulation rate evaluation**

$$CR = \left(\frac{\text{Cumulated solid mass going down through the plane}}{\text{Solid sample mass}}\right) \times \frac{1}{\text{Simulation\_time}}$$

N° Simulation	Solid	Density (kg/m^3)	Mean diameter D_{50} (µm)	Initial Volume in jet cup (cm^3)	CR (s^{-1})	CR_1 tracer 1 (s^{-1})	CR_2 tracer 2 (s^{-1})
1	FCC 100 g	1 450	70	V	**0.26**	**0.25**	**0.28**
2	Ilmenite 100 g	4 500	130	V/3	**0.68**	**0.65**	**0.69**
3	Ilmenite 310 g	4 500	130	V	**0.24**	**0.25**	**0.27**

- **Experimental tests are carried out at constant initial solid volume**

Testing Procedure	Apparatus	New attrition index	Air Jet Velocity	Comparison of solids with different properties	Characterization of the solid resistance

Characterisation of the solid mechanical resistance

- **Jet cup tests carried out for several testing durations**

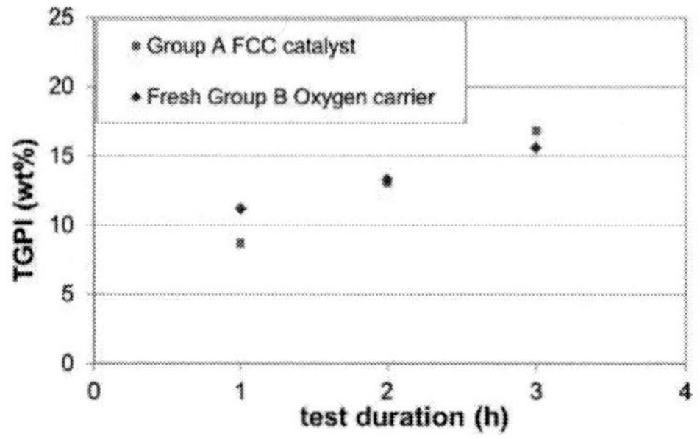

Testing Procedure	Apparatus	New attrition index	Air Jet Velocity	Comparison of solids with different properties	Characterization of the solid resistance

Characterisation of the solid mechanical resistance

- **Jet cup tests carried out for several testing durations**

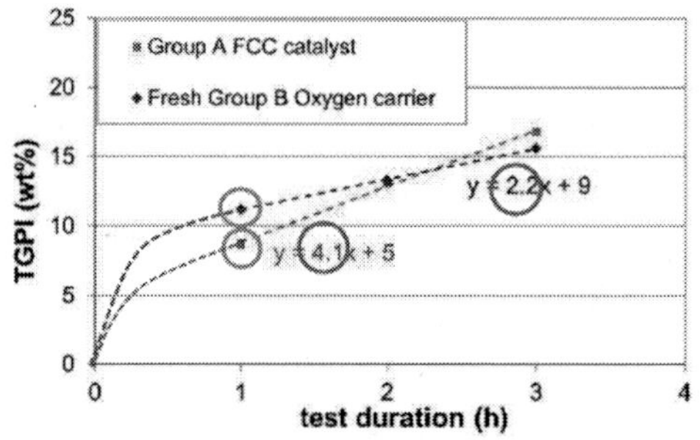

- **Attrition increase is linear over the tests durations**
- **Resistance to attrition is characterized by:**
  - **Attrition increase rate versus test duration**
  - **Attrition value for the first test duration**
- **Tests for several durations carried out on the same solid sample**

Outline

- Scientific context and objectives
- Jet cup attrition testing method development
- Use of the attrition testing method for the European Project « Success »
- Conclusions and perspectives

SUCESS Project

- European funded project
- Objectives
  - Scale-up production of synthetic oxygen carriers for the CLC process
  - Evaluation of the oxygen carriers in pilot plants
  - Economical evaluation of the process
- 16 partners
  - Universities: TUV (Vienna Austria), Chalmers (Swenden), INPT-IMFT (France), ...
  - Research Centers: IFPEN, SINTEF, SINTEF ER, VITO
  - Companies: EDF, Total, SGS, Johnson Matthey, Bertsch, EURO SUPPORT
  - 9 countries represented
- Budget: 10 millions of euros
- One of IFPEN objectives: Evaluate the resistance to attrition of the two oxygen carriers developped by the partners (Vito and Johnson Matthey)

SUCESS Project

SUCCESS

SEVENTH FRAMEWORK PROGRAMME

- **Material tested**
  - **Two synthetic fresh oxygen carriers**

Origin	Name used in this presentation	Mean Diameter D50 (m)	Particle Bulk Density (kg/m3)
Fresh Perovskite supplied by Vito	Fresh OC n°1 (Perovskite)	148	1 590
Fresh CuO/Al2O3 supplied by Johnson Matthey	Fresh OC n°2 (CuO/Al2O3)	155	980

  - **Two « used » synthetic materials supplied by university of Chalmers**

Origin	Name used in this presentation	Mean Diameter D50 (m)	Particle Bulk Density (kg/m3)
Perovskite tested for 100 hours in pilot (composition similar to the fresh Perovskite)	Used OC n°1 (Perovskite)	169	1840
Fe/Al2O3 tested for 30 hours in pilot (bad attrition performances)	Used OC n°2 (Fe/Al2O3)	136	1410

- **Testing program**
  - **2 test durations for each material: 2 and 3 hours**
  - **Each condition for each solid repeated four times**

SUCESS Project

SUCCESS

SEVENTH FRAMEWORK PROGRAMME

- **Results**

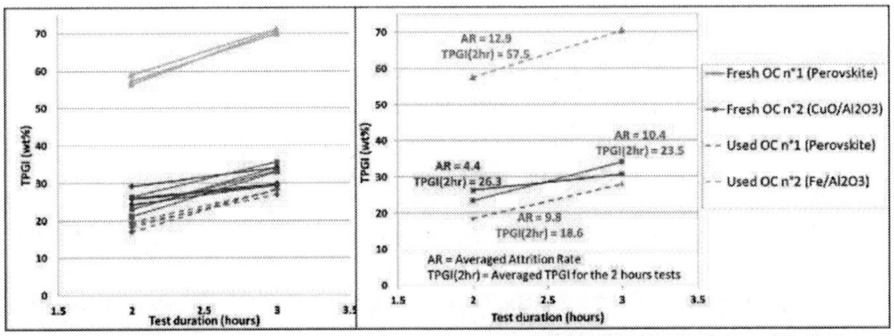

	Fresh OC n°1 (Perovskite)	Fresh OC n°2 (CuO/Al2O3)	Used OC n°1 (Perovskite)	Used OC n°2 (Fe/Al2O3)
Averaged Attrition rate (value / error +/-)	10.4 / 1.1	4.4 / 0.8	9.8 / 1.6	12.9 / 0.9
Averaged TPGI (2hrs) (value / error +/-)	23.5 / 2.2	26.3 / 2.1	18.6 / 1.1	57.5 / 1.1

SUCESS Project

- **Results**

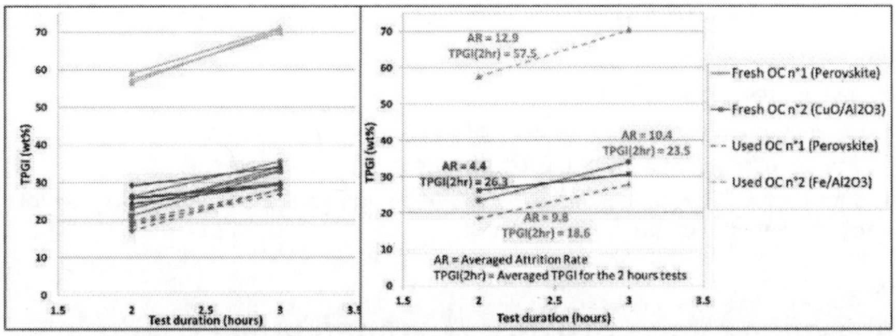

  - **Bad attrition performances for the used OC n°2. Result expected because of the bad performances in the Chamers university pilot.**
  - **Fresh OC n°2 has a better attrition resistance compared to the Fresh OC n°1**
  - **Perovskite material seems resistant to the oxydo-reduction cycles imposed during the 100 hours pilot plant testing (to be confirmed after longer tests on the IFPEN 10kW pilot)**

Outline

- **Scientific context and objectives**
- **Jet cup attrition testing method development**
- **Use of the attrition testing method for the European Project « Success »**
- **Conclusions and perspectives**

Conclusion

- **Development of attrition testing method to compare materials with different properties**
  - New attrition index (TPGI)
  - test with the same initial volume of particles from CFD results
  - attritrion characterized with the TPGI increase rate versus time.
  - Limitations: CFD simulations to be validated, attrition rate highly dependant on the PSD measurement method

- **Application for the SUCESS Europen project**
  - Fresh OC n°2 has a better attrition resistance compared to the Fresh OC n°1
  - Perovskite material seems resistant to the oxydo-reduction cycles imposed during the 100 hours pilot plant testing (to be confirmed after longer tests on the IFPEN 10kW pilot)
  - Attrition testing will be carried out again after passing the two fresh materials in the IFPEN 10 kW plant

Perspectives

- **Method developed allows to compare the materials but do not allow us to predict attrition at industrial scale**

Perspectives

- **Method developed allows to compare the materials but do not allow us to predict attrition at industrial scale**
- **Connect attrition in Jet Cup with the main attrition sources in the process**

Gas injectors

Cyclones

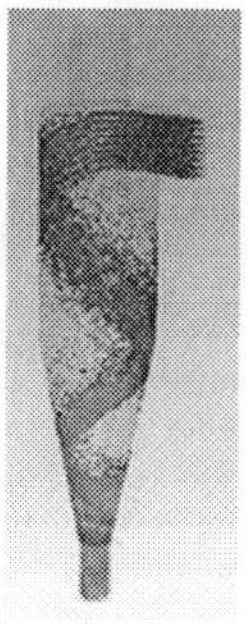

Perspectives

- **Method developed allows to compare the materials but do not allow us to predict attrition at industrial scale**
- **Connect attrition in Jet Cup with the main attrition sources in the process**
  - **Use of CFD/DEM to compare mechanical stress in attrition tests with the main process sources of attrition**
- **Use of a population balance model to predict attrition at industrial scale using results from attrition testing at lab scale**
  - **Population Balance: Joachim Werther, Ernst-Ulrich Hartge, A population balance model of the particle inventory in a fluidized-bed reactor/regenerator system, Powder Technology 148 (2004) 113– 122.**
- **PHD thesis on attrition in partnership with the university of Leeds (Professor Ghadiri, PHD student Fabio Fulchini)**

OPTICAL PARTICLE CHARACTERIZATION

DR. PAUL FREUD
DR. PHIL PLANTZ
MICHAEL TRAINER

MICROTRAC INC.

A MEMBER OF NIKKISO

Microtrac
Total Solutions in Particle Characterization

OPTICAL PARTICLE CHARACTERIZATION

- DYNAMIC LIGHT SCATTERING
  1 Nanometer to 5 Microns
- LASER DIFFRACTION - STATIC SCATTERING
  10 Nanometers to 3 Millimeters
- DYNAMIC IMAGE ANALYSIS
  5 Microns to Centimeters

A MEMBER OF NIKKISO

Microtrac
Total Solutions in Particle Characterization

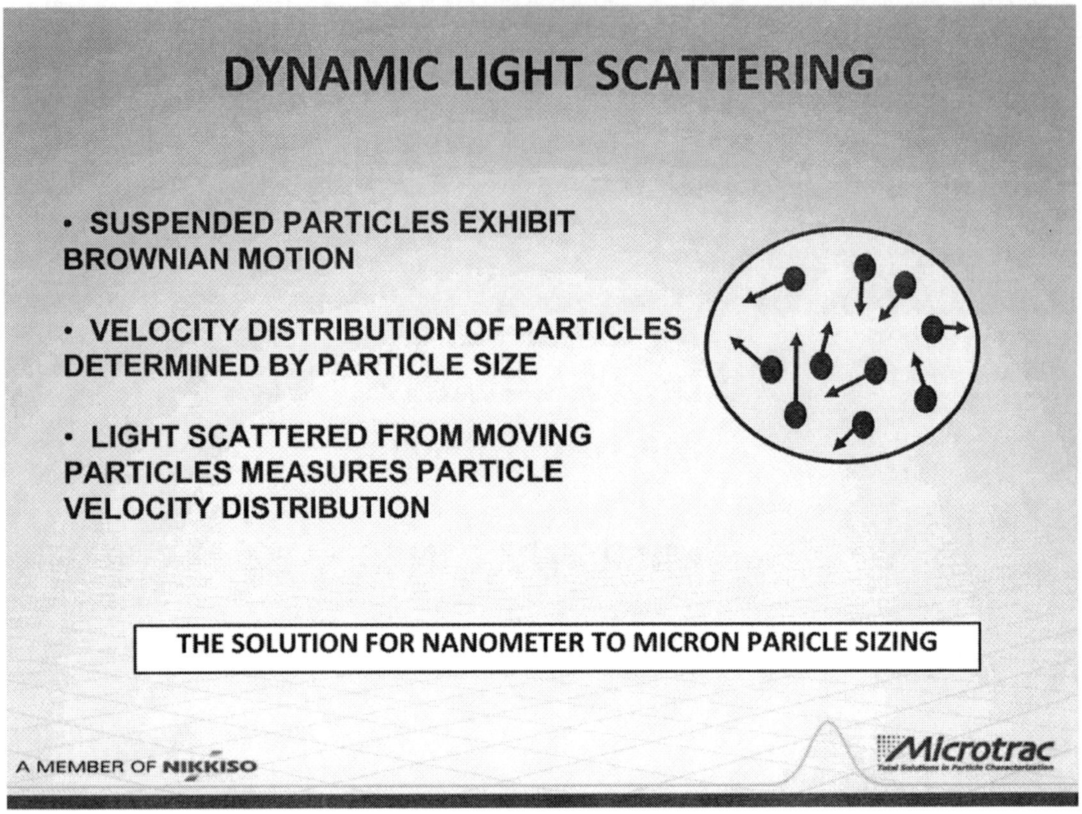
DYNAMIC LIGHT SCATTERING
• SUSPENDED PARTICLES EXHIBIT BROWNIAN MOTION
• VELOCITY DISTRIBUTION OF PARTICLES DETERMINED BY PARTICLE SIZE
• LIGHT SCATTERED FROM MOVING PARTICLES MEASURES PARTICLE VELOCITY DISTRIBUTION
THE SOLUTION FOR NANOMETER TO MICRON PARICLE SIZING
A MEMBER OF NIKKISO
Microtrac

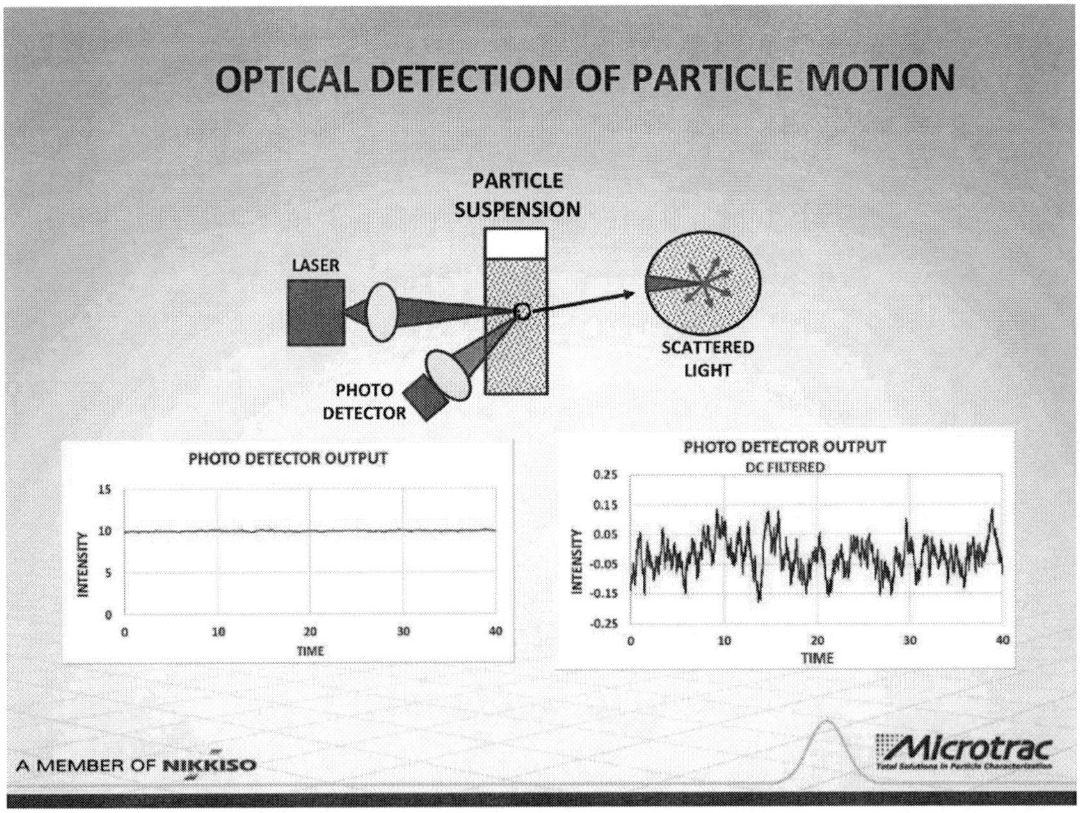
OPTICAL DETECTION OF PARTICLE MOTION
PARTICLE SUSPENSION
LASER
SCATTERED LIGHT
PHOTO DETECTOR
PHOTO DETECTOR OUTPUT
INTENSITY
TIME
PHOTO DETECTOR OUTPUT
DC FILTERED
INTENSITY
TIME
A MEMBER OF NIKKISO
Microtrac

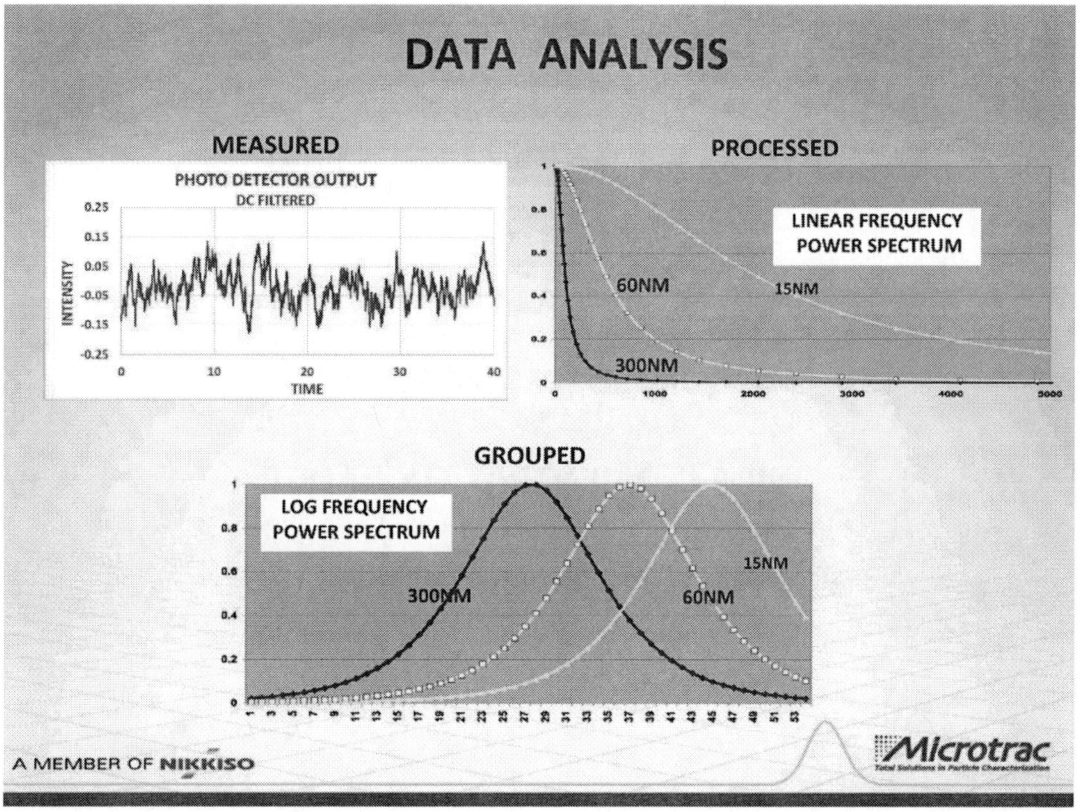
DATA ANALYSIS
MEASURED
PHOTO DETECTOR OUTPUT
DC FILTERED
INTENSITY
TIME
PROCESSED
LINEAR FREQUENCY
POWER SPECTRUM
60NM
15NM
300NM
GROUPED
LOG FREQUENCY
POWER SPECTRUM
300NM
60NM
15NM
A MEMBER OF NIKKISO
Microtrac

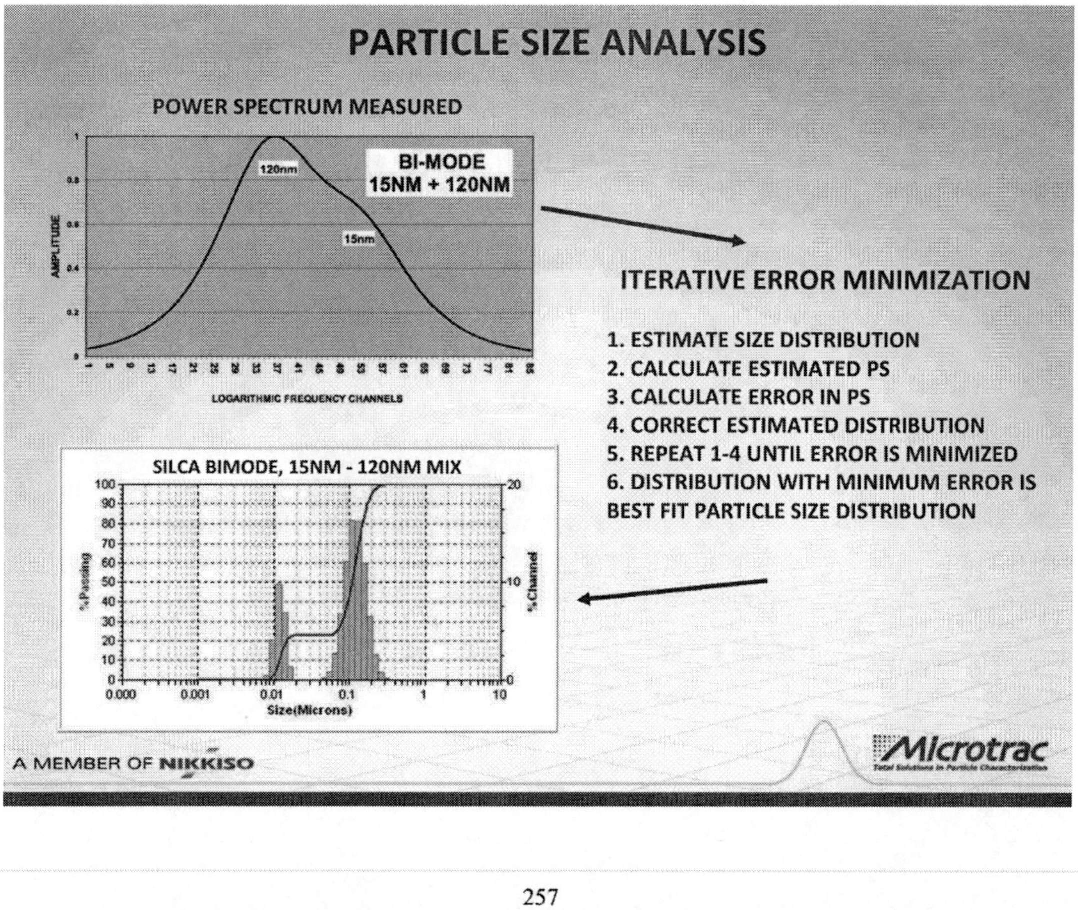
PARTICLE SIZE ANALYSIS
POWER SPECTRUM MEASURED
BI-MODE
15NM + 120NM
120nm
15nm
AMPLITUDE
LOGARITHMIC FREQUENCY CHANNELS
ITERATIVE ERROR MINIMIZATION
1. ESTIMATE SIZE DISTRIBUTION
2. CALCULATE ESTIMATED PS
3. CALCULATE ERROR IN PS
4. CORRECT ESTIMATED DISTRIBUTION
5. REPEAT 1-4 UNTIL ERROR IS MINIMIZED
6. DISTRIBUTION WITH MINIMUM ERROR IS
BEST FIT PARTICLE SIZE DISTRIBUTION
SILCA BIMODE, 15NM - 120NM MIX
%Passing
%Channel
Size(Microns)
A MEMBER OF NIKKISO
Microtrac

DYNAMIC LIGHT SCATTERING ATTRIBUTES AND LIMITATIONS

SIZE RANGE .8nm TO 6.5⌠m

LIQUID SUSPENSION ONLY

WIDE CONCENTRATION RANGE

VOLUME CONCENTRATION OUTPUT

REQUIRES ZERO SAMPLE FLOW

A MEMBER OF NIKKISO

Microtrac

LASER DIFFRACTION PARTICLE CHARACTERIZATION

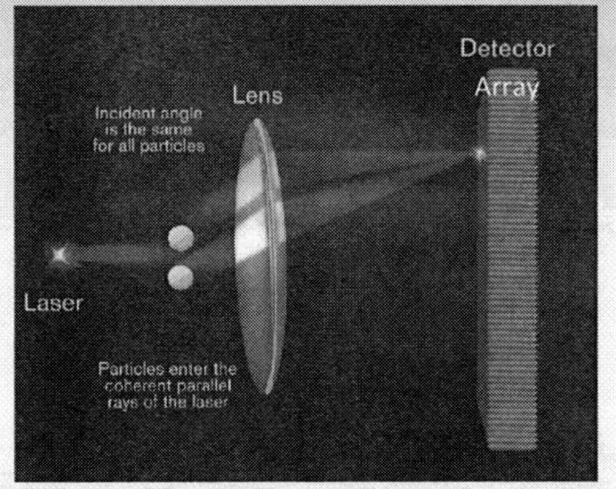

SINGLE LASER - SINGLE DETECTOR ARRAY
1MICRON TO 1000 MICRON

A MEMBER OF NIKKISO

Microtrac

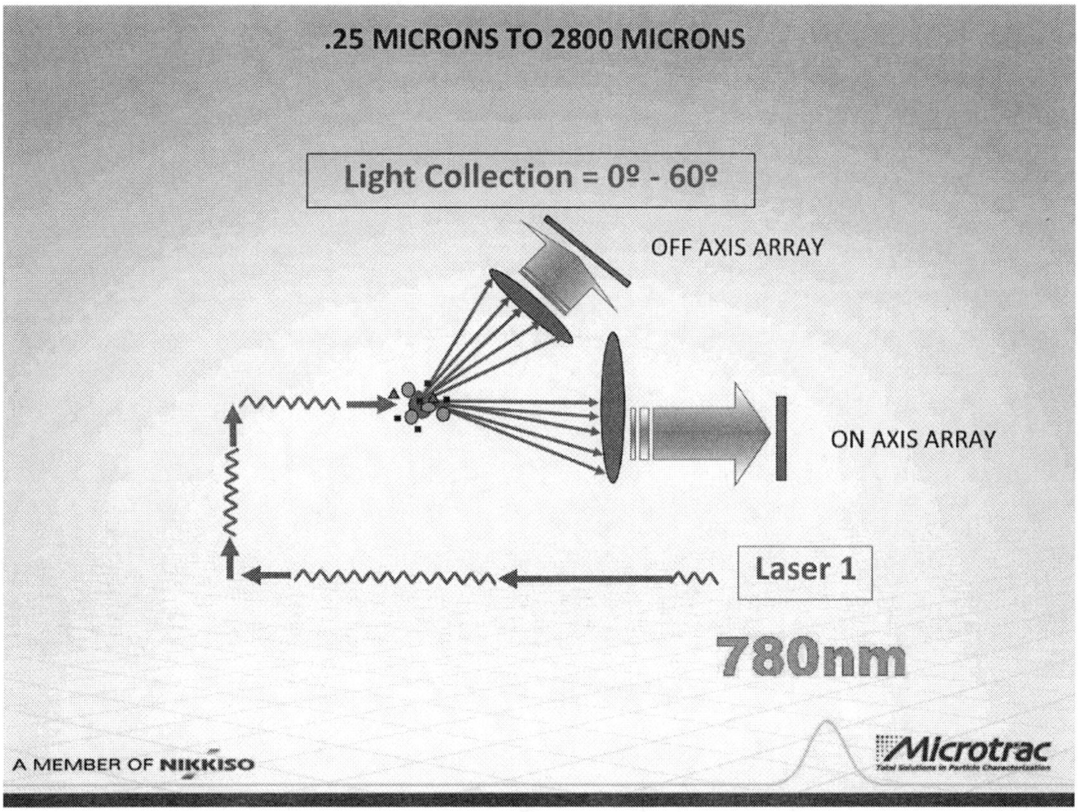
.25 MICRONS TO 2800 MICRONS
Light Collection = 0º - 60º
OFF AXIS ARRAY
ON AXIS ARRAY
Laser 1
780nm
A MEMBER OF NIKKISO
Microtrac

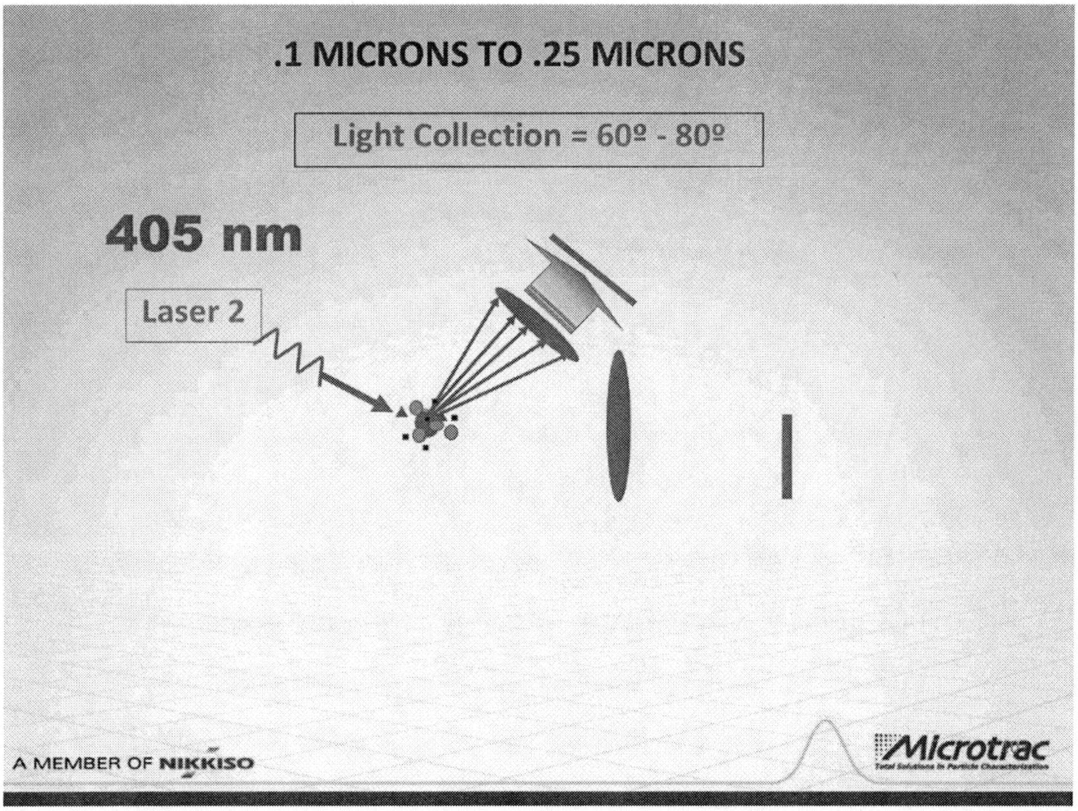
.1 MICRONS TO .25 MICRONS
Light Collection = 60º - 80º
405 nm
Laser 2
A MEMBER OF NIKKISO
Microtrac

.01 MICRONS TO .1 MICRONS

Light Collection = 80º - 165º

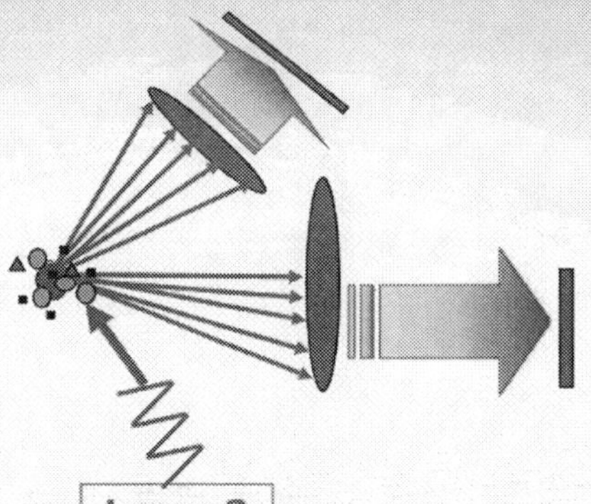

Laser 3

405 nm

A MEMBER OF NIKKISO

Microtrac

LARGE PARTICLES SCATTER TO SMALL ANGLES

SMALL PARTICLES SCATTER TO LARGE ANGLES

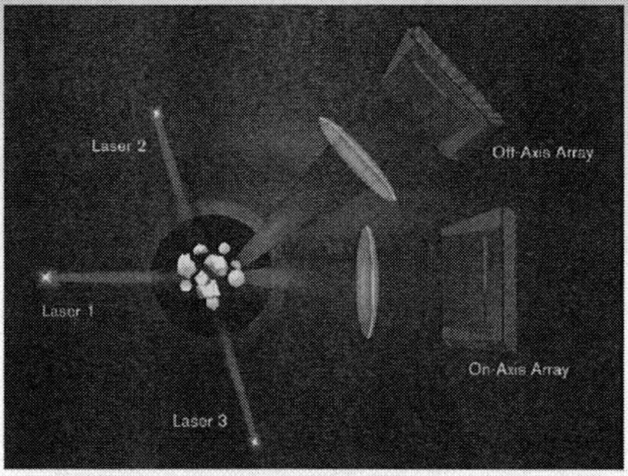

3 LASERS — 2 DETECTOR ARRAYS — 2 WAVELENGTHS

0 TO 165 DEG

10NM TO 2800

A MEMBER OF NIKKISO

Microtrac

SCATTERING TO SIZE ANALYSIS

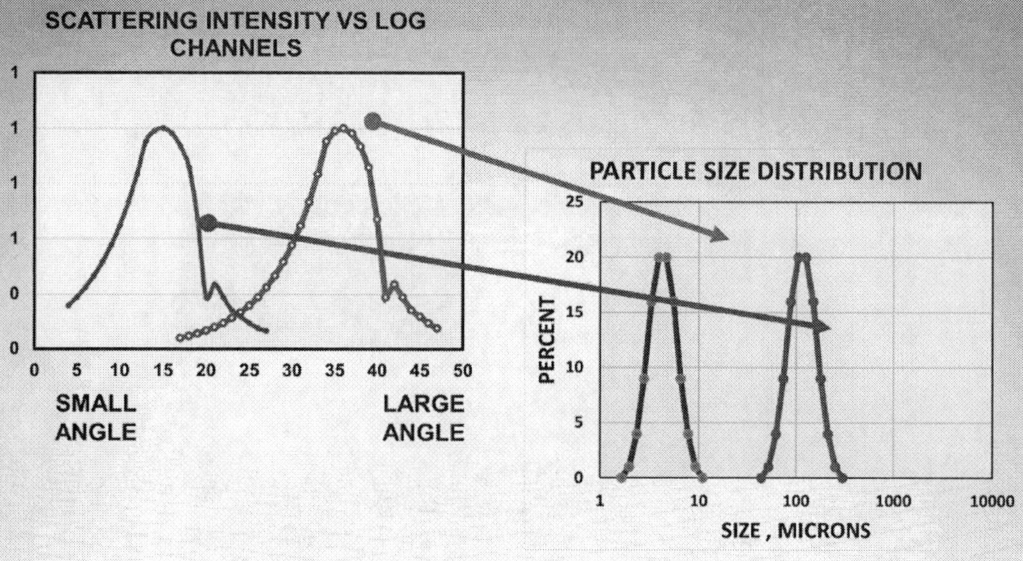

A MEMBER OF NIKKISO

Microtrac
Total Solutions in Particle Characterization

LASER DIFFRACTION
ATTRIBUTES AND LIMITATIONS

10 nm to 2800 ⌠ m SIZE RANGE

WET OR DRY FLOWING SAMPLE DELIVERY
LIQUID SUSPENSION or DRY POWDER

NARROW SAMPLE CONCENTRATION RANGE

VOLUME CONCENTRATION OUTPUT

A MEMBER OF NIKKISO

Microtrac
Total Solutions in Particle Characterization

DYNAMIC IMAGING
PARTICLE CHARACTERIZATION
5 MICRONS TO 5 CENTIMETERS

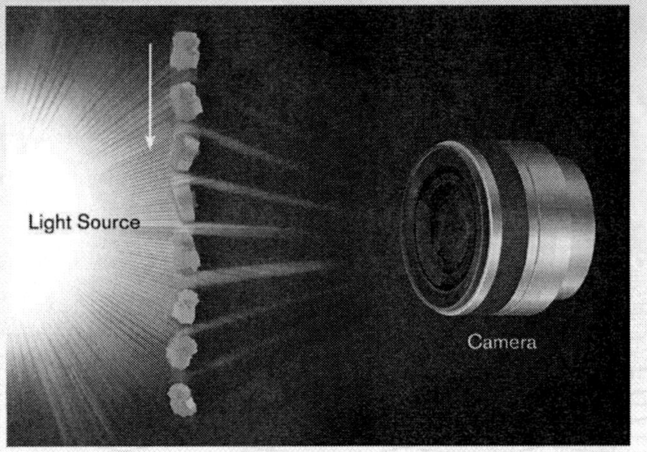

A MEMBER OF NIKKISO

Microtrac

Measurement Principle

Particle detection: The software counts the pixels covered by the particle and uses the size of the pixels to calculate the morphological parameters.

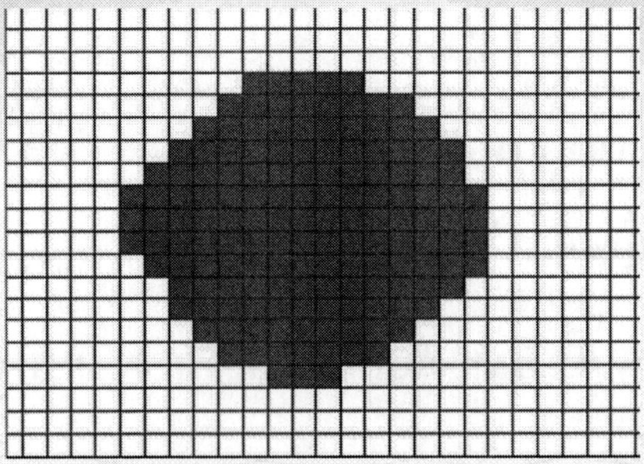

A MEMBER OF NIKKISO

Microtrac

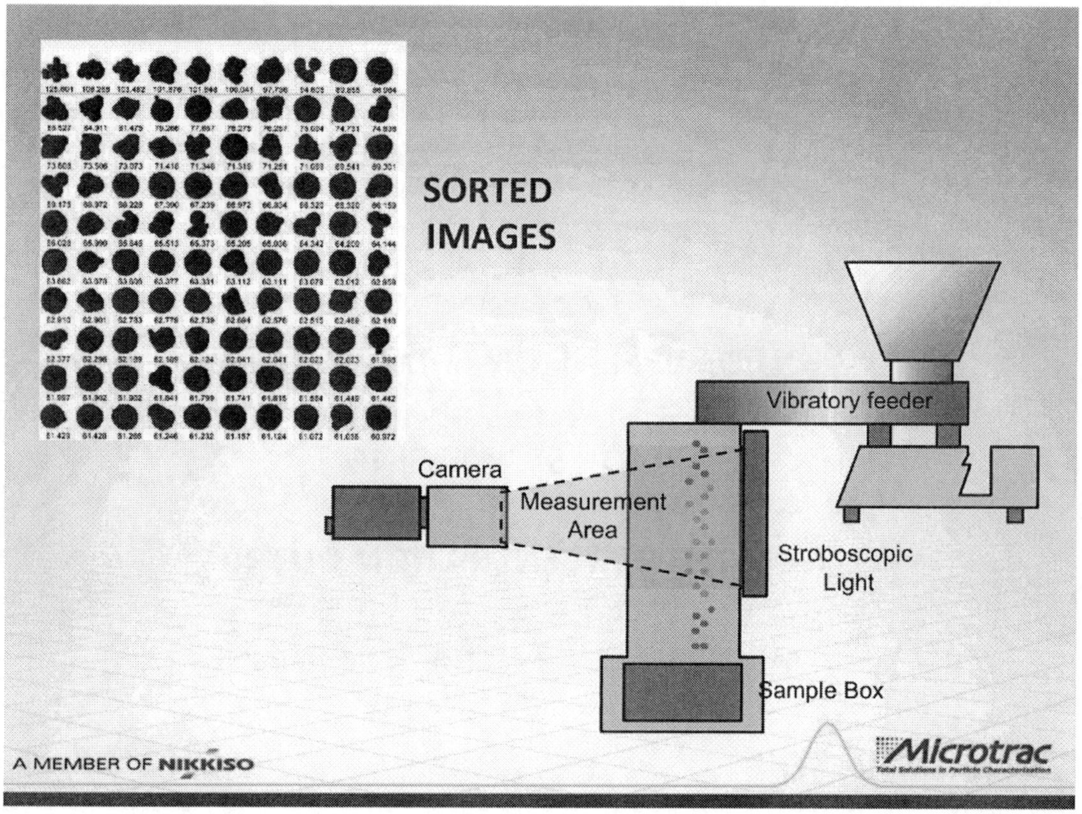
SORTED
IMAGES
Vibratory feeder
Camera
Measurement
Area
Stroboscopic
Light
Sample Box
A MEMBER OF NIKKISO
Microtrac

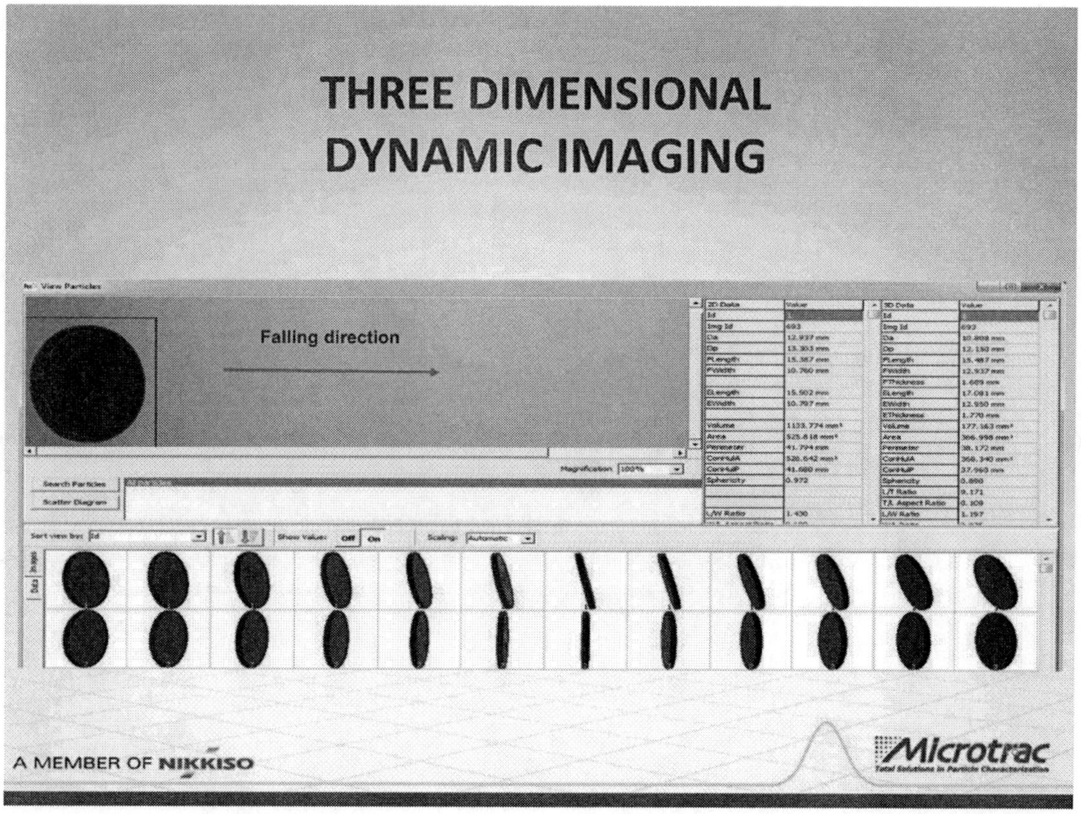
THREE DIMENSIONAL
DYNAMIC IMAGING
Falling direction
A MEMBER OF NIKKISO
Microtrac

DYNAMIC IMAGE ANALYSIS
ATTRIBUTES AND LIMITATIONS

5 ⌠m to 5 cm SIZE RANGE

LIQUID SUSPENSION or DRY POWDER

2D OR 3D IMAGING

NUMBER CONCENTRATION OUTPUT

NARROW CONCENTRATION RANGE

COMBINATION
LASER DIFFRACTION +DYNAMIC IMAGING

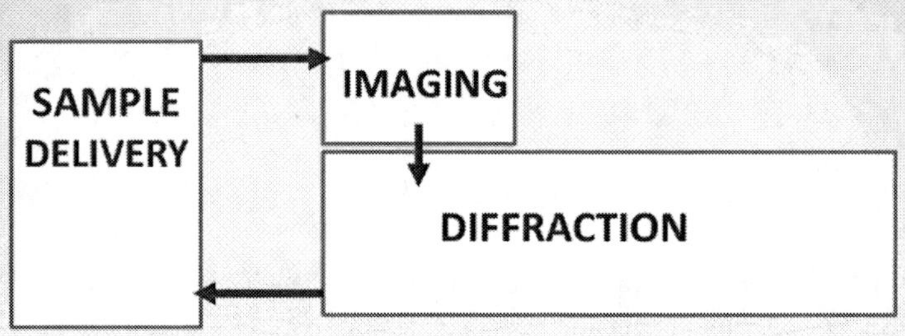

LASER DIFFRACTION
+
DYNAMIC IMAGING
ATTRIBUTES AND LIMITATIONS

ADD SHAPE TO DIFFRACTION DISTRIBUTION

COMBINE VOLUME CONC. OF SMALL NUMBER CONC. OF LARGE

IMPROVE ACCURACY OF LARGE

A MEMBER OF NIKKISO

Microtrac

THANK YOU FOR THE OPPORTUNITY TO DISCUSS OPTICAL PARTICLE CHARACTERIZATION

A MEMBER OF NIKKISO

Microtrac

Breaking up is hard to do: On the De-agglomeration of Cohesive Particles

Hrenya, C. M., University of Colorado at Boulder

Originally presented on: 4/13/2016 10:45:00 - 11:15:00

Cohesion between solids can arise from numerous sources. In this work, we consider cohesion arising from van der Waals forces, liquid coatings, and humidity. Experiments and DEM simulations are used to explore the role of cohesion on fluidization, and particularly on de-agglomeration.

ARIZONA STATE UNIVERSITY

Could Granule Morphology be the Key to Understanding Breakage Behavior?

Dr. Heather Emady
School for Engineering of Matter, Transport and Energy
Arizona State University

Frontiers in Particle Science & Technology:
Mitigation & Application of Particle Attrition
Houston, TX
April 13, 2016

Acknowledgements:
- FPST 2016 organizers
- Tianxiang Gao, PhD student

SCIENCE foundation ARIZONA INNOVATION AT WORK

ASU

ARIZONA STATE UNIVERSITY

Particulate Process and Product Design

Quantitative relationships between granule properties and formulation and process parameters are needed.

What are the desired particulate product characteristics?

Size, size distribution, bulk and tapped density, shape, porosity, strength, uniformity

How can the process be optimized to achieve the product characteristics?

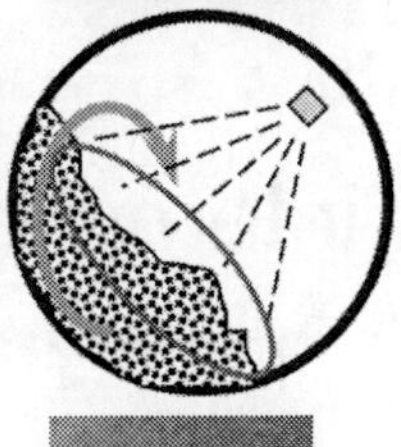

Rotation rate, shear rate, fill level, temperature, fluid addition method and rate

How can the feed formulation be optimized to achieve the product characteristics?

Particles: Size, size distribution, density, shape, porosity, strength, composition
Fluid: Surface tension, viscosity
Particle/fluid: Contact angle

ASU 2

Outline

1. Granule Characterization
2. Wet Granulation
3. Single Drop Granule Formation
   - Mechanisms
   - Granule Morphology
4. A Preliminary Granule Breakage Study
5. Summary & Future Directions

Characterization of Granules

- Size & size distribution
  - Most common
- Bulk density
- Porosity & internal structure
  - X-Ray Microtomography
- Shape
  - Typically 2D

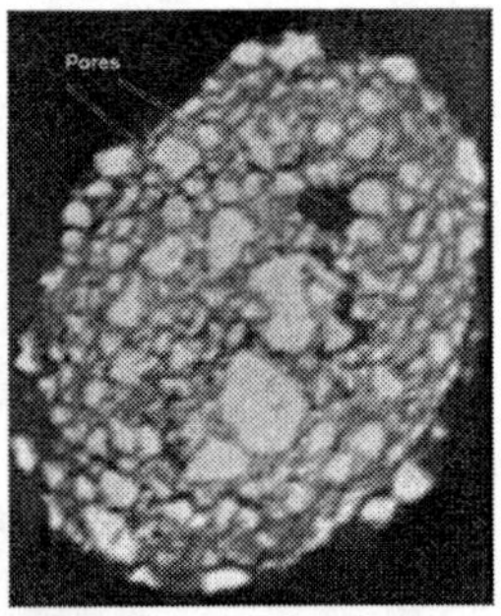

Rahmanian N, et al. Powder Technology. 2009;192(2):184-94.

ARIZONA STATE UNIVERSITY

Subsequent Performance Metrics of Granules that may be Dependent on Shape

- Strength and attrition resistance
- Packing (e.g., into tablet dies)
- Disintegration
- Dissolution

ASU 5

ARIZONA STATE UNIVERSITY

Wet Granulation Process

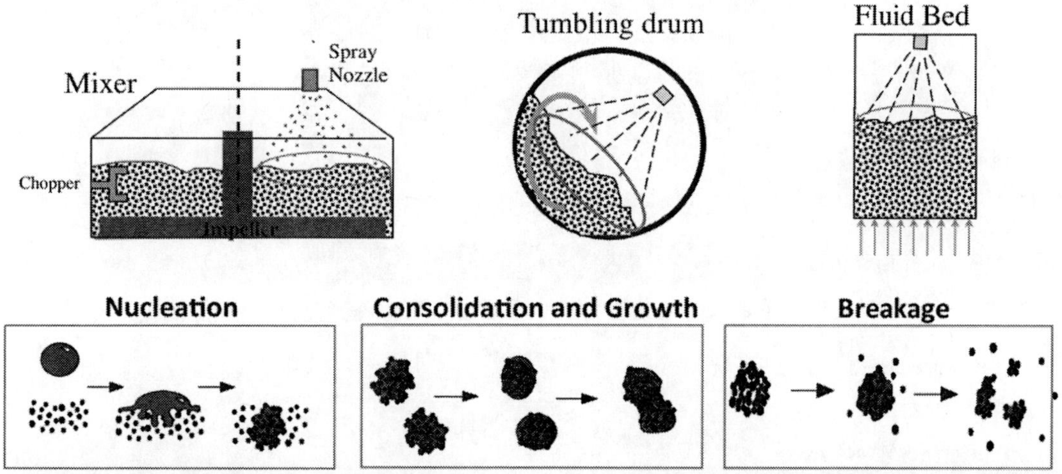

Achieve better control by separating rate processes

ASU 6

Granule Breakage

Parameters that would influence the breakage behavior:

1. ***Granule morphology***
2. *Internal structure and porosity*
3. *Granulation methods*
4. ***Particle and binder types***

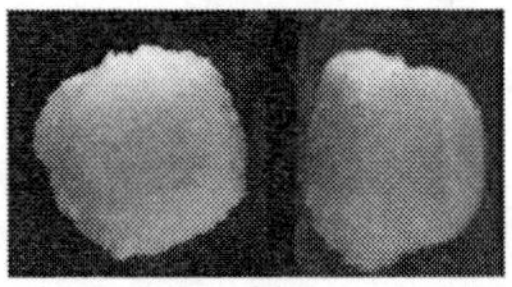
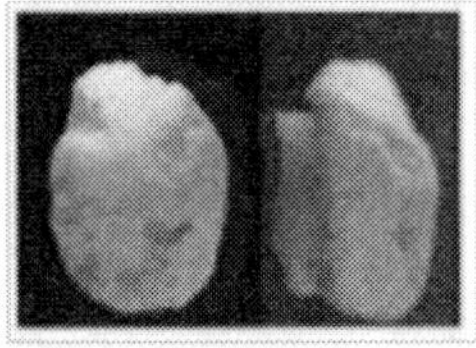
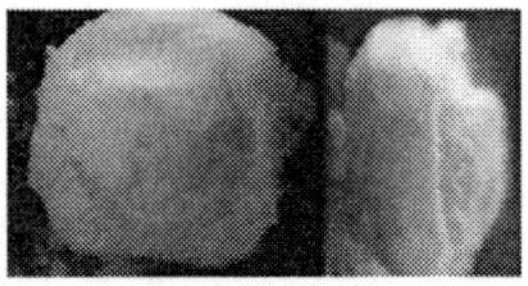

ARIZONA STATE UNIVERSITY

Granule Formation Mechanisms & Morphology

Finer particles < 10 μm
(high porosity)

Tunneling

Round Shape

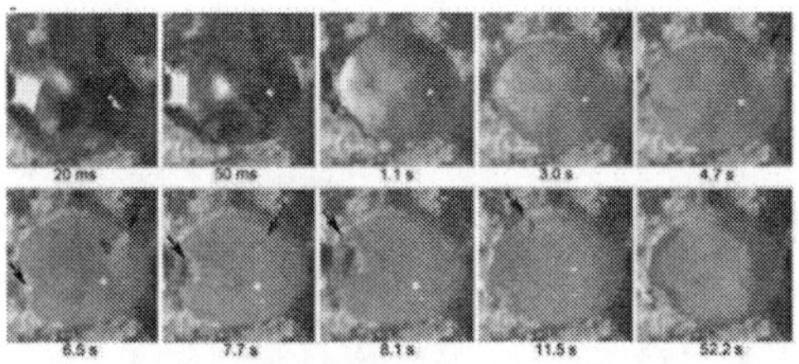

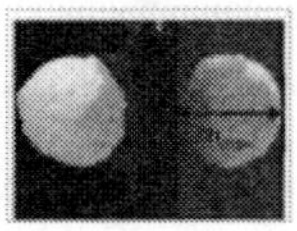

Coarser particles > 10 μm
(low porosity)

Spreading

Flat Disk Shape

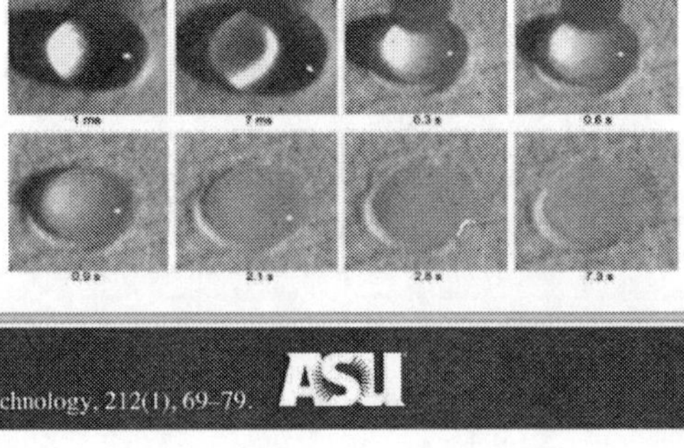

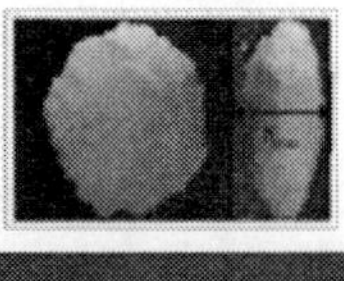

Emady, H. N. et al. (2011). Powder Technology, 212(1), 69–79.

Granule Formation Mechanism Regime Map

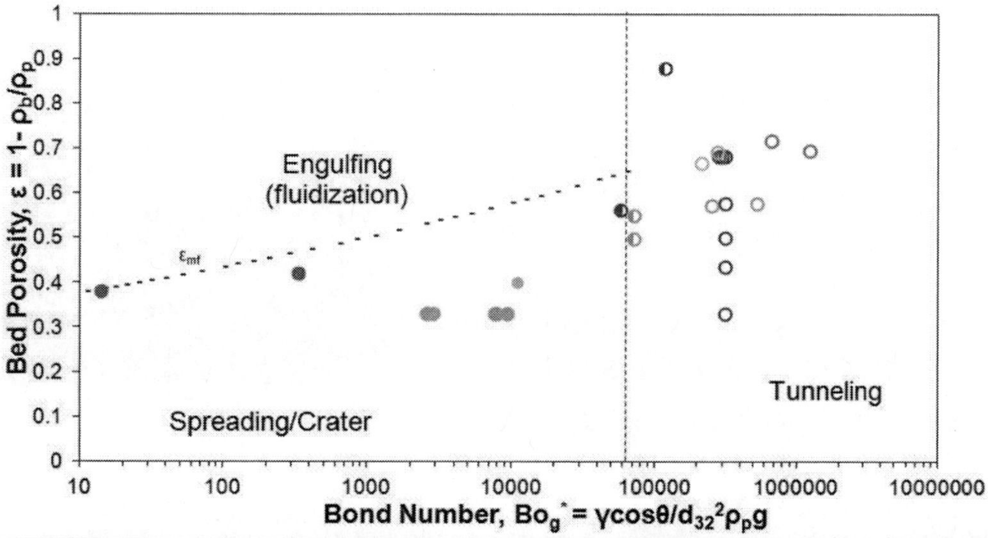

ARIZONA STATE UNIVERSITY

A Preliminary Granule Breakage Study

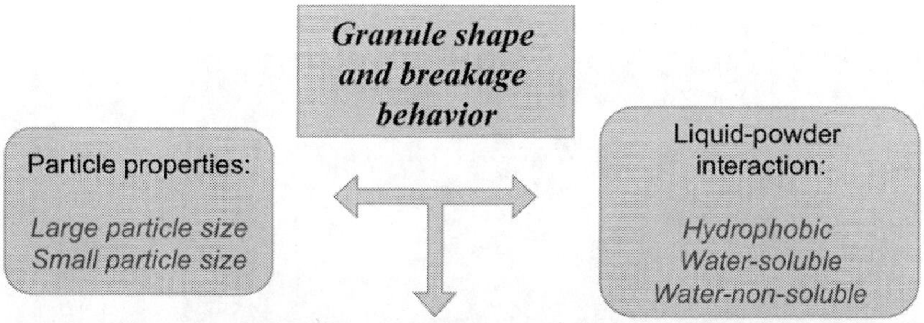

Particle & Binder Type	Hydrophobic	Water-soluble		Water-non-soluble
Powder of large size *(~100-200 μm)*	Compap L APAP	Compap L APAP	Sugar	Sugar
Powder of small size *(~1-10 μm)*	Micronized APAP	Micronized APAP	Powdered Sugar	Powdered Sugar
Drop Liquid *(~3 mm)*	APAP Saturated Water Solution	DI water	DI Water	Sugar Saturated Water Solution

ARIZONA STATE UNIVERSITY

Experimental Procedure

- 8 experiments are conducted
- Petri dish is prepared with powder
- Single drops are released on the powder bed, and a high speed camera records the drop impact and penetration
- Videos are analyzed to determine the formation mechanism
- Granules are extracted to characterize the morphology and breakage behavior

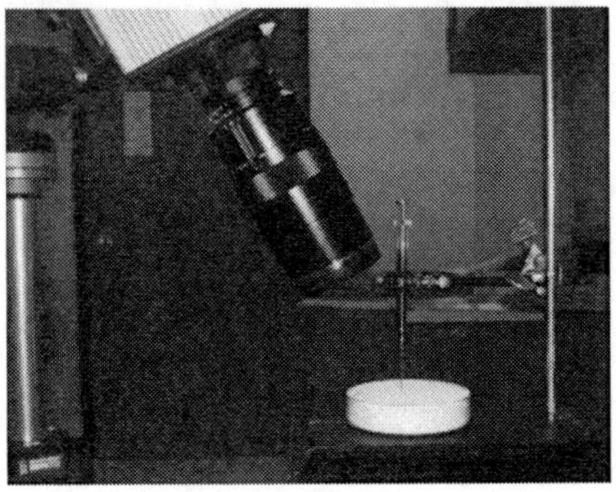

ASU 11

ARIZONA STATE UNIVERSITY

Granule Morphology Characterization

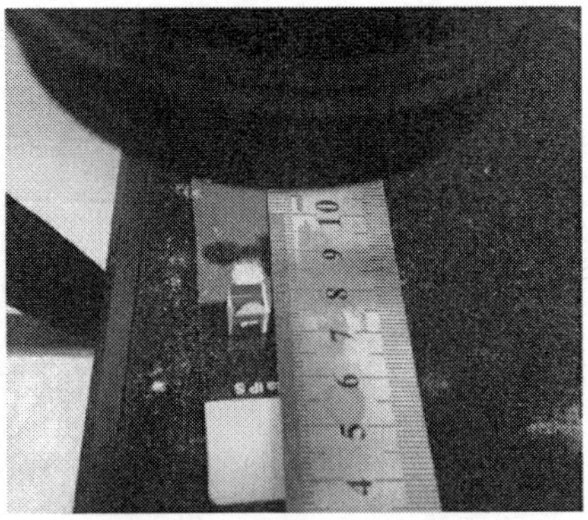

Two prisms are put together by long side of each other to get the top and side view of granules.

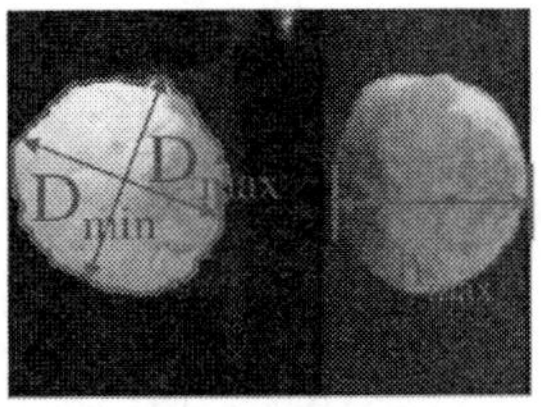

$$H.A.R. = \frac{D_{max}}{D_{min}}$$

$$V.A.R. = \frac{d_{area}}{h_{max}}$$

ASU 12

Granule Breakage Behavior Characterization

- A sieve with 2.00 mm mesh is used to characterize the breakage behavior.
- Granules are distributed inside the sieve.
- Allow a fall from 3 cm height for 3 times and record the remaining total weight.
- Repeat 4 times. There should be decreasing weights.

Granule Formation Mechanism - Spreading

APAP & Sugar

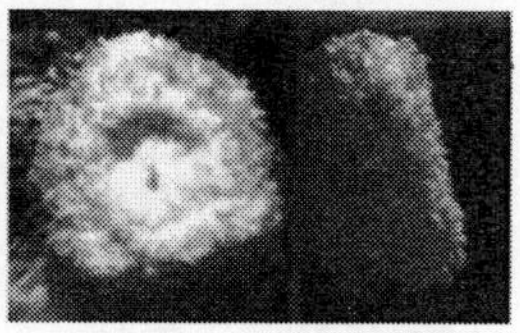

ARIZONA STATE UNIVERSITY

Granule Formation Mechanism - Spreading

APAP

Sugar

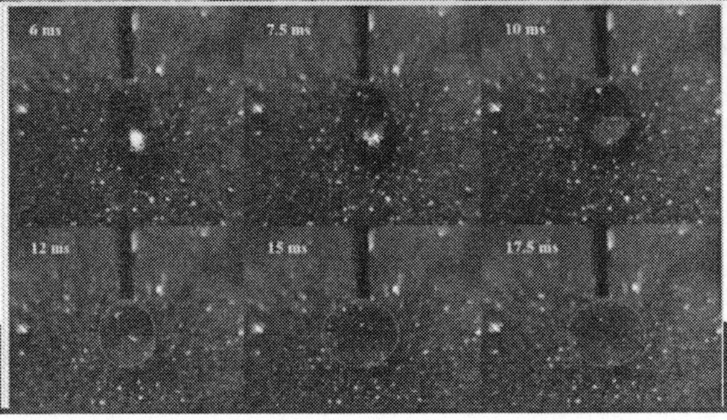

15

ARIZONA STATE UNIVERSITY

Granule Formation Mechanism - Tunneling

Micro-APAP & Powdered Sugar

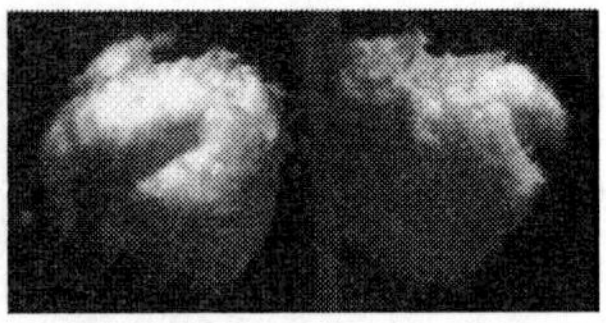

16

Granule Formation Mechanism - Tunneling

Micro-APAP

Powdered Sugar

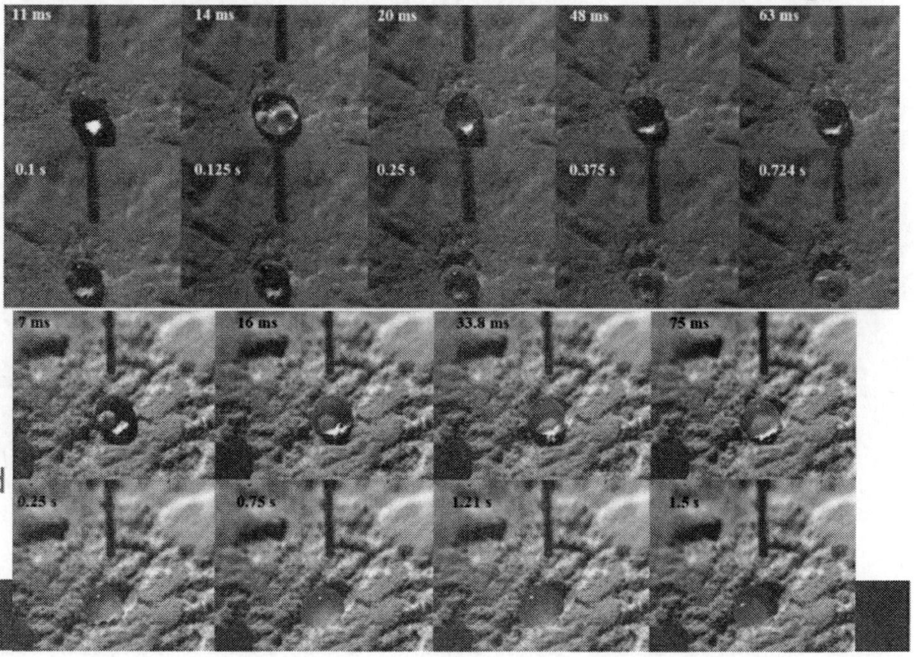

Granule Morphology Results

Experiments	APAP +Saturated APAP	APAP + DI water	Sugar + DI water	Sugar + Saturated sugar water solution
Granules				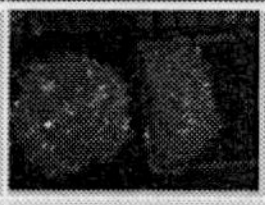
Morphology parameters	Size: d_a (mm): 5.24±0.53 h_{max} (mm):2.89±0.22 Shape: H.A.R: 1.06±0.06 V.A.R: 1.81±0.18	Size: d_a (mm): 5.52±0.71 h_{max} (mm):3.21±0.47 Shape: H.A.R: 1.04±0.08 V.A.R: 1.72±0.37	Size: d_a (mm): 8.04±2.79 h_{max} (mm)::3.21±0.56 Shape: H.A.R: 1.13±0.25 V.A.R: 1.41±0.38	Size: d_a (mm): 5.88±0.87 h_{max} (mm): 3.7±1.51 Shape: H.A.R: 1.09±0.05 V.A.R: 1.83±0.63
Experiments	Micro-APAP +Saturated APAP	Micro-APAP +DI water	Powdered Sugar + DI water	Powdered Sugar + Saturated sugar water solution
Granules	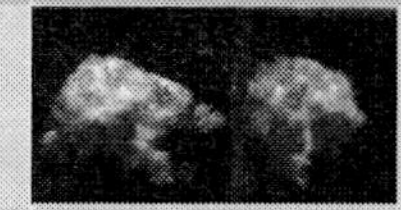	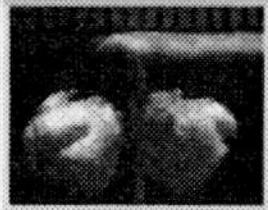		
Morphology parameters	Size: d_a (mm): 3.51±0.29 h_{max} (mm): 3.00±0.89 Shape: H.A.R: 1.04±0.05 V.A.R: 1.18±0.46	Size: d_a (mm): 3.50±0.85 h_{max} (mm): 2.81±0.64 Shape: H.A.R: 1.04±0.18 V.A.R: 1.25±0.06	Size: d_a (mm): 5.55±1.48 h_{max} (mm): 4.54±2.25 Shape: H.A.R: 1.13±0.04 V.A.R: 1.23±0.29	Size: d_a (mm): 4.31±1.27 h_{max} (mm): 4.13±0.62 Shape: H.A.R: 1.06±0.06 V.A.R: 1.04±0.15

Arizona State University

Attrition Results

	Formation Mechanism	Number of granules	Net weight of granules (g) after the given number of drops					Attrition percentage %
			0	1	2	3	4	
APAP +Saturated APAP	Spreading	17	0.411	0.399	0.391	0.382	0.372	9.49
Micro-APAP +Saturated APAP	Tunneling	19	0.107	0.088	0.086	0.085	0.081	17.8
APAP + DI water	Spreading	17	0.453	0.392	0.374	0.365	0.360	20.5
Micro-APAP +DI water	Tunneling	15	0.108	0.097	0.092	0.092	0.089	17.6
Sugar + DI water	Spreading	17	5.411	4.901	4.532	4.215	3.917	27.6
Powdered Sugar + DI water	Tunneling	6	0.41	0.376	0.368	0.37	0.368	10.2
Sugar + Saturated sugar water solution	Spreading	18	1.22	1.189	1.17	1.16	1.15	5.74
Powdered Sugar + Saturated sugar water solution	Tunneling	16	0.636	0.62	0.612	0.609	0.604	5.03

ASU 19

Arizona State University

Breakage Behavior Trend: APAP

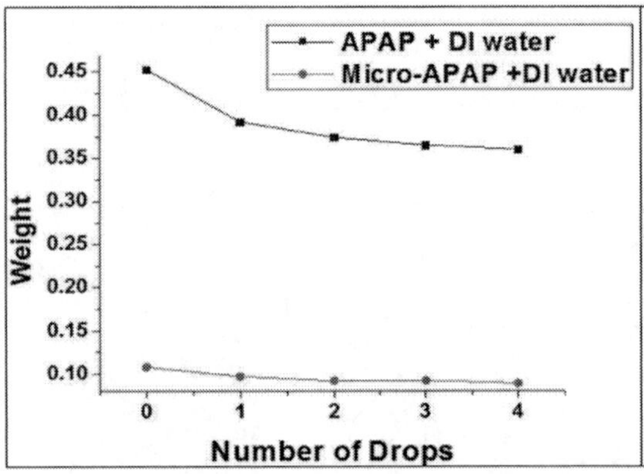

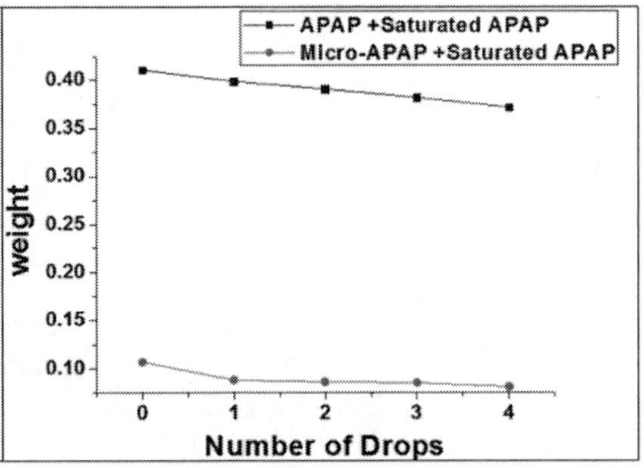

ASU 20

ARIZONA STATE UNIVERSITY

Breakage Behavior Trend: Sugar

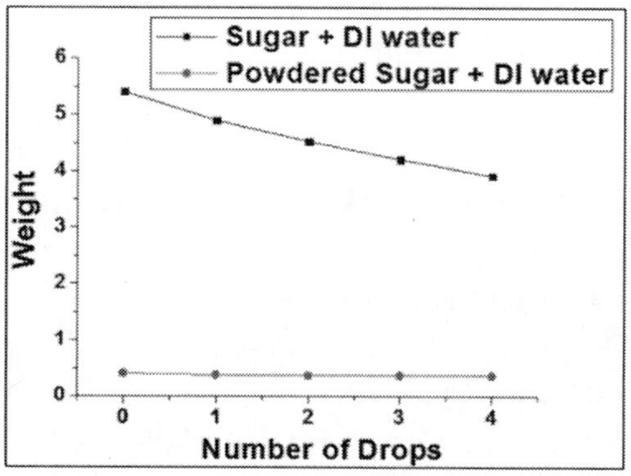

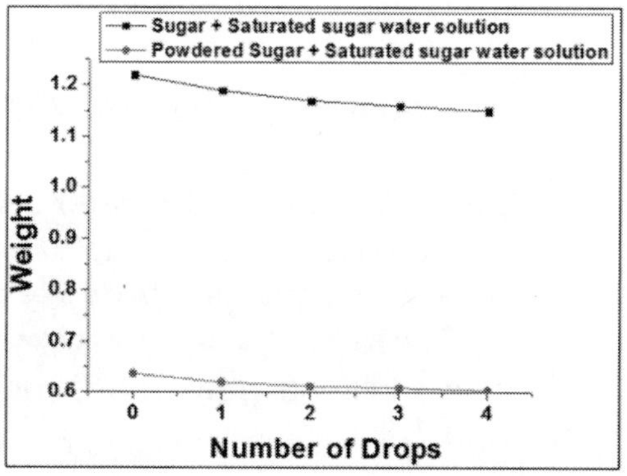

ASU 21

ARIZONA STATE UNIVERSITY

Summary

- Single drop formed granules have narrow size distributions, but the shape depends on the formation mechanism.
  - *Tunneling*, *Spreading*, and *Crater Formation* are the different granule formation mechanisms, which can be predicted by a regime map.
- Particle size has significant influence on the formation mechanism, granule morphology as well as attrition behavior.
  - Small particle size results in tunneling formation, rounder granules and **low attrition percentage**.
  - Large particle size results in spreading formation, plate-like shape and **high attrition percentage**.
- Hydrophobicity and solubility are complex and their influence on shape and attrition behavior needs to be further investigated.

ASU 22

ARIZONA STATE UNIVERSITY

Future Directions

The end goal is to create a link between the formulation properties, the granule formation mechanism, the granule morphology and internal structure, and granule performance metrics such as breakage and attrition behavior.

- Characterize the formulation to place these results on the regime map:
  - Liquid properties: viscosity, surface tension
  - Powder properties: particle size distribution, particle density, bulk density
  - Powder-liquid properties: contact angle
- Test additional powders of varying properties.
- Perform mechanical testing on granules using a more sophisticated technique.
- Characterize the internal structure of granules.

ASU 23

ARIZONA STATE UNIVERSITY

Thank You!

ASU 24

On the Origin of High Powder Cohesion after Milling: Micro-Scale Examination and Fundamental Approach to Reduced Cohesion

Davé, R., New Jersey Institute of Technology

Originally presented on: 4/13/2016 12:45:00 - 13:15:00

Micronization of powders leads to their very poor flow, packing, and dispersion, not to mention increased sticking tendency. Most of these problems are usually attributed to fine particle size, which results in higher cohesion as compared to particle weight. This is usually represented via granular Bond number, which increases with size reduction. However, in many cases, there is an additional factor, which further contributes to higher cohesion. This factor is related to the altered surface characteristics arising from micronization. Such behavior is usually prevalent in crystalline materials, where milling leads to exposing of higher surface energy sites or defects. This additional factor adds to the poor flow and handling of micronized powders. Unfortunately, this aspect is poorly investigated in literature and hence, poorly understood. In this work, fundamental investigation is conducted to shed light on this problem as well as its mitigation. Towards that goal, simultaneous micronization and surface coating of crystalline particles with nano silica is carried out to. The purpose is to examine the role of surface energetics and the influence of dry coating on both creation of nano-scale surface roughness and passivation/stabilization of high surface energy sites, and hence reduced cohesion. Pharmaceutical powders were used as test materials and a fluid energy mill was used to micronize the particles to several sizes under 30 µm with or without simultaneous nano-silica coating. Powder flow property and dispersibility were characterized using FT4 powder tester and Rodos/Helos laser diffraction particle sizer. Surface energy was characterized using a next generation Inverse Gas Chromatography instrument. Total surface energy as well as Lifshitz –van der Waals (LW) dispersion component of surface energy were measured as a function of milling intensity. In addition, surface energy heterogeneity was also examined, revealing the micro-scale changes that occur during milling of crystalline materials. Impact of dry coating with nano-silica on surface energetics and powder cohesion was investigated. It was found that dry coating leads to reduction in cohesion and granular Bond number, hence improved flowability due to both reduced LW dispersive component of surface energy and creating nano-scale surface roughness.

Attrition of Limestone Particles Related to the Fluidized Bed Capture of SO_2 and CO_2

John Grace and Jim Lim,
University of British Columbia,
Vancouver, Canada

Frontiers in Particle Science and Technology,
Houston, April 2016

Equipment deployed

<u>Attrition Instruments</u>:

1. Modified ASTM Air Jet Apparatus 
2. High-Temperature Target Impact 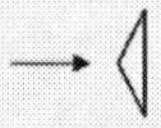
3. Miniature Circulating Fluidized Bed 

<u>Reactivity Measurements</u>:

Dual-Environment Thermogravimetric Reactor

All are custom-made and able to operate up to ~900°C.

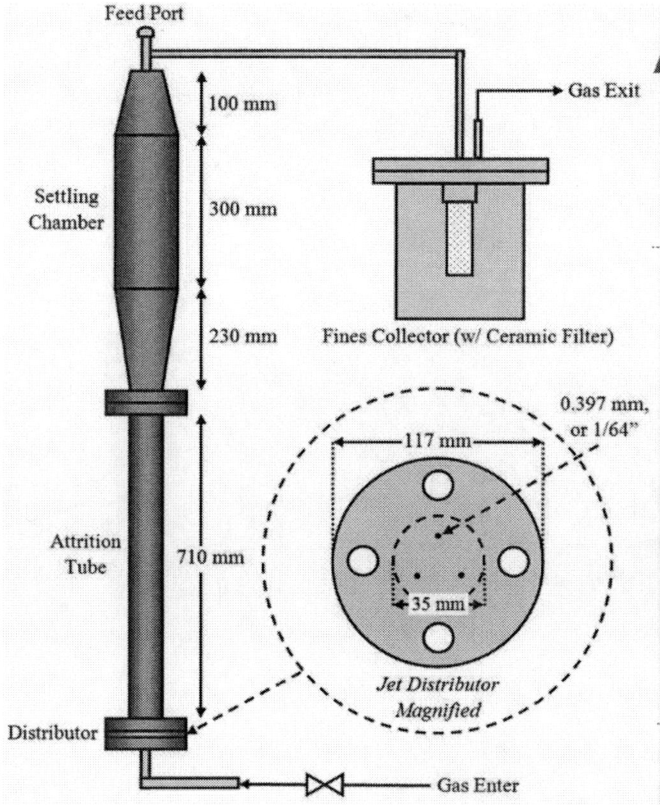

Air Jet Apparatus ASTM D5757

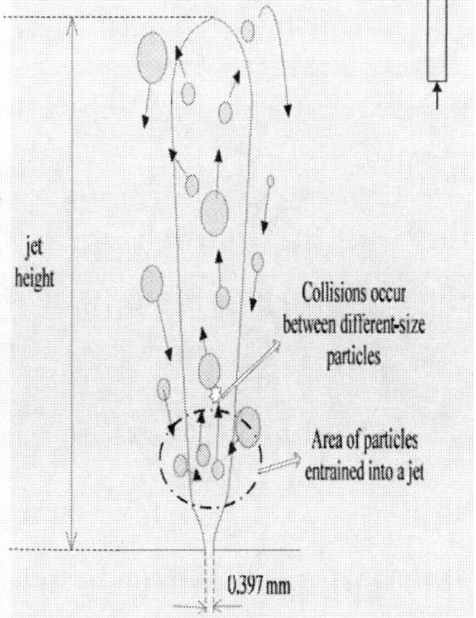

Air Jet ASTM Apparatus

Temp (°C)	Fluidizing Gas	Gas Flow Rate (L/min) @ 0°C, 1 atm	Sup. Gas Velocity (m/s)	Initial Mean Diameter (μm)	Test duration (h)
0	Air, 30-40% RH	10.0	0.173	10 - 180	5

$$\text{Air Jet Index } (AJI) = \frac{Mass\ of\ Fines\ after\ 5\,h\ test\ run}{Original\ Mass\ of\ Sample} \times 100\%$$

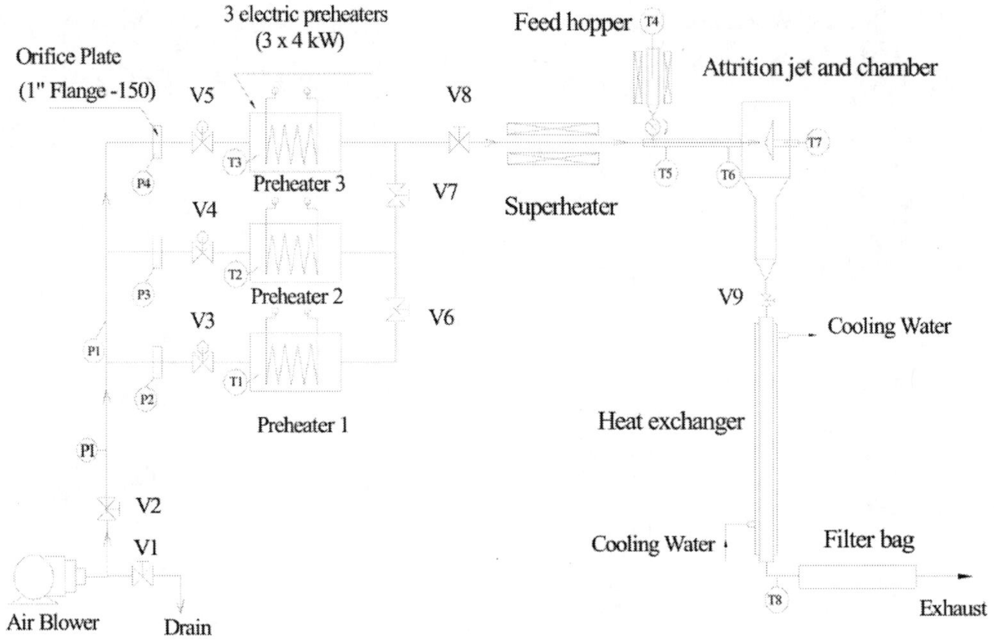

High-Temperature Target Attrition Apparatus

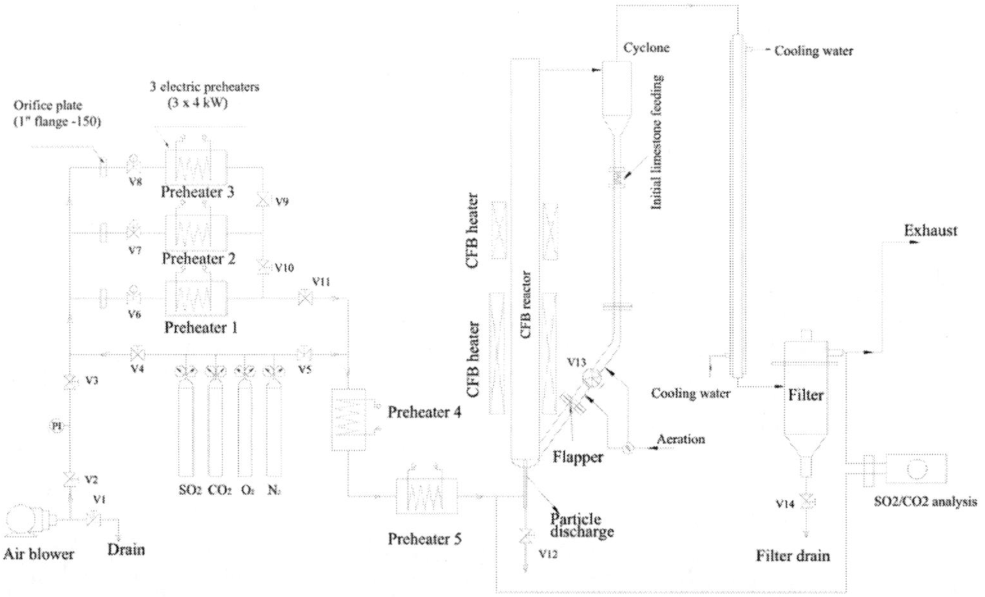

Small CFB facility for continuous attrition experiments

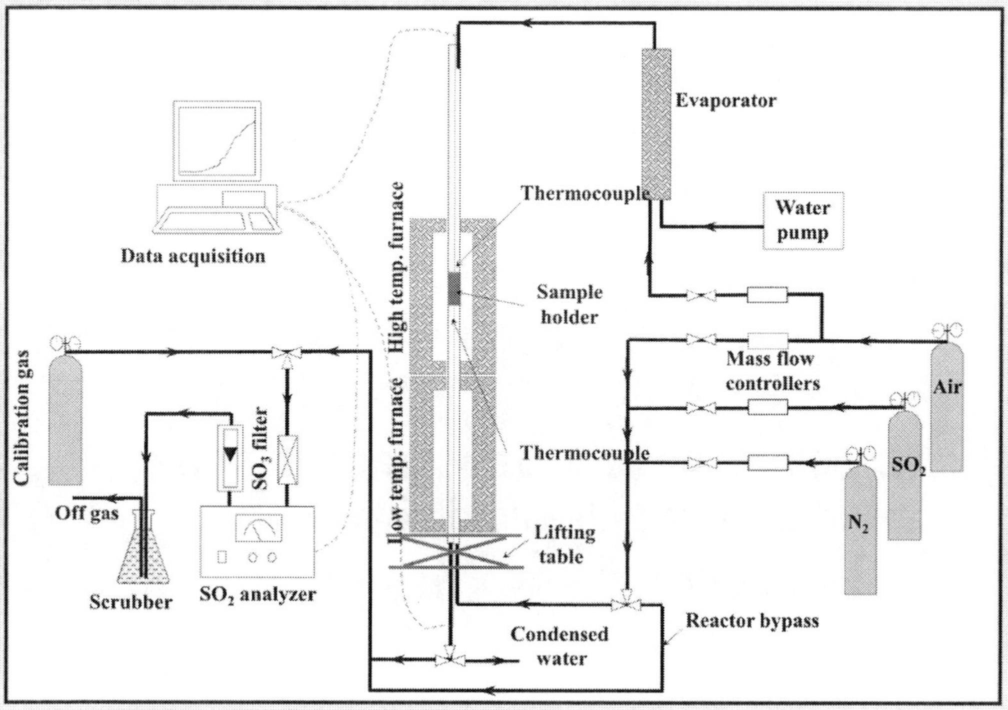

Dual-Environment Thermogravimetric Reactor

SO_2 Capture vs CO_2 Capture

SO_2 Capture	CO_2 Capture
Product accumulates on particle outer surfaces, larger volume blocks pores.	Carbonate product replaces what was originally there. Less blockage.
Capture is irreversible: ($CaO + SO_2 + ½O_2 \rightarrow CaSO_4$)	Calcination/carbonation cycling should be reversible: ($CaO + CO_2 \rightleftharpoons CaCO_3$)
Calcium utilization is key. Attrition is helpful, within limits.	Number of cycles is key. Attrition is largely undesirable.

Mechanisms of Attrition in Fluid Bed SO_2/CO_2 Capture

1. Reaction induced: due to Calcination gas release, pore stuffing.
2. Impact, due to impacts with other particles, heat transfer surfaces & cyclones.
3. Abrasion as particles rub against one another during the fluidization.

Thermal decrepitation could also be a factor, but is believed to be minor.

21 Worldwide Limestones Studied

Sample name (given by supplier)	Origin
Sormony	Canada
Hangzhou	China
Dolomite	USA
Faxe Grainstone	Denmark
South African	S. Africa
Jianshan	China
Havelock	Canada
Strassburg	USA
Lucerne Valley	USA
Faxe Bryozo	Denmark
Derbyshire	UK
Texada Island	Canada
British Coal	UK
Tex-Mex	USA
Luscar	Canada
Calpo	Canada
Michigan Calcite	USA
Clam Shells	Canada
Faxe Coral	Denmark
Faxe Wackestone	Denmark
Michigan Dolomite	USA

Pores and Fractures in Calcined Limestones

Sulphur X-ray Mappings of "Fully" Sulphated Limestones

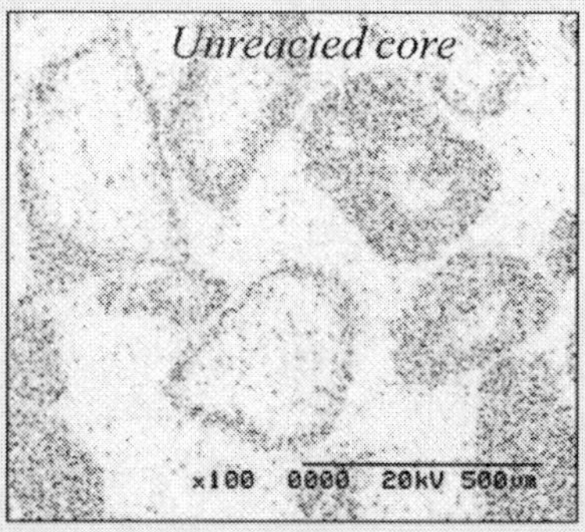

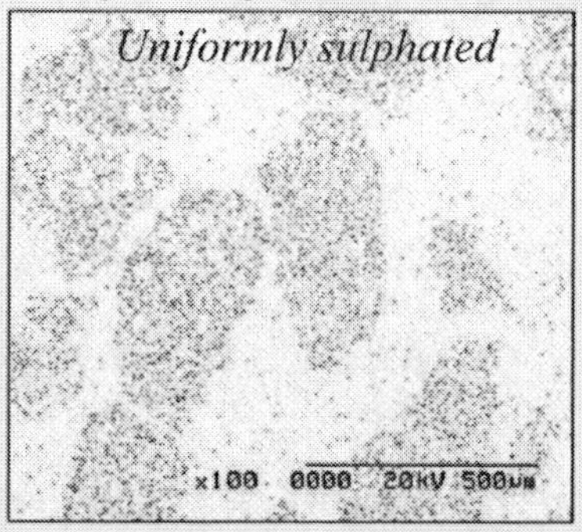

Lesson:

Limestones differ.

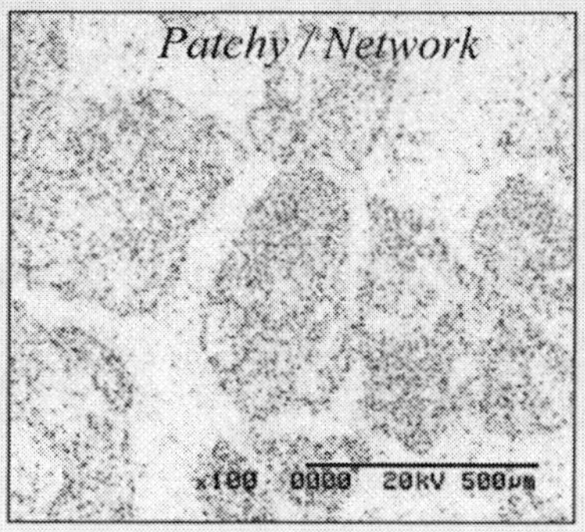

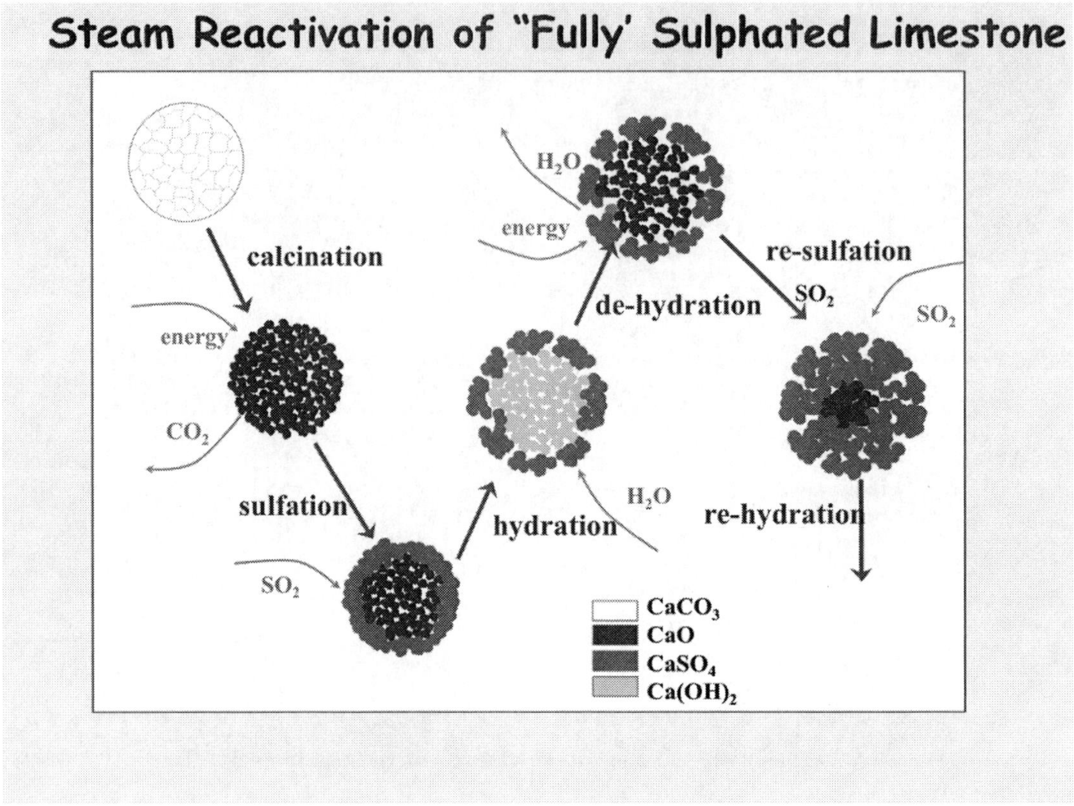
Steam Reactivation of "Fully' Sulphated Limestone
calcination
energy
CO_2
sulfation
SO_2
hydration
H_2O
de-hydration
H_2O
energy
re-sulfation
SO_2
SO_2
re-hydration
$CaCO_3$
CaO
$CaSO_4$
$Ca(OH)_2$

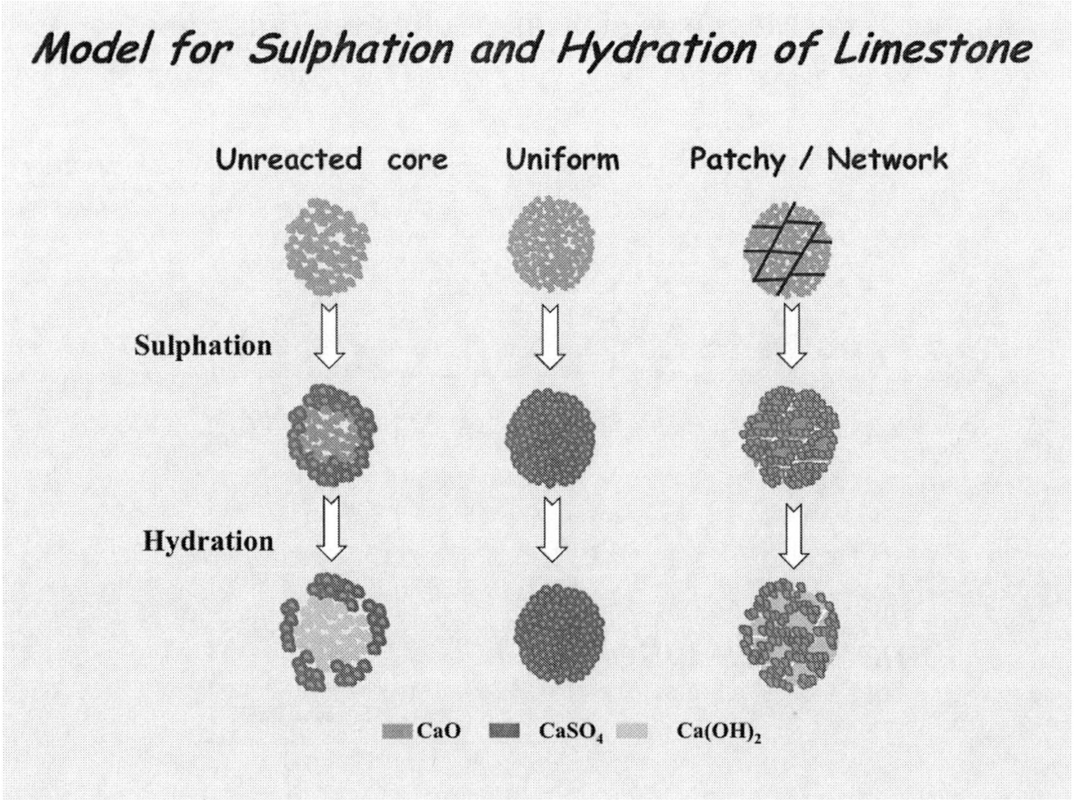
Model for Sulphation and Hydration of Limestone
Unreacted core
Uniform
Patchy / Network
Sulphation
Hydration
CaO
$CaSO_4$
$Ca(OH)_2$

Fracture Formation During Hydration

Unreacted core & Uniformly

Patchy / Network

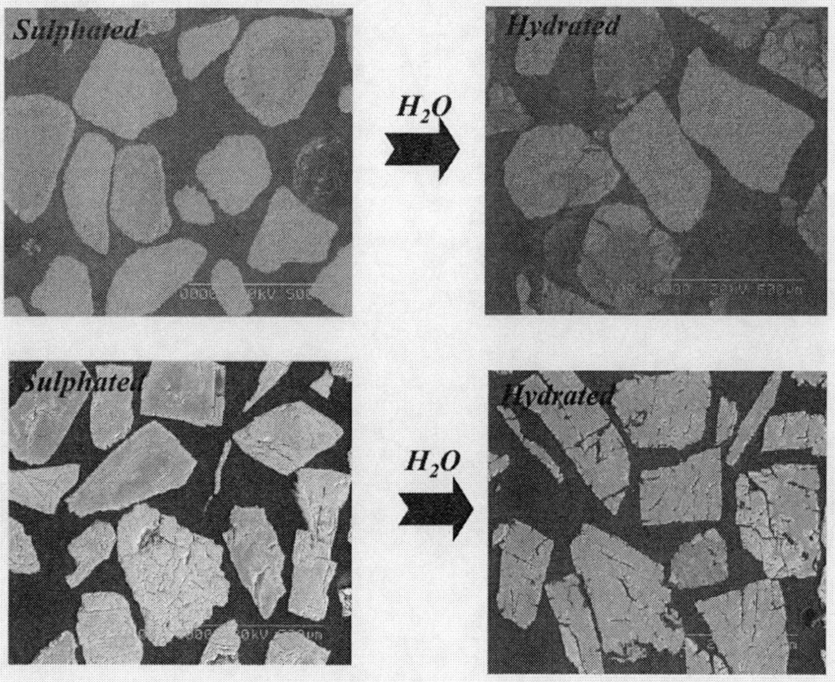

Effect of Reaction on Attrition

Attrition resistance:

$CaSO_4 > CaCO_3 > Ca(OH)_2 \approx CaO$

$CaSO_4$	$CaCO_3$	$Ca(OH)_2$	CaO
8.7 $\times 10^{-23}$	6.1 $\times 10^{-23}$	5.5 $\times 10^{-23}$	2.8 $\times 10^{-23}$

cm^3/molecule (Molar volume)

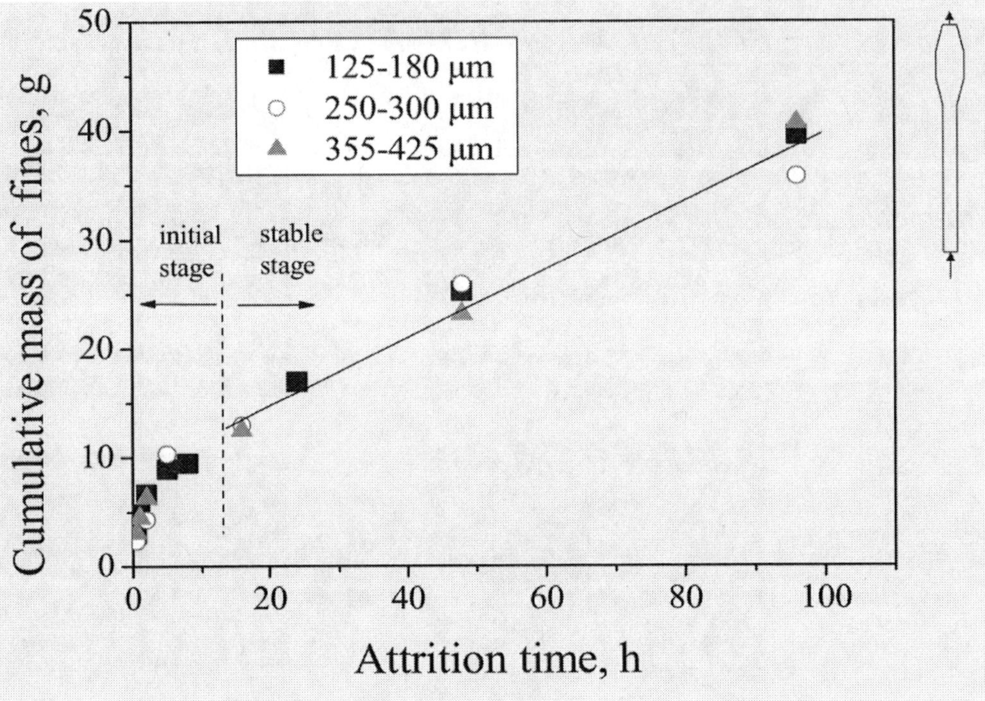

Linear relationship after an initial stage. Xiao et al., 2012

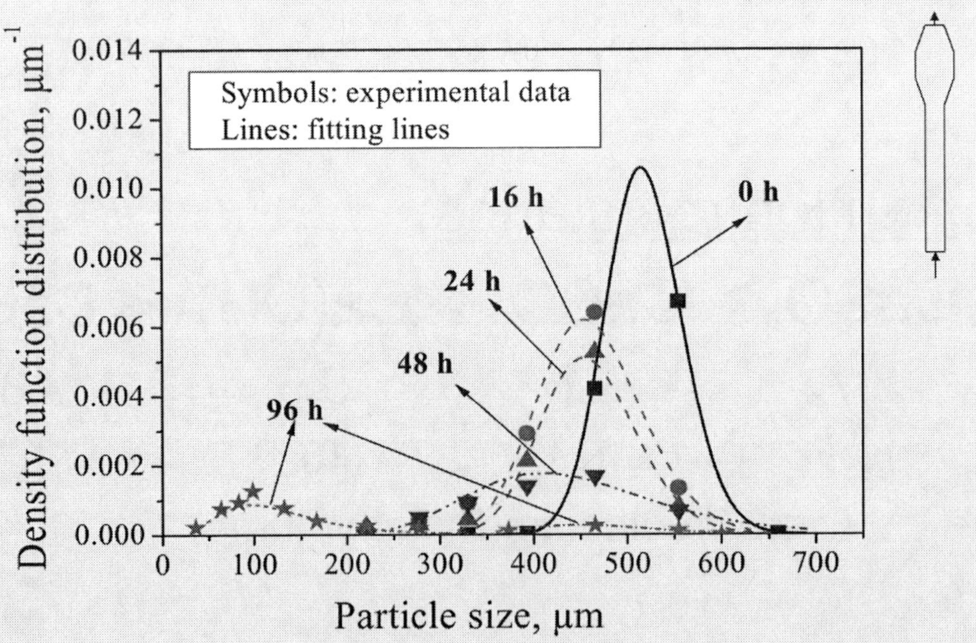

Particle size distributions of limestone particles after different attrition times.
Note the progressive change due to impact attrition.
Gang et al., 2014

SEM Image of agglomerates made up of attrition-generated fines

Gang et al., 2014

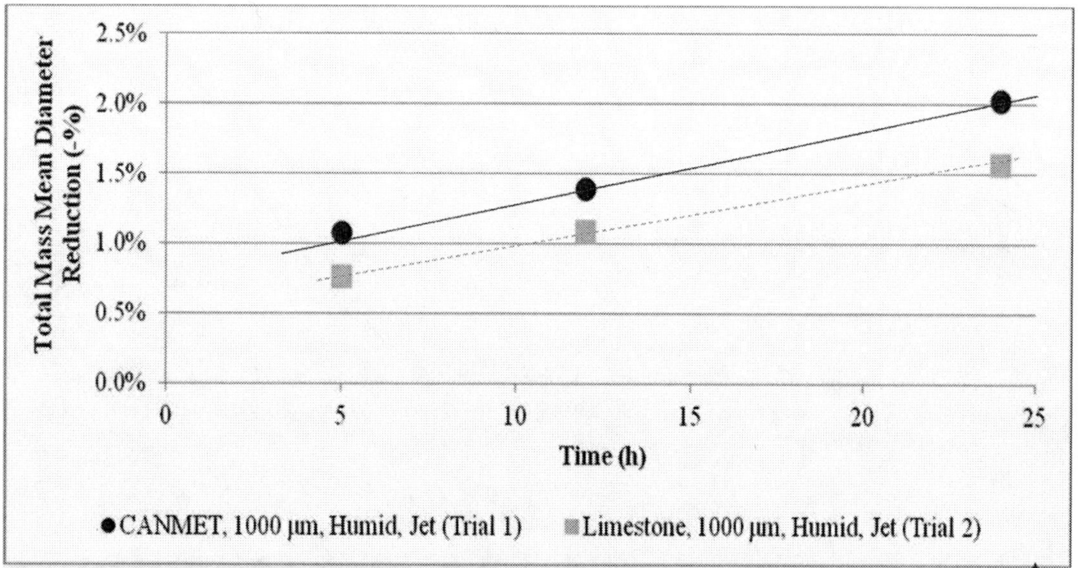

Total mass mean diameter reduction vs. time for uncalcined jet attrition lasting up to 24 h at 20°C.

Knight et al., 2014

SEM images of calcined CANMET pellets after 0 h (left) and 5 h (right) for jet attrition in humid air at 20°C

Knight et al., 2014

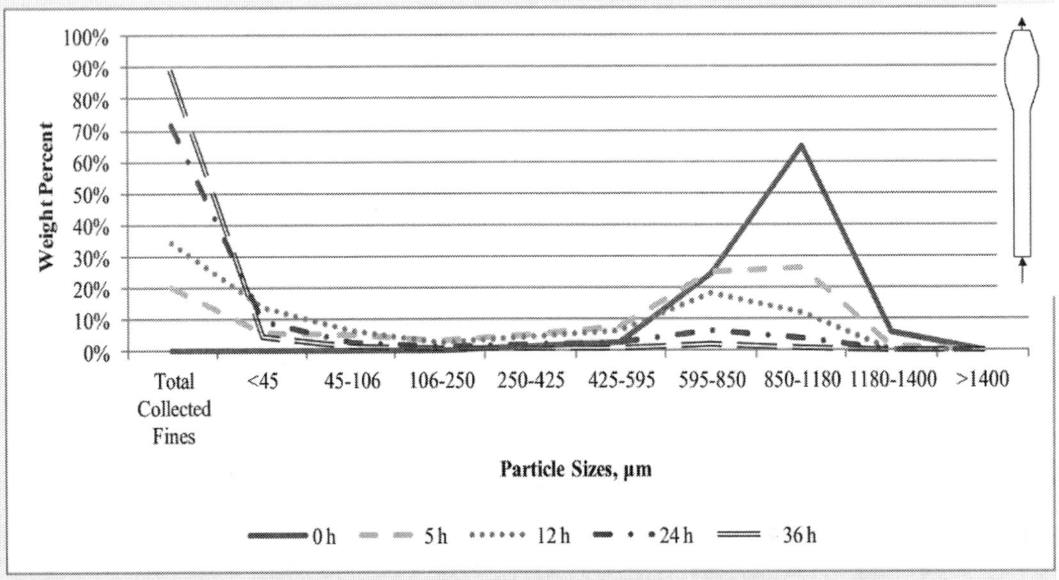

Differential particle size distribution for crushed Canmet pellets at 20°C, humid jet. The bimodal distributions suggest that abrasion is important.

Knight et al., 2014

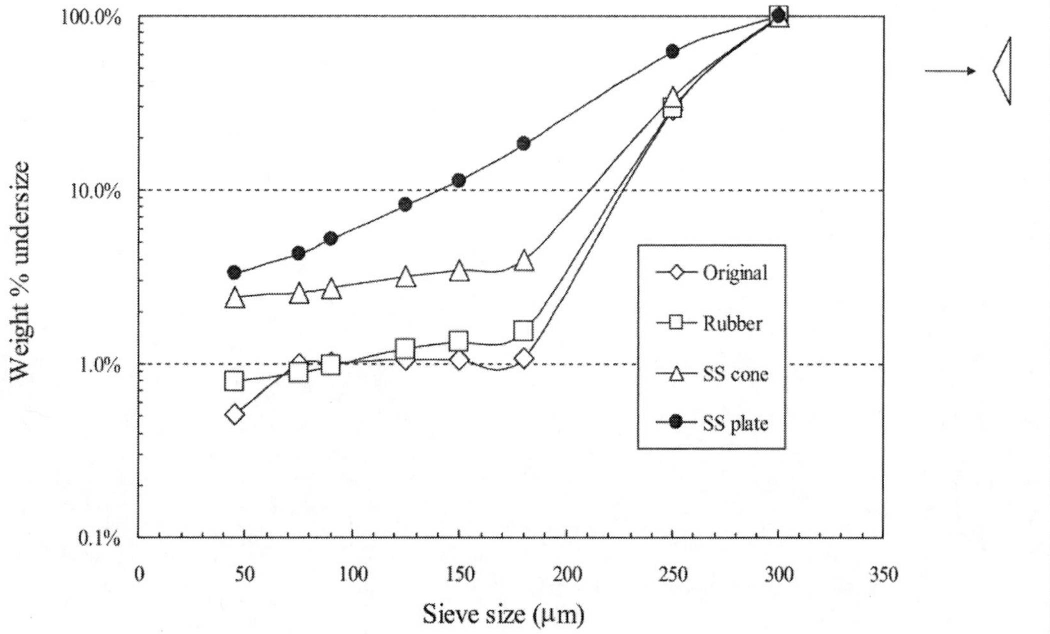

Particle size distributions for three different target and 250-300 μm limestone particle. Impact particle velocity = 74 m/s.

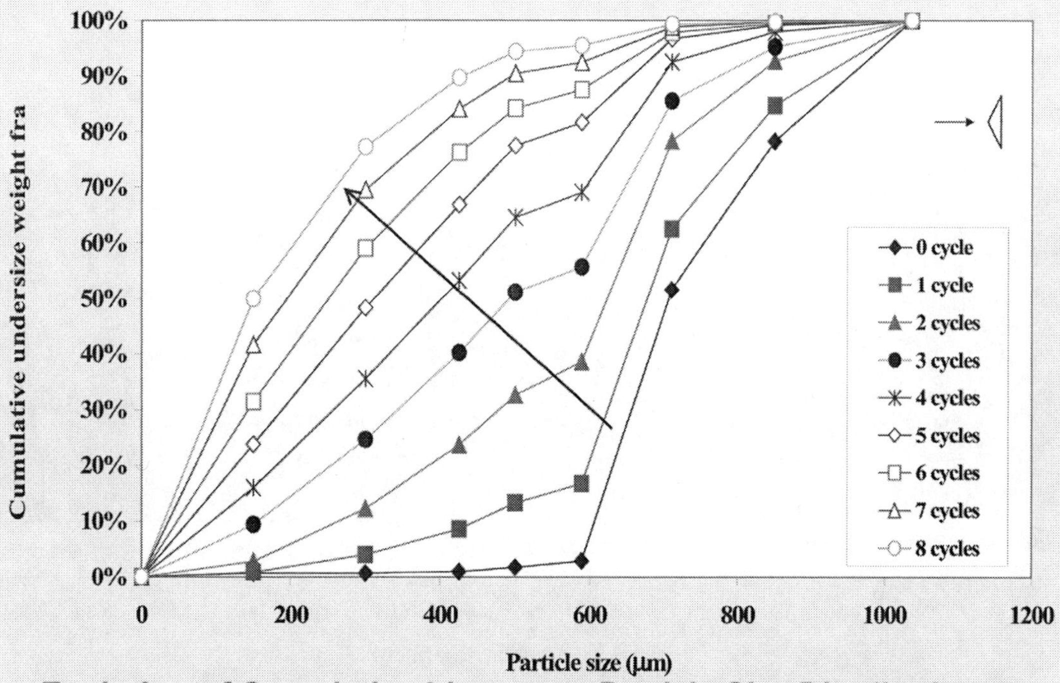

Evolution of Cumulative Limestone Particle Size Distribution with Repeated Impacts at 55 m/s and 25°C

Chen et al., 2007

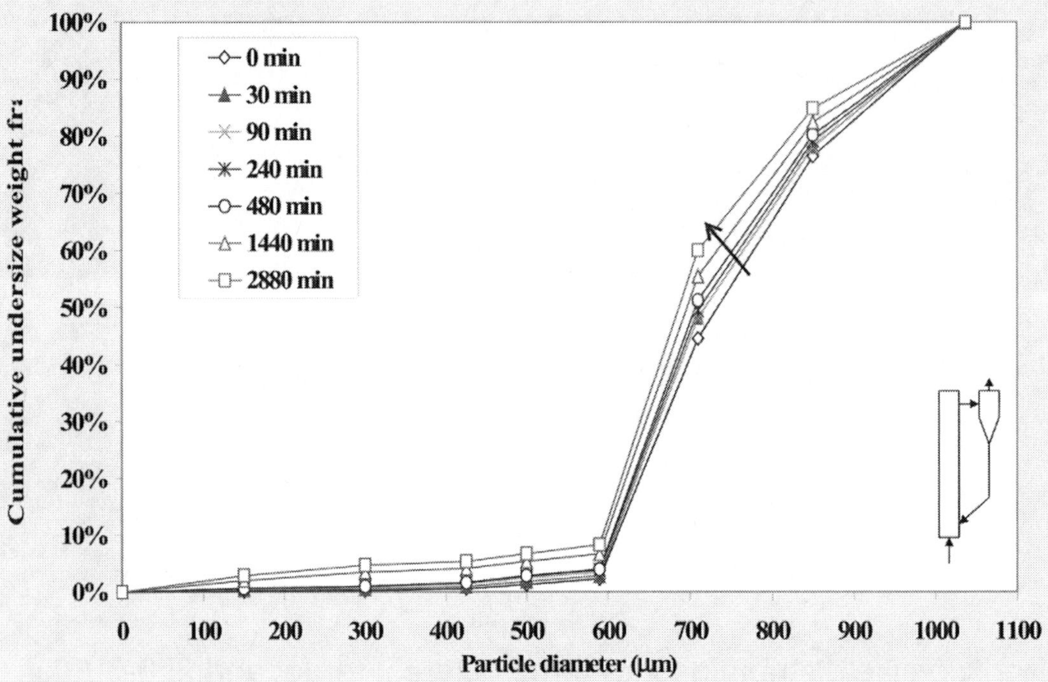

Cumulative particle size distribution for limestone CFB continuous attrition at 25°C and a sup. gas velocity of 5.6 m/s. Chen et al., 2008

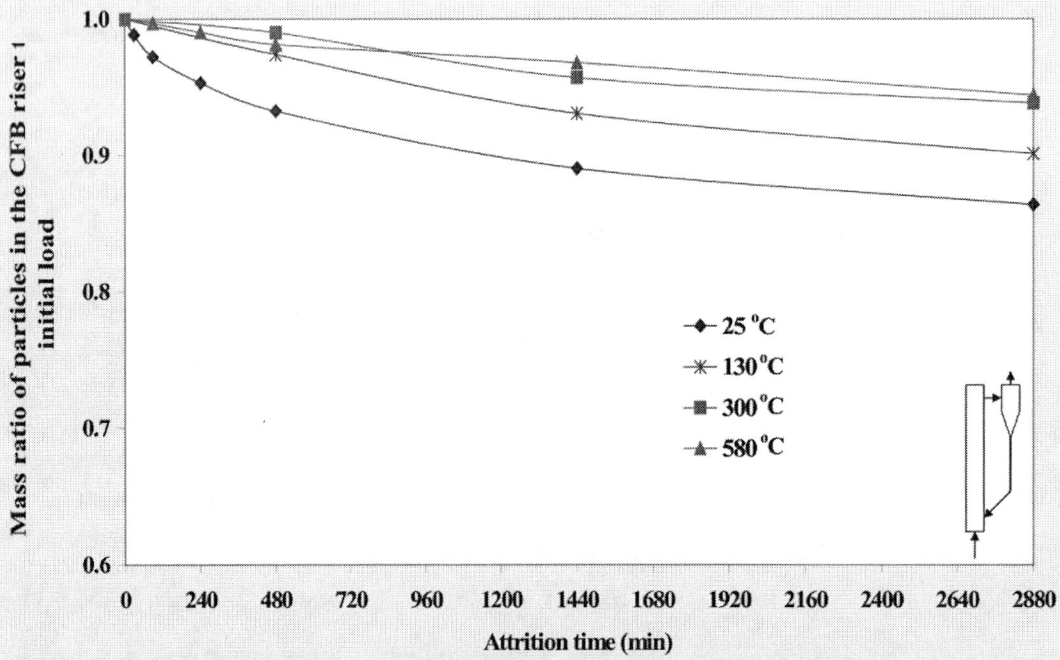

Mass change in CFB riser as a function of attrition time for continuous attrition at four temperatures and superficial gas velocity = 5.6 m/s. Note substantially less attrition as temperature increased. Chen et al., 2008

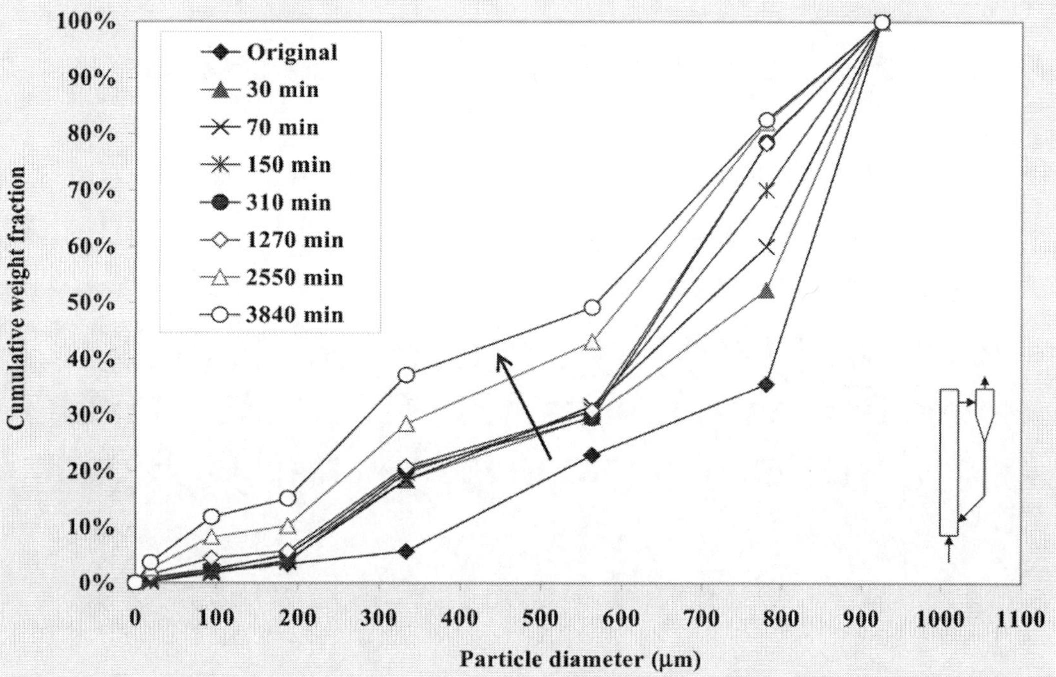

Cumulative particle size distribution for limestone after 64 h of attrition with calcination, sulfation in particle impact chamber. Temperature 850°C; sup. gas velocity 5.3 m/s; SO_2 feed concentration: 1150 ppm.

Chen et al., 2008

Attrition Model:

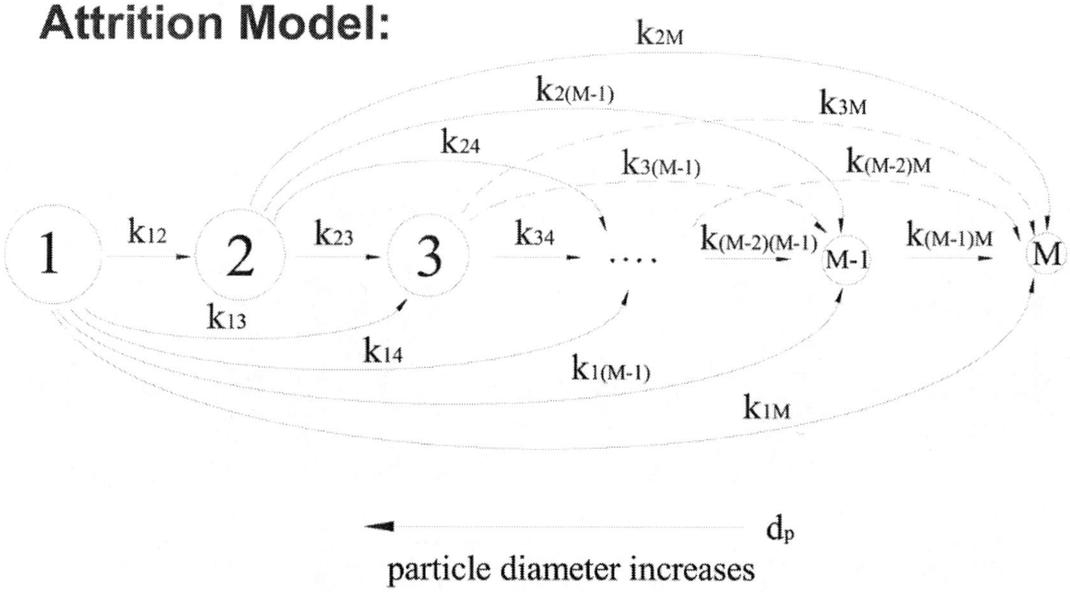

Particle impact exchange steps for different particle size intervals

Chen et al., 2011

Attrition Rate Constants

As a first approximation, the various *k* values can all be assumed to be equal if the particle size intervals are similar.

$k = k_o \exp\{-E/RT\} \times U_g^2$ gives good fits with experimental results, with $E \approx -2.35$ kJ/mol for Thames limestone in the small CFB unit.

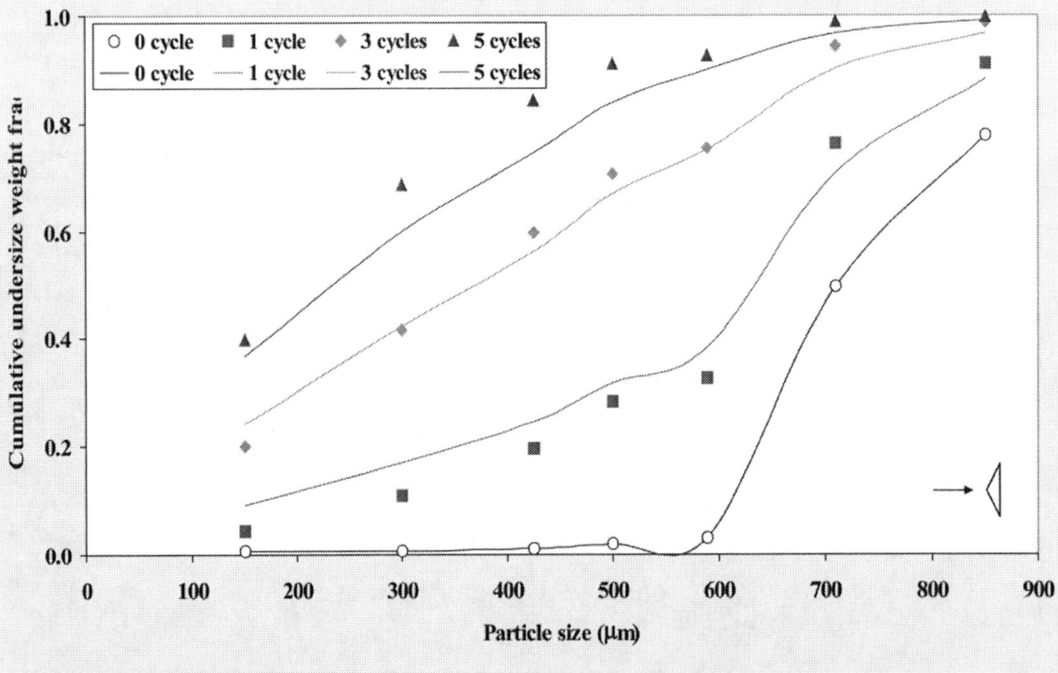

Comparison of experimental and predicted particle size distribution for limestone impact attrition. Temperature=31°C, pressure=1 atm, gas velocity = 76 m/s. (Points represent experimental data; lines are model predictions.) **Chen et al., 2011**

Some Findings/Conclusions

- The three attrition testing methods give results which are complementary.
- There is considerable variability among different limestones.
- There is a linkage between micro-cracks, which improve SO_2 and CO_2 capture, and attrition.
- Testing at room temperature tends to over-estimate the degree of attrition.
- Impacts, abrasion and reactions all influence the rate and extent of attrition.
- A simple model can predict the PSD evolution.

Acknowledgements

Funding:

Natural Sciences and Engineering Research Council of Canada,
Carbon Management Canada,
Alstom Power

People:

Zhongxiang Chen, Andrew Knight, Yan Li, Gang Xiao, Naoko Ellis

Application of Attrition-Resistant Materials for Novel Clean Energy Technologies

Kataria, A., RTI International

Originally presented on: 4/13/2016 13:45:00 - 14:15:00

At RTI's Energy Technology Division, we are developing cleaner energy solutions, a number of which utilize fluidized bed reactors (FBR) at their core. RTI has been developing process-specific fluidizable materials for a variety of technology applications including coal gasification, biomass conversion, CO2 capture and utilization, and syngas cleanup and conversion. We have developed significant expertise in designing highly attrition-resistant particles while imparting chemical reactivity and scaled up the manufacturing up to 100 ton scales. This presentation will share our experience and expertise in developing attrition-resistant sorbents for two applications: a) high-temperature syngas desulfurization, and b) CO2 capture from coal-derived flue gas. The ZnO sorbent employed for high-temperature desulfurization process is inherently a hard material and the critical research challenge was to optimize the balance between reactivity (lowering hardness) and attrition resistance. On the contrary, the PEI-based CO2 sorbents are supported on fluidizable silica support shells which are soft. RTI has addressed this limitation by designing a process that minimizes attrition.

Predicting and Controlling Attrition of Active Pharmaceutical Ingredients during Agitated Drying

Remy, B., Bristol-Myers Squibb

Originally presented on: 4/13/2016 14:15:00 - 14:45:00

Scale-up of agitated drying processes to minimize particle size changes in Active Pharmaceutical Ingredients (API) can be challenging. Particle attrition problems due to agitated drying are often discovered upon the initial scale-up from the lab to the plant. Traditional laboratory drying equipment has not successfully reproduced the degree of attrition observed at scale. Factors such as API characteristics, batch size, dryer geometry, residual solvent content, and the defined agitation protocol can impact the extent of attrition during drying. The development of new scale-down tools can provided a platform to study the mechanism for particle attrition during API manufacturing in a material sparing way. This work characterizes the propensity for particle attrition to occur by employing a laboratory agitated vessel that has been modified to measure impeller torque and reproduce the range of compressive and shearing forces observed during scale-up. A workflow for assessing risk of API attrition at manufacturing scale is described. The application of this methodology for several different APIs will be highlighted.

AUTHOR INDEX

AIChE
120 Wall Street, FL 23
New York, NY 10005-4020

ISBN 978-1-5108-2259-7

CURRAN ASSOCIATES INC.
proceedings.com